Hazardous Materials Medicine

Hazardous Materials Medicine

Treating the Chemically Injured Patient

Richard Stilp

Armando Bevelacqua

Library of Congress Cataloging-in-Publication Data applied for:

Hardback ISBN: 9781119663928

Cover Design: Wiley
Cover Image: Courtesy of Chris Hawley

SKY10044247_031323

Contents

Author Biographies and Acknowledgments

Richard's Biographical Information

Richard Stilp, PM, RN, MA. Richard completed his fire service career in April of 2018 as the Fire Chief of St Cloud Fire Rescue in Central Florida. He began his career in the fire service in 1976 on Orlando Fire Department where he rose in rank to District Chief and served as a paramedic from 1978 until his retirement. During his tenure on Orlando Fire Department, he worked on his off-duty days as an Emergency Department Registered Nurse at several hospitals in Central Florida and taught Paramedic School at Valencia College. Currently, he is the owner of Emergency Management Systems, Inc. consulting company, serves as the Vice President of the Florida HazMat Symposium, and a member of the NFPA Hazardous Materials Technical Committee.

After retirement from Orlando FD in 1999, he served as the Corporate Director for Safety and Security for Orlando Regional Healthcare Hospitals, Executive Director of the Central Florida Fire Academy, and Regional Hazardous Materials Coordinator for the Urban Area Security Initiative for Region V.

As an author, Richard has authored a number of technical/text books related to hazardous materials response and terrorism preparedness. In addition, he has published a number of articles for both fire and nursing professional publications.

Richard holds an Associate in Arts, an Associate in Science in Nursing, a Bachelors of Arts in Business, and a Masters of Arts in Management. He is a Fire Instructor III and a Master Exercise Practitioner (MEP) through FEMA.

Acknowledgments

After a lifelong career in the fire service, it is difficult to thank every person who had a significant impact in my work life and the completion of this project.

First and foremost, to my writing partner and friend Armando "Toby" Bevelacqua. From the day we were put together on Orlando Rescue One, we immediately began collaborating on different projects. Those projects have continued for 40 or so years. Although technically my friend, Toby is more like my brother. We may disagree and even fight over our differing thoughts or styles but in the end, we come to neutral ground. I will always love this guy. Thank you for being a great friend and fellow collaborator.

Lieutenant Gary Bass, my first real lieutenant who showed me the importance of continuous training to be the best I could be. "Don't just practice until you get it right, practice until you can't get it wrong!"

District Chief Leo Wright who always stood up for the firefighter paramedics in the early days when they were viewed more like a red-headed step child instead of part of the team. He always saw the value in the concept of a heavy rescue that was specially equipped and staffed with well-trained firefighter/paramedics who were dedicated to all special operations within the department. Leo was never afraid to make a decision and his command presence was second to none.

Joe Landerville, a friend from a different department, who directed me to the first IAFF hazmat train the trainer program and started my career in hazardous materials training that has carried me across the country and beyond. The IAFF hazmat leadership recognized the value of the medical side of hazmat response and provided me the opportunity to both develop and teach hazmat programs across the county. It is also the only job that I was ever fired from, but that is another story. Politics makes strange bedfellow, and destroys them as well.

Doctor Robert Duplis who supported some crazy ideas and backed the hazmat medical project from the beginning. Although you are gone, you will never be forgotten.

Doctor Todd Husty who authored the foreword for this book. Dr. Husty has always kept the health and safety of his paramedics above all. He supports all aspects of this program and has been a friend for many years. As a medical director, there is no one finer!

Although I love the field of hazardous materials response, I have always focused on the medical aspects of hazardous materials. This is different than most of the other hazmat instructors teaching on a national level. Many of those who teach other hazmat subjects have embraced my focus and supported me along the way. They have always had the insight to recognize that a knowledgeable hazmat technician with medical/toxicological background is an asset to any team. Those friends and great hazmat instructors include (in alphabetical order):

Bob Bradley
Chris Hawley
Mike Hildebrand
Joe Leonard
Greg Noll
John O'Gorman
Bob Royal
Greg Socks
Jason Waterfield

And finally, to Rick Edinger who had the confidence to place me on the NFPA Hazardous Materials Committee so I can participate in the development of national hazardous materials standards and continue to participate in the advancement of this field on a national level.

Armando's Biographical Information

Armando Bevelacqua PM, BS, is a 37 plus year veteran of the fire service. He is the recipient of the 2010 "In the Zone Award," the "Level A Award" for leadership service and support in education of the hazardous materials first response community, and the "Dieter Heinz 2016 Instructor of the Year Award." He retired from the City of Orlando Fire Department, Florida, where he served as the Chief of Special Operations, Homeland Security, and

Emergency Medical Service Transport. Armando also teaches at local colleges, instructing Fire and EMS classes. He writes freelance, publishing articles and educational textbooks. He has published with topics on report writing for EMS responders and a chemistry book geared for the first responder. He has presented nationally on several issues in the disciplines of technical rescue, EMS, hazardous materials, and management. Armando lectures to fire departments throughout North America and Europe. He is an adjunct instructor through the Department of Defense as well as with several federal agencies involved with force protection.

Chief Bevelacqua serves on several federal, state, and local committees. He held membership to the Inter-Agency Board (IAB) for training and exercise development discussing issues affecting USAR and HazMat deployment, and training as it relates to terrorism. He serves as a member of the NFPA 470 (472, 1072, 473) and 475 Technical Committees and the ASTM standards development committee for emergency response. Chief Bevelacqua has assisted in the development of standards and protocols such as the Rocky Mountain Poison Control for the development of standardized medical protocol for the WMD event and for the State Department for WMD training embassy delegates.

His latest endeavor is to create educational videos and comics for the first response community. Educating new and seasoned responders to the ever-advancing technologies that are entering the first response arena.

Acknowledgments

It has been a long while since I have thought about the day that I stepped into my first emergency responder-related course. I truly did not understand the ramifications of what would become a lifelong career. Not sure what draws individuals into this journey but I am so glad to have been a part of the early days of EMS and my career as a firefighter and eventually as a chief officer.

We sometimes do not consider those that came before us, or stood shoulder to shoulder with us, how they have had an impact on our lives and career, and to that end is a short list of those that have had strong influences on my career. These are just a few because there have been so many: Dr Eugene Nagel who taught segments of my then EMT II class, Dr. Armando Garcia ACLS Instructor, and emergency room physician, Dr. Barth Green who taught us how to properly support a spinal injury, Captain Randy Boaz City of Miami Fire, Chief Bernie Tillson Boca Raton Fire, Chief Ed Hesse Orlando Fire, Terry McGowan, John "Sal" Saltalamacchia, and Oscar Grimes to name a few. But what these individuals did for me is they made me not only want to be a better medic and in some cases forced me to be a better responder as a whole.

I would also like to thank some of my HazMat brothers and sisters that over the years have made me think about responses in a different light. Many responders have influenced some of my work over the years or have contributed to the work contained herein such as Kristina "Krissy" Kreutzer PhD. Her work with Risk Assessment and how to look at hazards made it simple to "see" toxicity. She would often challenge me in the classroom, until I told her that she should be in the front of the classroom. A few years later she was and delivered her concepts of hazard assessment to the fire service. Thank you Krissy for your work and service. We lost Krissy a few years ago may you RIP my friend. I miss your response challenge questions.

Another hazmat friend was Ludwig Benner. Often, I would call or email Ludie, to give me his thoughts on a subject. Although he would always remind me that he was not a responder, he always had an interesting twist on a process, concept, or idea, many hours where spend talking about hazmat with him. During the writing of this book, I would discuss concepts and ideas about basic hazmat principles and concepts of response. Ludwig passed during the editing process of this manuscript; I miss our discussions tremendously my friend.

Thank you, Chris Hawley, for helping me with writer's block, our discussion in the hallway at the Florida conference was exceedingly valuable. Chief Bob Royall for his support on just about anything I come up with (just do not call him out when you are the key note – although the deer in the headlight look was priceless). Mike Hildebrand and Greg Noll for their inspiration to write and get involved. Thank you for the friendship. Christina Baxter for her work on PPE and contribution to this book in Chapter 7. I think she learns from us as much as we have from her. Mike Callan for his down to earth approach to hazmat response, his concept of safe, unsafe, and dangerous inspired the chemical time line, thanks Mike for all you do and all your help over the years! Robert Cruthis thank you sir for your additions to the Triage piece and your review of the SOP and of course Al Valerioti for his CAMEO contributions. Thank you sir.

Over the years countless responders have called, emailed, approached me at conferences and during class, asking questions. All these questions have been much appreciated, as many times my response to the responder was "I don't know but let me find out." To all of these people a sincere thank you as you also helped me learn and grow within this discipline. To all the responders that have actually used the information contained herein, thank you for learning a new approach toward prehospital care. Continue your learning, keep asking questions!

Of course, Rick Stilp my partner on Heavy Rescue 1 during this journey from the beginning to date and the coauthor of this book and many other works. Our conflicts and discourse on many topics along with the heated discussions have definitely made me consider other methods and approaches to different problems and ideas. Thank you for the lifelong friendship.

And lastly, my wife Mary Michelle she puts up with some of my most esoteric discussions (which has in some cases led to some very interesting projects), but most important her support on this project. The long nights of research while writing this edition, sometimes during a vacation which would occur because of something I would read or a dot I wanted to connect, thank you for allowing me to think through the process. Additionally, thanks for allowing me to verbalize my frustration at times, but most important allowing to do my work as a whole.

Additional Recognition

Jointly, the authors would like to thank those who have played a big part in this particular project:

Thanks to Chris Pfaff for supplying excellent hazardous materials related photographs for this text.

Thanks Chris Hawley, our good friend for his contribution for the cover photo and so much more over the years.

Butch Loudermilk who has always been a behind the scene supporter. He always has something great to say and has provided the authors with great review of the subjects in this book.

Robert Cruthis who has, on many occasions, taught with the authors and has continuously provided great insight and enthusiasm for this project and many more.

Don Guillette, the most talented paramedic instructors we have ever met. Don is known in Florida as the "Capnography King." He is a great cheerleader for this program and has attended more of the authors' classes than any other student over the years.

Foreword

I teach a lot and have done so for a long time. Emergencies, including prehospital emergency medicine, is my specialty. But sometimes what I teach is more basic.

I teach that a true professional knows what they know and knows what they do not know. A true professional embrace that concept without shame but actually, with pride. That is because a true professional may not know everything but we know where to find the answers.

And I can say with authority that I do not know enough about hazmat! I mean, I really do not know enough about hazmat!! But I know where to find out!!! Right here in this book!

But it is more than just a resource. This book should be a reminder that although, we who practice emergency medicine in a prehospital or in-hospital setting may not know everything there is to know about hazmat. We need to recognize and remember that our patients and the scenes we run on may have a toxic source.

I teach all levels of practitioners to the concept of a differential diagnosis. After the chief complaint/presentation, the history of the present problem or illness (OPQRST, etc.), and some observation and/or physical examination, we should have a list of likely suspected conditions that are causing our patient's difficulties. That is how we will ultimately arrive at a final diagnosis. And yes, paramedics make presumptive diagnoses or else you cannot use any medications in your drug box. The problem is, in the heat of battle or in the middle of the night, or with highly anxious and distraught family members around, we sometimes forget those things that should be on our list, on our differential.

An important basic medical concept is that if a patient is exhibiting signs and symptoms that are not consistent with diseases and conditions that we are familiar with, we should investigate what substance may be causing the problem. Frequently, these are side effects of medications. I have made a lot of house payments based on side effects of medications! But this is how we also spot that toxic problem. Someone who is turning blue and cannot breathe for apparently no reason is pretty suspicious.

But it is more than that. I am always impressed with how complicated it gets out there doing prehospital emergency medicine in uncontrolled environments. It can be dangerous. It worries me for everyone but it is even worse since my middle kid became one of my paramedic firefighters. When you have skills testing/scenarios, the first words out of your mouth is "scene safety." But are you really thinking that as you are absorbing the pesticide through your own skin or you are running into a subway filled with sarin? Do we really check our situational awareness at each scene and with each patient? Your world is dangerous enough. Please be careful and remember your hazmat lessons.

I am so sure that I do not know everything about hazmat, that one of my assistant medical directors is an occupational medicine and a hazmat specialist, Dr. Stan Haimes. As good as he is, he still does not know everything! That is why we have protocols that include calling the poison center and that is why we need Rick and Armando. Thank you for the resource that you have created but thank you even more for the reminder of the importance of hazardous materials in our profession.

Todd M. Husty, DO
Office of the Medical Director
City of Lake Mary, City of Longwood,
City of Maitland, Orlando/Sanford Airport,
City of Oviedo, City of Sanford,
County of Seminole,
Seminole County SWAT, City of Winter Springs SWAT

Preface

There is a never-ending argument concerning the provision of medical care. Is the practice of medicine an art or a science? Physicians have argued this for over a century. And of course, there is no real answer. Both can be debated. What is apparent is that the provision of medical care at all levels, from basic EMT to skilled surgeons requires the learning and understanding of anatomy and even more importantly, physiology. It takes both understanding the science to comprehend the intricate workings of the body systems combined with the art of intuition to know when a patient is in medical trouble. The field of Hazardous Materials Medicine is no different.

Unfortunately, there are numerous reports and case studies that have revealed that a healthcare provider failed to understand the seriousness of a chemical exposure only to see the patient's condition deteriorate. Victims have been misdiagnosed and not treated with the care necessary to make them well or to reduce the physiologic damage that will continue if the treatment is not effective. Many times, these exposures and the resulting injuries require the initial care of first responders who are dispatched to the scene of the incident. These field EMTs and Paramedics are generally very well trained in the recognition of medical, cardiac, and trauma-related issues. But their training and experience is lacking when it comes to chemical exposures. Then, to make matters worse, emergency department providers also may lack the knowledge and experience to quickly and efficiently treat the victims of chemical exposures.

This book is a culmination of over 40 years of effort by the authors. Upon development of fire department-based hazardous materials response teams in the 1980s, the authors realized the need for additional education and training in case one of the hazardous materials team members inadvertently got exposed during an emergency response. The effort started by developing training that was specific to the chemicals found within their city, especially in industry. This initial list of chemicals was somewhat short.

It did not take long to realize two important factors that would change the overall objective. First, the main interstate highway and a major railroad, both going east and west through their city carried almost every imaginable chemical used in the state and that interstate highway has a long history of being the most dangerous highway in the state. This broadened the list substantially. Second, and possibly more importantly, realizing that the majority of chemical exposures were not taking place at hazardous materials emergency scenes. Instead, they were taking place in homes and small businesses within the community. These exposures normally did not cause injuries to a group of people but, most of the time, it involved only one or two victims.

The goal of the initial training was to teach not only the recognition of various signs and symptoms associated with chemical exposures, but also to educate the medical responders about the physiology associated with the exposure. When specific treatments were taught, the students also had to understand the physiology of the therapy involved. With buy-in from the department and the medical director, a program was developed and presented. Word of this new program reached outside of the authors' department and soon surrounding departments were requesting the training. The program grew from there. To the author's knowledge, it was the first hazardous materials medical program offered in the United States and today, the longest standing program in the country. All other programs in the country were based on the work previously done by these authors.

An earlier book, Emergency Medical Response to Hazardous Materials Incidents, was published to place the hazardous materials medical program in a more formal package. Although this book was basic, it was well ahead of its time. Many paramedics stated that they had never heard of a program to teach prehospital responders how to treat chemical exposures. Unfortunately, many did not care. These exposures did not happen often and some paramedics were very happy only treating the symptoms that were initially noticed without understanding why they were occurring. But times have changed and now the thirst for knowledge has increased and the occurrences of chemical exposures have multiplied.

Although there have been other publications on this subject, none contain the depth and detail this text contains. It has always been the desire of the authors to present information that is complete and detailed but in a format that is understandable to the reader. This has always been a challenge but one that has been taken very seriously. This text represents a body of work that is relevant, understandable, important, and ready for those eager to expand their knowledge and skills in providing medical care to those exposed to hazardous agents.

1

HazMat Medicine and the HazMat Medic

Introduction

The late 1950s and into the early 1960s brought innovation to prehospital care as we know it today. These were the "wild west" days of a new concept called EMS (Emergency Medical Services). Many early pioneers of EMS had the vision to recognize that the lifesaving efforts typically done in the hospital setting could be accomplished in the field.

Because of the death of his daughter, Dr. Safar trained and organized one of the first prehospital EMS services in the country. While working in Pittsburg, he rigorously trained a group of ambulance attendants over nine months and in 1968, put these highly trained prehospital care providers to work. These ambulance attendants operated out of Pittsburgh's Presbyterian and Mercy Hospitals. They were instructed to manage airways, start IVs, deliver babies, and interpreted EKGs. They were known as the Freedom House Ambulance Crew. They were indeed the first civilian-trained paramedics in the country.

In 1974, the University of Pittsburg received a grant from the U.S. Department of Transportation to create a curriculum for nationwide emergency medical services. Dr. Safar and Dr. Nancy Caroline developed the standards for education, training, ambulance design, and specialized equipment.

A few innovations that occurred simultaneously contributed to a new standard for prehospital medical care. Dr. Karl Edmark perfected the heart defibrillator, giving rise to the Seattle Fire Department Medic One, a rescue program in 1970. At the same time, the revolutionary "LifePac" was being developed by Dr. Peter Safar from Pittsburg. Dr. Safar and Dr. James Elam developed the understanding and application of what was called the ABCs (Airway, Breathing, and Circulation) of CPR (Cardiopulmonary Resuscitation). All of these developments contributed to the foundations of EMS and the new emerging field of prehospital care.

Meanwhile in Florida, an electrical engineer-turned anesthesiologist, Dr. Eugene Nagel was looking at the fire department as a community resource to provide initial prehospital care. He convinced the Miami Fire Department, because of their strategic locations within the community, to have specially trained firefighters with these new lifesaving techniques and equipment to respond to medical emergencies. Radio communication equipment was provided so that the field providers could talk directly to a physician in the hospital and receive critical guidance at a medical emergency. In 1964, Dr. Nagel became the medical control officer (Medical Director) for the City of Miami Fire Department.

Hazardous Materials Medicine: Treating the Chemically Injured Patient, First Edition.
Richard Stilp and Armando Bevelacqua.
© 2023 John Wiley & Sons, Inc. Published 2023 by John Wiley & Sons, Inc.

Other leaders in this field were Walt Stoy, PhD, for his contribution of education materials and systems design, and Fire Chief Jim Page, a Battalion Chief with the Los Angeles County Fire Department and founder of JEMS (Journal of Emergency Medical Services) Magazine. Chief Page paved the way for development of EMS as a discipline. He also served as a technical consultant to the TV show "Emergency." All had a hand in developing field personnel providing advanced prehospital care that later held the title of "paramedic."

In the 1970s, the television show named "Emergency" started to play on Saturday evenings. The show featured emergency calls of all types but focused on emergency medical calls. The two lead actors portrayed the paramedic/firefighters Jonny Gage and Roy Desoto. These two treated everything from trauma patients involved in a car accident and rattle snake bites to cardiac arrest. This show was a great propelling force to bring advanced prehospital life support to every community in the United States. The viewing audience witnessed what could be accomplished by well-trained and skilled paramedics. The public's expectation that ambulance attendants could save lives before the patient reached the hospital increased across the nation and raised the bar for prehospital medical care.

Both of the authors of this book became paramedics in the 1970s. In the early days of EMS, many agencies struggled to get enough paramedics trained to provide these advanced skills. Training standards were not consistent across the nation and largely depended on what the local physicians wanted their prehospital providers to know and what skills to perform. A paramedic in the City of Chicago was different than a paramedic trained in Miami. In fact, in some areas of the country, paramedics were called EMT IIs and not paramedics at all.

As this medical specialty grew in popularity, new radio systems were being introduced that allowed prehospital providers the opportunity to talk to an emergency room physician and send an EKG tracing over the radio waves. Cardiac defibrillators became more portable. Suddenly, the role of the advanced life support provider became more technical and required more knowledge to provide the best prehospital care possible.

At almost the same time, the field of hazardous materials response began to take root. There were more and more industrial accidents involving chemicals occurring both in facilities and along transportation routes. Highway crashes and train derailments often made the news because of the fires, explosions, injuries, and deaths that they caused related to the hazardous cargo.

Just as fire departments welcomed the new field of emergency medicine, they also embraced the field of hazardous materials emergency response. It would seem logical that there would be a blending between the two specialties that had found their way into the fire service. But that has not been true in the past. Seldom was there a focus of hazardous materials exposures being a subject of emergency medical training.

The paramedic curriculum has always included exposure to carbon monoxide and sometimes even went as far as teaching the pathophysiology of pesticide poisoning, but that was about it. Those EMS agencies, whether they are fire department based or not, found themselves responding to industrial, farm, highway, and railroad incidents where victims were contaminated and sick from chemical exposures. Many times, these victims were being swept up from their location and transported to the hospital without the thought of decontamination, causing injury to both the prehospital care providers and the hospital staff.

But worse than that was the injury suffered by the victim. Even today, there existed a great lack of knowledge of chemical exposure injury by both the prehospital providers and the emergency department staff. This has led to victims suffering and even dying because of a lack of appropriate care. None of these providers received adequate training to treat a victim suffering from acute chemical exposure and the resulting injury. Instead of embracing this field, it was pushed aside because it was either too technical, required too much training, or just did not happen enough to be concerned about.

The regional poison control centers across the nation were put in place to assist with these situations. They are fairly quick to respond and will send critical information to the hospitals to assist in the care of these patients. Many times, the centers are contacted by the emergency physicians, but unfortunately, too often this important step is missed by the physician, and needed information is not obtained.

Rapidly identifying the injury and providing appropriate treatment for the chemical injury will reduce the recovery time and will save lives but is often an afterthought. Although the regional poison control centers provide factual and detailed information, any expected delays in the information will ultimately delay the treatment and may lead to a poor outcome or long recovery time. Basic knowledge of chemical exposures and an understanding of how physiology may be affected is critical to providing quick and efficient treatment.

The premise for this book and the subject of chemical exposure and injury has been a career-long venture for the authors. The field of hazmat medicine has been their passion for many years. Their focus has always been to train both prehospital personnel and hospital care providers to a level where they can confidently assess and treat the victims of chemical exposure. The authors have learned from real-life experience, reviewing case studies, and conducting research to train health care providers to a level of care expected by their customers (see Figure 1.1).

Case Study – Sarin Attack in the Tokyo Subway

On March 20, 1995, the religious-based terrorist organization call the Aum Shinrikyo, under the direction of their leader Shoko Asahara, conducted an attack in the political district of Tokyo. Plastic bags filled with Sarin (a viscous liquid that vaporizes) were used to deploy the nerve agent poison that affected 15 different subway stations and trains. At the time, this was the largest terrorist attack ever carried out. In total, over 6000 victims were affected and 19 died.

Shortly after the Sarin was released into the subway cars, emergency calls began to come into the call center. In total, 1364 EMTs and 131 ambulances responded to 15 different

Figure 1.1 All emergency responders/receivers should understand the consequences of chemical exposure and injury. Rapid diagnosis and treatment is necessary for positive outcome and rapid recovery.

subway stations. At the time, the EMTs working in Tokyo were advanced practice EMTs and were trained in advanced airways and IV therapy but were required to receive direct orders from a physician to do either. Because of the ongoing disaster, physicians were not available to provide those orders and as a result, only one patient received an advanced airway and IV. No other advanced life support was conducted outside of the hospital.

Although some of the victims died or had more serious signs and symptoms that required care in the hospital, most of the 6000 only complained of eye pain, dim vision, and headache related to miosis caused from low-level exposure. Those treated in the hospital were related to breathing, somatic, and cardiac issues.

The hazardous materials response team was managed within the Tokyo Metropolitan Fire Department (TMFD). The team had advanced capabilities including infrared gas analyzer and gas chromatograph-mass spectrometer (GCMS). The police department has similar equipment but only had select individuals trained to use the equipment and did not maintain a hazardous materials response team.

The first call for assistance came into the dispatch center at about 8 a.m. At about 10 a.m., the TMFD identified the chemical offender as acetonitrile (methyl cyanide). This analysis was incorrect. At about 11 a.m., the police department correctly identified the material as sarin using their own GCMS.

The after-action reports completed well after the incident found several major points for concern. First, there was a lack of training and equipment available for a chemical response including the absence of prehospital decontamination which led to the secondary contamination and injury to both emergency responders and hospital staff. 135 EMT alone developed acute symptoms and received medical treatment at the hospital. Second, the EMTs were hindered in the care of the victims because, although they were trained and equipped, the EMTs were unable to provide advanced care unless given direct permission by a physician but all physicians were overwhelmed in the hospital and unable to provide the required permission. Finally, it was determined that triage in chemical disasters, including nerve agents, is of limited use since delayed symptomatology that can be consequential and normal data points were developed for trauma patients.

When this incident took place in 1995, the hazmat medical program provided in this book was well underway. This incident provided the authors with addition motivation to prepare for all kinds of chemical events. In fact, the authors of this book took many lessons learned from terrorist activities around the world and authored the book "Terrorism Handbook for Operational Responders." This became the first Terrorism-based book for emergency responders in the United States and was published well before the terrorist attack on September 11, 2001 and the Anthrax incidents that followed. Now, almost 30 years later, this program addresses all of the shortcomings that were found in Tokyo and other significant chemical/biological incidents. Since that time, there have been many chemical incidents that have caused extensive injuries and deaths. All have provided evaluation points to improve the information found in this text. It has always been the author's goal to prepare and teach hazardous materials medical information so the effects from chemical incidents of all types are reduced.

History

In the 1980s, when we first envisioned this field of study and began developing and presenting coursework to our fellow paramedics, we found that some criticized us, believing that treating chemically injured patients was more of a Ghostbusters operation and there are not enough of these patients to truly make a difference. Others recognized the need for the train-ing, embraced the additional knowledge and skills, and adopted much of what they learned into their responses.

With greater training and knowledge of chemical incidents, prehospital responders began recognizing that there were many more chemically injured patients than they had previously thought. Many of the incidents were reported to dispatch centers as a general illness or injury not as a report of chemical exposure.

The authors found that many incidents that were dispatched as normal EMS responses were really related to chemical exposures. One example involved two different cases involv-ing elderly women who lost consciousness during a hair-dying procedure. In both of these cases, a hair dye was placed on the victim's head, and after the dying procedure was complete and the victim's hair was being washed in a salon sink, they lost consciousness. These inci-dents occurred only three days apart involving the same beautician. When the second inci-dent took place, an investigation revealed that the hair dye, made outside of the United States. contained a high nitrite concentration. The nitrite concentration combined with the elderly patients with poor circulation contributed to the loss of blood pressure and loss of consciousness. After years of presenting these cases to students, many have reported back that they have experienced similar emergency calls involving hair dye.

Another incident that has been misdiagnosed for years, especially in Florida, is the emer-gency call for assistance to a patient with heat stroke. Typically, when emergency services arrive, they find a patient unconscious after working outside on the lawn. The patient's skin is not significantly hot, and their blood pressure is abnormally low. In some of these cases, there is evidence of fertilizer stuck on the patient's legs from the knees to the ankles, which is often ignored. The EMS responder starts treatment for a heat-related injury not under-standing that the loss of consciousness is not related to heat exposure.

Fertilizers are rated according to its content and contain three numbers. The first number is the percentage of Nitrogen, the second is the percentage of Phosphorus, and the third the percentage of Potassium. Those fertilizers with a high first number are especially toxic. For example, 20-3-3 contains 20% nitrogen. The nitrogen is often sodium nitrite or similar chemical.

In many of these cases where homeowner have spread high nitrogen fertilizer without protecting their sweaty skin, the exposure causes a complete loss of vascular tone resulting in a loss of blood pressure and consciousness. This is misdiagnosed as a heat-related injury instead of a chemical exposure leading to mistreatment. The authors have presented these cases during classes over the years. Again, many students have returned to us stating that this scenario has happened to them as well, but because of their recent hazmat education, they have been able to diagnose and treat appropriately.

The decision to take on the task of developing a hazardous materials toxicology-based program led us to have many spirited discussions with our medical director, personal conver-sations with other health care professionals. and evaluating our own experiences. Using our teaching and emergency response experiences, we created a program that would benefit our department, our team, and ultimately brought the approval of our medical director. We knew that we were venturing into an area of study that had not been attempted before in the prehospital community. Little did we know that it would eventually become a national pro-gram of study.

We did not enter into this endeavor without appropriate backgrounds. By the late 1980s, both of us already had more than a decade of experience serving as field paramedics.

Both instructed paramedic school at the local college and had taught advanced classes in ACLS, pharmacology, pathophysiology, trauma life support, and were serving as hazmat technicians on our department's hazmat team. With the blessing of both our department and our medical director Dr. Robert Duplis, we started down this path of developing the first hazmat Advanced Life Support program and course in the nation.

Over the years of actual scene response, teaching hazmat medicine, writing, and researching, we have had the opportunity to generate many operational procedures. Throughout those years, numerous departments and agencies have requested our policies, procedures, and our medical SOPs to use in their agencies. Even State agencies have incorporated our policies and procedures into their medical systems. In one case, we were requested to participate in the development of a national evidence-based protocol for response to victims of weapons of mass destruction and become reviewers of national curricula (Figure 1.2).

That leads into the present day and this text. Although there have been several other books published in this area of study, including an earlier book published by the authors of this book, this one is the most complete text available. This text represents lessons learned, years of research, review of many case studies, and years of teaching and guiding other agencies on the development of hazardous materials medical programs. We have always believed that if you want to truly understand and function at a high level then, you must teach. We have learned a lot over the past 40 years of response and instruction. A good deal of this learning, from our perspective, came from our students and their experiences. Through them, we have learned and matured in our thoughts and response concepts.

We understand and teach what we know works, both in the field and in the emergency department. From the beginning, we have focused on what can be accomplished in the field. We begin with a quality assessment to form a differential diagnosis, then follow that with appropriate treatments based on the assessment findings and, when possible, determination of the offending chemical.

Through the response and review of many emergency incidents and exercise design and evaluations, we identified both the extent and limitations of field practice. Of course, the emergency departments will always have the ability to take treatment further than a field paramedic can. For example, many toxidromes could be addressed in this text, but the application of treatment of these toxidromes is beyond the capability of field advanced life support trained personnel and many hospital emergency departments. Some require extensive

Figure 1.2 The hazardous materials medical technician program emblem has changed through the years. These emblems represent the program from 1990 to present.

laboratory testing and long-term care or chelation therapies. These do not have a place in the field or the emergency department.

There exists a fallacy that training for chemically exposed patients should only be provided to those paramedics that support the hazardous materials teams. This is misleading as most chemical exposures take place with individuals not associated with a hazardous materials incident. There also exists the thought that emergency department medical providers do not need the training because resources are at their fingertips to guide them in the treatment process. After several experiences which resulted in unnecessary death, we have found that this is also not true. Both prehospital care providers and hospital emergency department medical providers need training and education on both assessment and treatment for the chemically injured patient.

For example, one of our case studies involves a 12-year-old girl who was exposed to 70% hydrofluoric acid. The father had purchased the acid for his pressure cleaning company because of its ability to remove rust stains from the stucco on buildings. His daughter knocked over a 1 gal jug wetting the grass with the acid. The 12-year-old then slipped in the wet grass, exposing both legs, buttocks, and back. The injured girl was taken to the hospital, where the father carried her into the hospital in his arms, handed her to a nurse, who subsequently handed her to a physician who placed the girl on the bed. By this time, the young 12-year-old was unconscious.

Complaints of burning arms by the father, nurse, and physician soon followed. No one in the Emergency Department understood the effects of hydrogen fluoride. They only understood that it burned their skin and eyes. Unfortunately, a bolus or two of calcium gluconate or calcium chloride may have saved the girl's life. A lack of awareness and training caused the misconception that hydrofluoric acid was similar to hydrochloric acid. Both cause injuries at the site of contact, but one greatly differs from the other in systemic toxicity. This demonstrates the importance of training in chemical exposures at all levels of the emergency medical chain of response.

Events

Although true hazardous material incidents do not occur every day, what is seen frequently is the personal misuse of chemicals both in the household and in the workplace, causing exposures. These are typically not classified as hazardous materials incidents. Most of the patient exposures that are treated by emergency responders do not come from a dispatched hazardous materials incident. Instead, the emergency response was dispatched as some kind of general medical call, and it was not until the EMS providers arrived and found that an exposure to a chemical was responsible for the resulting symptoms.

Unfortunately, most emergency responders do not have appropriate training to deal with these patients. Not only do these patients often present a secondary hazard to the health care provider, but their recovery is dependent on how they are treated immediately after the exposure. Many times, the lack of understanding and the continued tissue damage related to the exposure causes patients to decompensate while in route or while awaiting lab results in the hospital emergency room.

This text was developed to provide both the knowledge and skill necessary to assess, diagnose, and treat an acutely chemically injured patient. In addition, the materials presented in this book will provide background, physiology, and logic to the assessment and treatment process.

In some incidents, it will be easy to determine the chemical and route of exposure to cause the injury. A chlorine leak at a water treatment facility is one of those examples. These are not difficult to determine, and responders are rarely led in the wrong direction. But, when an employee of a chrome plating company is found unconscious in the shop, and there is no clear cause for the incident, a detailed assessment must be completed to determine if the patient's illness is due to exposure or a medical cause. Today, emergency responders and emergency department nurses, and physicians have equipment immediately available that can help with the determination of the cause. And if not, through a detailed focused assessment, the care provider can be led into a differential diagnosis strategy.

Situational Assessment Continuum

Through many years of evaluating case studies and participating in emergency responses related to chemical exposures, we have developed a Situational Assessment Continuum (SAC) model to follow in both assessing the patient and the scene. This model has three components, (as seen in Figure 1.3), that can be evaluated on an emergency scene. Contained in each of these components is critical information that will assist the responder in developing a diagnosis and treatment plan. The three components are:

Patient presentation (Exposure and toxidromes)
Event conditions (Scene evaluation)
Scene assessment (Hazard identification)

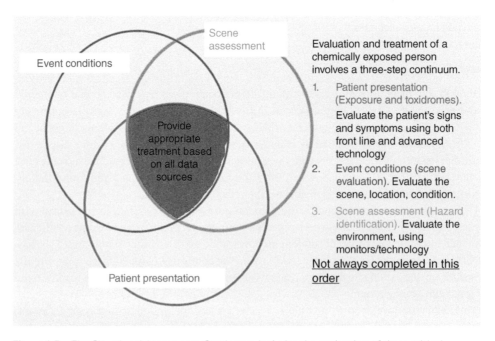

Figure 1.3 The Situational Assessment Continuum includes the evaluation of three critical components after a chemical exposure.

Every scene and situation is different, and it is realized that not all three components are always available. Take for example, the patient who walks into an emergency room or fire station and states that he/she has been exposed to some chemical and is now having difficulty breathing. In this case, there are no components of the environment or scene to evaluate to assist in determining what kind of chemical may be involved. But can a health care provider still provide care? Absolutely, but now they are treating symptoms, evaluating the pathophysiology, and providing treatment that is guided by the known physical changes that are taking place.

The process involves three different areas of assessment to develop a full picture of the patient's state of exposure. There is no set order of this process, so information can be gathered as the assessment in each area is completed. The more information that is gathered, the more confidence a medical provider will have when treating the patient. When detailed information is received, the medical provider must be able to quickly narrow down the differential diagnosis and, in many cases, determine the exact chemical and to what degree the patient is exposed/injured. Described below are the three areas of assessment.

Patient Presentation

Although these components can be completed in any order or concurrently, the patient assessment is usually done first. The only other consideration that needs to take priority over the patient assessment would be to ensure scene safety. On a hazardous materials scene, this is an extremely important step. In a later chapter, the hazmat patient assessment, or as it is referred to in this text, the "toxidrome exam," (page 15) will be covered in detail. A good patient evaluation is paramount to providing appropriate and efficient treatment.

Event Conditions (Scene Evaluation and Size-up)

Evaluating the event involves the initial size-up of the scene. It includes any information that is gathered on dispatch, en route, and what you see when you arrive. This may include information provided by bystanders and what is witnessed upon arrivals such as smoke, chemical clouds, containers present, and type of industry. In hazmat team terms, this is referred to as "recognition and identification." In addition to the typical hazmat size-up are those external factors such as weather, personal accounts, and the activity occurring at the time of the incident. Gaining information from just the event will assist the responder in some initial thoughts on the types of chemicals that may be involved.

Scene Assessment (Hazard Identification)

This assessment is a detailed evaluation of the location of the initial injury. This includes the use of air monitoring equipment to determine the level of concentration that is present. Identifying the state of matter (i.e. solids, liquids, vapors, and gases), which gives further clues on the potential route of entry. This assessment may include air monitoring, collecting samples for analysis, or more detailed studies, depending on the circumstances. However, observation of the scene with the associated symptomology and event location all play a role in understanding the toxidrome.

Summary

Hazardous materials emergency medicine has improved over the last four decades. Advanced equipment that can be used to assist in the assessment of chemically injured patients that was previously available only in hospital ICUs are now found in the prehospital setting and used by both emergency responders and emergency department personnel. But, to use these tools to the fullest capability takes training, understanding of the diagnostic equipment, and an in-depth understanding of the physiology of the poisoning.

In addition to the improvements of equipment, there have also been great improvements in the educational base to become a paramedic. The paramedic textbooks of the 1970s included just a few hundred pages of text. This is pale in comparison to today's paramedic text of 2500 pages. Just a few years ago, nurses who were working in emergency rooms were only required to obtain a diploma or associate's degree are now required to be certified emergency nurses, and many are BS degree nurses or hold higher degrees. Emergency department physicians are now required to complete a residency in emergency medicine and are board-certified emergency physicians. The educational and skill requirements have been greatly improved since we started this endeavor and will continue to improve in future years. Many of those who, in the past, thought that hazardous materials medicine was an exercise in futility have now seen the advantage of this area of study in both paramedicine and emergency department care.

The true power of this text is that an instructor with a reasonably good EMS background and with a bit of hazardous materials response background combined with the information available in this text, can present the coursework in this area of study to prehospital and hospital emergency care providers.

We envision that the subject of hazmat medicine will become a permanent part of the national paramedic curriculum and certainly be a subject for emergency medicine in medical school. This is the goal that we have strived for, over so many years of teaching, researching, and delivering this content.

2

Exposures

Introduction

So often emergency response calls come in as medical emergencies only for the responder to arrive and find that the medical symptoms are caused from a chemical exposure. Many of the incidents that the authors have had experience with have been dispatched as typical medical calls such as severe eye pain only to find chemical paint remover spilled in the eyes. An emergency response to a "heatstroke" finds a patient unconscious in the front yard from fertilizer exposure. Shortness of breath response becomes a chloramine inhalation injury because the homeowner mixed bleach with ammonia. It is easy to justify toxicologic training for every emergency response paramedic and hospital emergency department staff.

Unfortunately, those seeking information concerning suicides easily find guidance on the internet on sites specifically meant to provide detailed instruction. These sites guide suicide seekers in the how-to for the use of cyanide, hydrogen sulfide, carbon monoxide, and sodium nitrates to complete their self-destruction. Even premade exit bags for simple asphyxiation are sold on these sites. Not to mention the self-treatments found at health food stores that contain cyanide derivatives to self-treat cancer.

In the past, nitrates were added to cologne in an effort to get a physiologic response when applied. Just imagine going on a date with someone you are infatuated with and then applying the new cologne that was just purchased. The nitrates cause vasodilation and a decrease in blood pressure. This leaves the wearer of the cologne with lightheadedness and a giddy feeling. The intent by the cologne manufacturer is to generate the feeling of love. Of course, this is only temporary chemical love. And a victim with a nitrate sensitivity would most certainly wreck his car on the way to picking up his date.

It is easy to see why emergency health-care providers must have an understanding of toxicology and the physiology that is altered by chemical exposures. Without this knowledge, a reasonable differential diagnosis cannot be reached and appropriate medical care cannot be provided.

Case Study – Derailment in South Carolina, a No-Notice Evacuation Event

On January 6, 2005, at 02:40 a.m., a Norfolk Southern freight train derailed and crashed in Graniteville, South Carolina. The train was hauling several rail cars filled with liquefied chlorine gas. At least one car was severely damaged leaking the toxic chlorine gas into the

Hazardous Materials Medicine: Treating the Chemically Injured Patient, First Edition.
Richard Stilp and Armando Bevelacqua.
© 2023 John Wiley & Sons, Inc. Published 2023 by John Wiley & Sons, Inc.

environment. The gas, which was heavier than air, created a toxic cloud that covered a largely populated area including a textile mill with 500 night-shift workers.

The regional poison control center was contacted by a person living near the train derailment who was complaining of a chemical odor and burning eyes. The poison control center contacted the local emergency department and requested to talk to the only on-duty emergency physician. The emergency physician was already overwhelmed with one critical patient, six patients with pulmonary edema, and over 100 awaiting treatments in the waiting room.

The on-scene safety officer advised the emergency department that the chemical involved was sodium nitrate. The poison control center was notified and researched the health effects from sodium nitrate and advised the hospital that the symptoms being seen at the hospital did not match sodium nitrate. Just 15 minutes later, the chemical was reported to be methanol.

An hour after the incident, the chemical was finally correctly identified as chlorine. The poison control center provided information for the correct treatment of chlorine exposure. The chemical leak resulted in the death of nine victims, 529 sought immediate care, 18 were treated by their personal physicians, and 5400 residents were evacuated.

This incident clearly identifies the need to recognize the signs and symptoms of specific exposures and provide the critical care necessary to treat the symptoms. A missed differential diagnosis may guide the health-care provider to inappropriate care. In this case, treating sodium nitrate would involve stabilizing blood pressure and treating methemoglobinemia instead of treating bronchiole constriction and chemically induced pulmonary edema. This miss-guidance would lead to patients becoming more ill and possibly dying from a treatable exposure.

Patient Presentation

In the effort to effectively assess the patient's condition, there must be a basic level of understanding of the anatomy and physiology of human systems and the ability to efficiently evaluate the patient's condition based on your physical findings.

When exposure to harmful chemicals takes place, it gains an opportunity to injure the body. This effort to injure the body takes place through one or more different courses, which are referred to as *routes of exposure* into the body. Toxicology books list many routes of entry, including intravenous, inhalation, intraperitoneal, intramuscular injection, subcutaneous injection, oral, and cutaneous exposure. This book only uses examples or routes commonly found in hazmat emergencies. Each section deals with the anatomy pertaining to the exposure and explains in detail the physiology involved in that exposure.

The general routes of entry into the body experienced by those exposed to hazardous materials (chemical exposure) are inhalation, absorption (through the skin and eyes), ingestion, and, to a much lesser degree, injection (as demonstrated in Figure 2.1).

Biohazard exposure through injection accounts for more reported injuries than any other route. Some would state that regardless of how the body contacts a chemical, all routes of entry into the body involve absorption. Some authorities list routes of entry as "routes of absorption." In the basic sense of the phrase, this statement is true. But, for reasons of simplicity, absorption is addressed as exposure to the skin and eyes.

Note: In this text, we will use the EMS designation of routes of exposure as Absorption, Inhalation, Ingestion, and Injection.

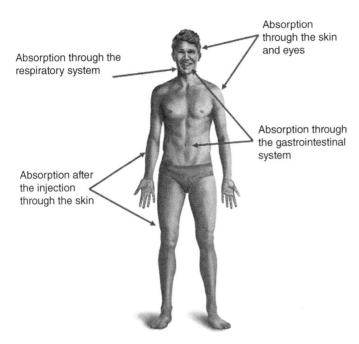

Absorption
through the skin
and eyes

Absorption through the
respiratory system

Absorption through
the gastrointestinal
system

Absorption after
the injection
through the skin

Figure 2.1 The routes of exposure (entry) include inhalation, ingestion, injection, and absorption through the skin and eyes.

Once a toxin is exposed to tissue, it may cause damage to that tissue. Other times the toxin may utilize that tissue as a means of gaining access into the body, where it seeks out an organ system completely dissociated from the point of entry. The organ system affected by exposure is then referred to as the *target organ*. Those chemicals that specifically target the nervous system are called neurotoxins. For example, organophosphates target their effects primarily on the parasympathetic nervous system. Solvents, another neurotoxin, target the central nervous system (CNS). Almost all organ systems can be affected by some type of toxin. The most common target organs are the nervous system, the cardiovascular system, the liver, and kidneys.

The effects from exposure may be felt almost immediately, causing acute signs and symptoms. Probably more often, the effects are noticed days, weeks, or years later, lasting for long periods. These chronic symptoms appear because of damage caused to an organ system by the invading toxin. Some types of cancer can be related to toxic exposure many years earlier. Asbestosis, a form of chronic lung disease and its associated cancer, mesothelioma, is linked to the inhalation of asbestos, often 10–20 years earlier.

The exposure to a hot environment and extremely cold materials will also be addressed in this text. Although these are not exposures typically talked about in a hazardous materials book, they are very important subjects and are explored in detail here in this text. A paramedic whose responsibilities include the overall well-being of a hazardous materials team needs to understand these conditions and the potential physiological impacts. A rapid assessment and appropriate treatment are critical for a positive outcome for these emergencies.

This chapter overviews the key points of anatomy and physiology as it relates to chemical and environmental exposure. The chapter will begin with a review of respiratory anatomy and physiology and progress through cardiovascular, neurologic, skin and eyes, and finally gastrointestinal system. The chapter is not intended to be a full lesson in anatomy and physiology but instead a review evaluating the important points that need to be understood to provide a good patient evaluation after chemical exposure.

The Toxidrome Exam

The toxidrome exam is used whenever a chemical exposure is suspected. It is not typically used for medical or trauma unless there is an associated exposure. The signs and symptoms after an exposure can be subtle or delayed and is intended to identify specific clues that occur after an exposure (see Figure 2.2).

- Overall Assessment
  - Vital signs including temperature, Masimo™/Rainbow to assess carboxyhemoglobin and methemoglobin levels.
- Eyes
  - Pupil size, reaction to light, iris detail, and opaqueness (haziness). Epithelial exam for sloughing, sclera blood circulation.
- Mucous Membrane
  - Color and moisture
- Respiratory Status
  - Rate, lung sounds, capnography, oximetry
- Cardiac Status
  - Heart rate, EKG to evaluate critical rhythms and QRS (sodium channel blockers) such as ST (hypocalcemia) and PR interval widening (slowed repolarization), ectopy.
- Skin
  - Color, moisture, temperature, burns, irritation
- Musculoskeletal
  - Tremors and/or rigidity, hyperreflexia, fasciculation
- Mental Status
  - Confusion, obtunded, aggressive, alert
  - General
  - Odor Present

Respiratory System

Overview

Respiration is defined as the inhalation of fresh air that is moved through air passages beginning at the nose and mouth and terminating in the alveoli. This fresh air fills the alveoli, and through diffusion, the oxygen is absorbed into the bloodstream, where it is carried to the cells to be used in the process of energy production (cellular metabolism). Once oxygen goes through cellular metabolism, carbon dioxide is produced as one of the waste products. Carbon dioxide is released into the bloodstream and is eventually diffused back into the alveoli and exhaled through the nose and mouth.

This definition gives a good overview of respiration, but the anatomy and physiology involve so much more detail. The medical provider must understand what is involved in the process of respiration to accurately assess and treat respiratory hazmat exposure.

Toxidrome exam	Asphyxiants Early	Asphyxiants Late	Irritants Early	Irritants Late	Hydrocarbons & Halogenated HC Early	Hydrocarbons & Halogenated HC Late	Corrosives	Cholinergics
Eyes - Pupil Cornea								
Dilated (mydriasis)							O	O
Constricted (miosis)								O
Corneal - Cloudy			O				O	
Opaque				O			O	
Lacrimation			O	O	O		O	O
Photophobia				O	O			
Blurred Vision					O			
Reddness/irritation	O	O	O		O		O	
Mucous Membranes								
Ashen	O				O			
Cyanotic		O				O		
Dry			O	O	O			
Moist								O
Respiratory Status								
Labored Breathing		O	O		O		O	
Tachypnea	O				O		O	
Bradypnea		O						O
Bronchorrhea							O	O
Bronchospasm							O	O
Lung sounds								
Wheezing/Stridor		O		O		O	O	O
Rales/Ronchi/PE		O		O		O	O	O
SaO$_2$ (below 90%)	O							
Cardiac Status								
Tachycardia	O		O		O			O
Bradycardia		O		O				O
Hypotension		O						
Hypertension	O				O			O
EKG interval widening		O					O	
Chest Pain		O				O		
Palpitations	O				O			
Skin Color and Temperature								
Reddening			O	O				
Ashen/Cyanotic		O			O			O
Cold/Clammy					O			O
Diaphoresis					O			O
Intense skin pain			O				O	
Gastrointestinal								
Diarrhea								O
Urination								O
Emesis	O	O	O	O	O			O
Salivation								O
Abdominal pain					O			O
Musculoskeletal								
Tremors/Rigidity								
Fasciculations/Seizures		O						O
Mental Status								
Confusion/Weakness					O			O
Altered mental status		O				O		O
Dizziness	O				O			
Headache	O				O			
Anxiety	O	O			O	O	O	
General-Odor present	O	O	O	O	O	O		O

Figure 2.2 The toxidrome exam is a detailed physical assessment concentrating on areas of the body commonly affected by a chemical exposure.

An evaluation and review will start at the top of the respiratory tree and work through the complete process of cellular respiration. Respiration is divided into external respiration and internal respiration. External respiration is the function that moves air containing oxygen from the outside environment to the alveoli. Internal respiration is the movement of oxygen from the alveolar tissue into the bloodstream, where it is transported to the cells. Once oxygen enters the cells, it is used in cellular respiration to develop energy for body function.

One of the waste products of cellular respiration is carbon dioxide. Upon production, carbon dioxide is moved into the bloodstream through several means and carried back to the lungs, where it is exhausted during exhalation.

Respiratory System Anatomy and Physiology

External Respiratory System

Air Entry and Filtration   Air is drawn into the body by decreasing the pressure in the thorax. This is accomplished through complex actions that involve tightening of the diaphragm and muscles of the abdomen while the intercostal muscles also constrict to bring the rib cage upward and outward. These actions cause an enlargement of the chest cavity and lungs that result in negative pressure in the chest (intrathoracic) cavity that causes air to flow in until the pressure is equalized with the outside atmosphere. The filtration process begins with the nose hairs and continues into the fine bronchioles (see Figure 2.3).

Nose Hairs   As the air enters the nose and mouth, a filtration process is underway to clean the air of impurities that may harm the alveoli. This filtration process starts with the nose. The mucous coated nose hairs provide some initial filtering by capturing larger particles that may be found in the air. Although not extremely efficient, nose hairs can filter particles as small as 7–10 microns in size.

Turbinates   The second step in the filtration process is the boney structures located in the back of the nasal cavity called turbinates. Turbinates are made up of three sets of bony

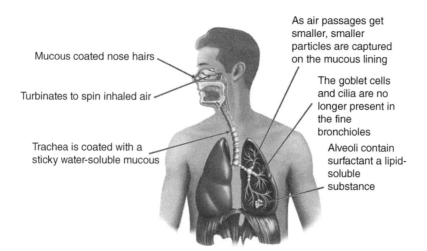

Mucous coated nose hairs

Turbinates to spin inhaled air

Trachea is coated with a
sticky water-soluble mucous

As air passages get smaller, smaller particles are captured on the mucous lining

The goblet cells and cilia are no longer present in the fine bronchioles

Alveoli contain surfactant a lipid-soluble substance

Figure 2.3   Movement of air from the nose and mouth to the alveoli. Various processes are in place to remove particles before reaching the alveoli.

protrusions on each side of the nasal passageway. These bony structures are surrounded by a richly vascular mucous membrane covering. Their function is to spin the air as it enters the trachea. This spinning effect causes a centrifugal force in the trachea spinning larger solid impurities against the outside wall of the trachea and lower airways.

The turbinates are very sensitive to irritation. Common allergies that cause nasal congestion are the result of swelling of the turbinates. A chemical irritant exposure to the turbinates causes swelling and nasal congestion. Severe nasal congestion will cause a person to start to mouth breath. In the case of mouth breathing, both the filtration accomplished by the nasal hairs and the spinning action of the air caused by the turbinates are lost.

Cilia and Mucous The next phase of filtration takes place in the airways. From the trachea to the bronchioles, the airways are lined with hair-like projections called cilia. Scattered throughout the cilia are goblet cells that secrete mucus. This mucus coats the airway walls and serves to trap this foreign material allowing cilia to sweep it to the top of the trachea, where it is swallowed down the esophagus or spit out.

Mucus found in the airways is a water-based material. This is important to know as the mucus' goal is to capture solid foreign materials, but it also combines/mixes with water-soluble gases and vapors. When water-soluble irritating or corrosive gases/vapors are introduced into the respiratory system, the result is the toxifying of the mucous that eventually will injure the underlying tissue. Figure 2.4 identifies each of these filtration processes as air makes its way to the alveoli.

Alveoli The terminal ends of the lungs contain the alveoli. Alveoli are tiny sacs surrounded by tissue that is richly circulated with capillary beds. Each alveolus is encapsulated with alveolar tissues that are only one to two cell walls deep. This tender tissue is intended to allow the oxygen that reaches the alveoli to easily enter the bloodstream through a process of diffusion (as demonstrated in Figure 2.5). Oxygen is also pushed into the bloodstream

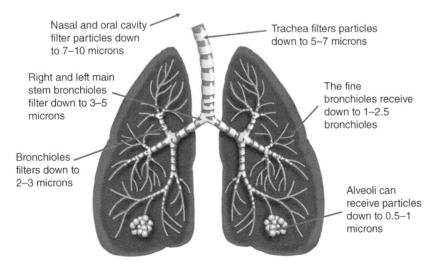

Figure 2.4 Filtration of air from entry to the alveoli. Water-soluble mucous coats the airway from the nasal cavity to the fine bronchioles allowing the absorption of water-soluble chemicals.

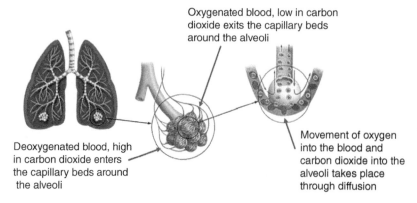

Oxygenated blood, low in carbon dioxide exits the capillary beds around the alveoli

Deoxygenated blood, high in carbon dioxide enters the capillary beds around the alveoli

Movement of oxygen into the blood and carbon dioxide into the alveoli takes place through diffusion

Figure 2.5   The movement of oxygen and carbon dioxide into and out of the blood take place at the alveolar capillary membrane.

under pressure. If it was possible to measure or count the number of alveoli, it is said that there are over 300,000,000 alveoli with more than $1000 \, ft^2$ of surface area.

Through normal respiration, oxygen reaches the alveoli at a concentration of 21% in the air. In a normal healthy person breathing at a rate of 16–24 times a minute, this amount allows the hemoglobin in the blood to saturate to the level of 96–99% as measured with a pulse oximeter. Increasing oxygen concentration in the alveoli will allow the hemoglobin to become fully saturated and even increase the concentration of oxygen circulating in the blood serum. By providing positive pressure ventilation (PPV) and maintaining that pressure through the use of CPAP with increased oxygen concentration, the blood can transport much more oxygen both combined in the hemoglobin and the circulating serum.

Surfactant   The alveoli contain a chemical called surfactant. Surfactant is a complex compound containing six lipids and four proteins and is produced in the lungs at the point of use. Non-water-soluble (lipid-soluble) gases/vapors are repelled by the water-soluble mucous lining the airways and are eventually inhaled into the alveoli. Surfactant is lipid-soluble and will readily mix with any lipid-soluble irritants, corrosive gases, or vapors that reach the alveoli.

Note: The lipid base surfactant found in the alveoli plays an important role of reducing the surface tension of the alveoli. This causes these small air sacs to be less able to collapse (more stable) when the pressure in the lungs is equal to the atmospheric pressure outside of the body. This occurs during every breath at the end of the respiratory cycle.

Alveoli compromised by a toxin can lead to several different issues. Injured surfactant (mixed with a chemical or diluted by infusing serum) will not provide decreased surface tension and will allow the alveoli to collapse when the alveolar pressure is equalized with the atmospheric pressure at the end of the respiratory cycle. Just like a fresh balloon, it takes a much higher pressure to reinflate the alveoli once it is collapsed. The collapsed alveoli lead to a condition called *atelectasis*. Atelectasis reduces the surface area of the lung tissue and decreases gas transport in the lungs.

Lung Pressures and Respiration   When evaluating the pressures in the lungs during respiration, it is important to understand when pressures are higher than, lower than, or equal to the outside atmosphere. Upon inhalation, the diaphragm tightens, the thoracic cavity size increases, and inhalation takes place. The intrathoracic pressure is lower than the outside atmosphere, and air rushes in. During this time, the alveoli are being stretched open by the decreased pressure in the thoracic cavity.

Once inhalation is complete, the diaphragm and intercostal muscles relax, and exhalation begins. During this activity, the pressure in the thorax and lungs increases and forces air out of the inflated alveoli and into the atmosphere. This process is continued until the pressure in the lungs is equal to the pressure on the outside of the body. When the pressure is equalized both inside and outside of the body, this is called the "end of the respiratory cycle, or end tidal."

Internal Respiratory System

Oxygen and Hemoglobin   Once in the bloodstream, oxygen is primarily transported on the hemoglobin of the red blood cells. There are 250 hemoglobin molecules per red blood cell; each can carry one molecule of oxygen. There are approximately five to six million red blood cells per cc of blood. Therefore, each cc of blood has the capability of carrying more than one billion molecules of oxygen.

To be more specific, oxygen (O^{-2}) is bound with the ferrous iron (Fe^{+2}) found in the hemoglobin of the blood. This bonded hemoglobin is called oxyhemoglobin. Although the majority of oxygen is carried by this process in the bloodstream, oxygen is also carried to a lesser degree in the blood serum (see Figure 2.6).

Other gases being carried in the bloodstream are carried in a solution of the blood serum as well. One of the waste products from cellular metabolism is carbon dioxide. There are three ways that carbon dioxide is transported back to the lungs/alveoli. The modes of transportation in the blood are:

1) Dissolved as a gas in the circulating blood
2) As bicarbonate in the blood
3) Bound to hemoglobin as carbaminohemoglobin

Although oxygen transportation is a simple process from the alveoli to the cells, carbon dioxide transportation is more complex.

Another example of gases being transported in the solution of the blood is nitrogen. Nitrogen makes up about 80% of the air we breathe; therefore, it is reasonable to expect that nitrogen crosses the alveolar–capillary membrane and enters the circulating blood volume. Most SCUBA divers are familiar with this because of the danger of breathing compressed air under the pressure of water as they make their dives. When a diver surfaces too fast from the

Figure 2.6   Oxygen is primarily carried on the red blood cell. Some oxygen is carried in the solution of the blood not attached to the red blood cell.

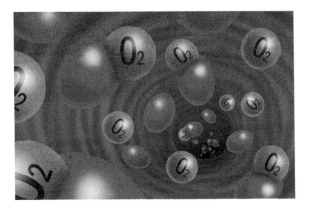

depth, they suffer from a medical condition called the "bends." The bends is a condition that takes place when a diver rises to the surface too quickly causing the formation of bubbles of nitrogen in the blood. The nitrogen bubbles obstruct blood circulation in the joints and sometimes the heart and brain.

Breathing Rate   The breathing rate is determined by four factors. These include voluntary, hypercapnic drive (elevated levels of CO_2), hypoxic drive (decreased levels of oxygen), and pH drive (decreased levels of pH or increased acidosis).

Voluntary   The conscious control of breathing is under the direction of an area of the brain known as the cerebral cortex. The cerebral cortex controls all voluntary muscle movement. Any chemical exposure that can depress the central cortex (in the CNS) will interfere with the person's ability to control respiration. CNS depressants such as hydrocarbons and opiates both can slow respirations to dangerous rates.

Hypercapnic Drive   The primary process that drives the involuntary breathing rate is carbon dioxide. As a person's metabolic rate increases, so does the production of carbon dioxide. This is detected by the rhythmicity center of the brainstem, and the respiratory rate is increased. This allows the lungs to release more carbon dioxide during exhalation.

Hypoxic Drive   The second involuntary influence on breathing rate comes from the concentration of oxygen in the blood. Normal blood oxygen level is between 80 and 100 mm/Hg (blood gas reading). Once the oxygen level drops below 60 mm/Hg, the peripheral chemoreceptors located at the bifurcations of the aortic arteries and the aortic arch stimulate an increase in depth and rate of respiration to increase the blood oxygen levels.

pH Drive   When pH changes in the blood, sensors located in the aortic and carotid bodies detect the changes, both the lungs and the kidneys have a part in controlling blood pH. An abnormally low blood pH (acidosis) causes an increased respiratory rate. Rapid breathing releases excessive carbon dioxide and thus reduces carbolic acid and increases pH, which in turn slows breathing.

Cellular Respiration   Cellular respiration is the process of taking fuel (sugar is the most prevalent) combined with oxygen to produce energy. Although the process is somewhat complex, this concept is the key to understanding the function of cellular respiration. Some physiology books classify cellular respiration as a combustion reaction that provides energy and produces heat, water, carbon dioxide, and a small amount of carbon monoxide.

This concept should sound familiar as your gasoline-powered vehicle uses a similar process. Your vehicle uses gasoline mixed with oxygen and applies it to a combustion process where energy is produced. As by-products, heat, water, carbon dioxide, and carbon monoxide are formed.

To be more specific, cellular respiration is a set of metabolic reactions that take place in the cells of the human body. These reactions convert biochemical energy from nutrients when oxidized with oxygen into adenosine triphosphate (ATP). The nutrients include sugar (carbohydrates), amino acids (fats), and fatty acids (proteins).

In total, three reactions take place in cellular respiration. These reactions consist of Glycolysis, Krebs Cycle (also called the Citric Acid Cycle), and Oxidative phosphorylation. Oxygen is used in each of the processes.

Glycolysis takes place in the cytoplasm of the cell. This process uses two ATPs at the beginning of the process and produces four ATPs during the second phase of the process rendering a net gain of two ATPs.

The next step is the Krebs Cycle which takes place in the mitochondria. The Krebs Cycle produces carbon dioxide as a waste product. The fact that there is a predictable amount of carbon dioxide produced during cellular respiration, it can be measured at exhalation, and the findings are used to determine the efficiency of cellular metabolism. The Krebs Cycle produces two ATPs.

The final step of cellular respiration is oxidative phosphorylation (also called the electron transport chain) which takes place on the membrane of the mitochondria. Oxidative phosphorylation produces the most energy, with the production of up to 34 ATPs. During this process, an enzyme, *cytochrome c oxidase,* is responsible for transferring O_2 in the electron transport chain. This process yields water (H_2O) as a by-product. This process is used to release energy in the form of ATPs and is by far the most effective producer of energy in the cellular respiration cycle.

It is important to understand that some chemicals can interfere with this process of producing ATP during aerobic cellular respiration. Cyanide, sulfide, azide, and carbon monoxide can bond with cytochrome c oxidase, effectively stopping the process of aerobic cellular respiration. These chemicals are called phosphorylation uncouplers as they bind to cytochrome c oxidase and inhibit the enzyme from functioning. When cytochrome c oxidase is not functioning, electron transport fails to progress, and the result is cellular asphyxiation.

A simplified look at the chemical reaction that takes place during cellular respiration is $C_6H_{12}O_6 + 6O_2 \rightarrow 6CO_2 + 6H_2O + \text{heat}$. In theory, this process can produce up to 38 ATPs.

These processes represent cellular metabolism. Cellular metabolism produces energy and as a by-product also produces heat, water, carbon dioxide, and small amounts of carbon monoxide. Interestingly, there is some comparison between cellular metabolism (recently called cellular combustion) and internal combustion found in an automotive engine. An automobile uses gasoline and oxygen to combust creating energy, heat, water, carbon dioxide, and carbon monoxide. A cell uses sugar and oxygen to produce energy, heat, water, carbon dioxide, and carbon monoxide. It is this similarity that has produced the use of the term, cellular combustion, to describe cellular metabolism (see Figure 2.7).

Figure 2.7   Respiration is required to produce cellular metabolism/combustion. Oxygen and fuel stimulate metabolism and produce energy, heat, carbon dioxide, water, and small amounts of carbon monoxide.

Chemical and Physical Form of Respiratory Exposure

Injury to the respiratory system results from the ability of a chemical to gain access into the respiratory system. The offending chemical must be in a form that allows access into the respiratory system. Gases, vapors, solid particles, and liquid aerosols are all easily inhaled.

Gases, Vapors, and Fumes

Gas is a chemical with a boiling point below normal room temperature. A vapor, which is similar to a gas or fume, is the molecular dispersion of a solid or liquid chemical within the air. Toxic gases and vapors readily enter the respiratory system because they easily mix with the atmospheric air and, depending on the concentration and physical properties, can injure all levels of the respiratory system.

Solid Particles

Solid particles are particles of a chemical floating within the surrounding air. When entering the respiratory system, the larger particles are filtered by the hairs in the nostrils. Others are trapped in the upper respiratory system and are rapidly swept out by the cilia. Very small particles (smaller than a micron) suspended in the air may have the ability to fall deeper in the smaller passageways and even in the alveoli. If the particle is larger than one micron but smaller than three microns, it can reach the fine bronchioles but not penetrate the alveoli. Some may be blown out during the exhalation phase. The particles that may become mixed with secretions of the passageways can break down into solutions and be absorbed through the lining and into the bloodstream. Others may become trapped forever, leading to pathologies years later. Figure 2.8 demonstrates particle sizes as compared to a strand of hair or one piece of beach sand.

Pneumoconiosis is a disease of the lung that develops due to the inhalation of dust particles. Asbestosis and silicosis are two examples of pneumoconiosis occupational diseases

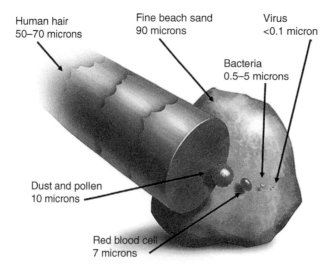

Human hair
50–70 microns

Fine beach sand
90 microns

Virus
<0.1 micron

Bacteria
0.5–5 microns

Dust and pollen
10 microns

Red blood cell
7 microns

Figure 2.8 Comparison of particle size.

caused by inhalation of asbestos or silica. Another example is coal dust (essentially very fine carbon particles) causes black lung in miners and similar diseases in firefighters who do not protect their respiratory system fighting fires. This condition is seen more often in wildland firefighters where efficient respiratory protection is unavailable.

Aerosols/Mists

Aerosols and mists are fine droplets of a liquid suspended within the surrounding air. Depending on their solubility in water, they are capable of injuring all levels of the respiratory system.

Concentration and Duration

The amount of air drawn into the lungs with each breath combined with the concentration of the chemical within the air both determine the dose of the toxin received. A person breathing rapidly and deeply during exercise receives a higher dose of a chemical than a sedentary person breathing lightly, regardless of the concentration in the air. An example of this principle is a child breathing toxic air. The child receives a much higher dose per pound of bodyweight because children breathe much faster than an adult in a given period of time.

Miners used this principle when they took canaries into the mines with them. The thought was that when the canary loses consciousness and hits the bottom of the cage, suffering the effects from toxic gas, the miners still had time to escape. Not only do canaries breathe much faster than humans, but their respiratory system is much more efficient, absorbing toxic gases much faster than the miners would.

Note: Doctor John Scott Haldane (1860–1939) developed the process of the "miner's canary." Canaries were carried into the mine and would fall ill long before the toxic gases affected the humans, allowing for a safe escape. Dr. Haldane will also be discussed later in the text because of his research into carbon monoxide poisoning and hyperbaric oxygen therapy.

With brief encounters at lower concentrations, a chemical may be trapped and evacuated by the respiratory system's natural functions, not causing any injury at all. At higher concentrations, the protective functions of the respiratory system may be overwhelmed, and certain injuries would occur. Low concentrations over a long period create a similarly devastating injury as seen in Figure 2.9. One reason this occurs is that eventually, the natural protective features of the respiratory system are overwhelmed, allowing the toxin access to deeper areas of the system.

The relationship between the concentration and duration during a toxic exposure is referred to as the dose–time relationship. To express the dose–time relationship in other terms, an exposure to a high concentration for a short period of time may be similar to that of a low concentration over a long period of time.

Types of Injuries Resulting from Chemical Exposure

Chemically Induced Bronchiole Constriction

It is commonly misstated that irritating chemicals produce bronchiole spasms. This, in most cases, is not a true statement. A bronchiole spasm is the constriction of the smooth muscles that surround the bronchioles. This occurs when there is an allergic or sensitivity reaction and not necessarily when an irritating chemical is inhaled. Instead, what takes place when

The dose time relationship

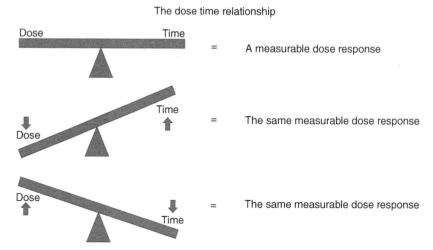

Dose   Time   =   A measurable dose response

Dose↓   Time↑   =   The same measurable dose response

Dose↑   Time↓   =   The same measurable dose response

Figure 2.9 A high dose of a chemical over a short amount of time can cause the same result as a low dose of a chemical over a long period of time.

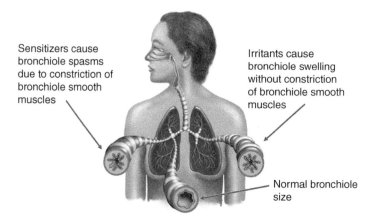

Sensitizers cause bronchiole spasms due to constriction of bronchiole smooth muscles

Irritants cause bronchiole swelling without constriction of bronchiole smooth muscles

Normal bronchiole size

Figure 2.10 Inhalation of a chemical irritant causes swelling of the bronchiole tissue while a sensitizer causes bronchiole spasms due to smooth muscular constriction.

an irritating chemical enters the respiratory tree is swelling and inflammation. The differences between these two processes are demonstrated in Figure 2.10.

Upon auscultation, both generate the sound of wheezing because both swelling and spasms cause constriction of the air passageways. When a chemical exposure takes place, and the result is shortness of breath and wheezing, it is impossible for an emergency medical responder to determine whether swelling, spasm, or both are involved. This is an important concept to recognize, so the emergency medical provider must treat both. By giving both a smooth muscle relaxer in combination with an anti-inflammatory drug, both conditions will be addressed.

Atelectasis and Disruption of Surfactant

To treat patients suffering from injured surfactant and atelectasis, the patient care provider must provide a couple of treatment procedures that will both reinflate the alveoli and keep them inflated. First, PPV is needed to reinflate the alveoli. In contrast to the regular patient-provided ventilations, the pressure is positive in the alveoli during both inhalation and exhalation. This may be enough pressure to inflate the alveoli. It is difficult to start inflation, but once the process opens the alveoli, it is much easier to keep it open.

The next process that needs to follow PPV is the use of Positive End Expiratory Pressure (PEEP). Reinflating collapsed alveoli is like blowing up a balloon for the first time. This can be accomplished using a PEEP valve attached to a PPV device such as a bag valve mask (BVM) or ventilator. Using these devices allows a care provider to provide Continuous Positive Airway Pressure (CPAP). Both can be provided and maintained using a ventilator with a manual PEEP setting. Utilizing a setting of 10–12 mm/Hg will maintain open alveoli.

Caution must be exercised as increased expiratory pressures can weaken blebs in the lungs and cause spontaneous pneumothorax, but this only occurs if there is a pre-existing weakness in the lungs. In addition, using CPAP also increases intrathoracic pressure and will increase the oxygen demands of the heart muscle and increase blood pressure. The care provider must be aware of all of these conditions when providing CPAP to patients.

Chemically Induced Pulmonary Edema (Noncardiogenic Pulmonary Edema)

The other condition that exists when irritating or corrosive chemicals reach the alveoli is a condition called Chemically Induced Pulmonary Edema. When the offending chemical reaches the alveoli, the cell walls become damaged and begin to leak blood serum into the alveoli. As the alveoli begin to fill with fluid, less and less oxygen is exchanged with the blood, and hypoxia ensues as the condition continues.

The lung sounds will mock that of cardiogenic pulmonary edema. Typically, rales will be heard on auscultation, indicating fluid invasion into the alveoli. Wheezing may also be heard due to the narrowing of the airways from the edema caused by the irritant passing through and injuring the upper airway tissue.

Unlike cardiogenic pulmonary edema, which is caused by increased pulmonary pressure, the normal treatments utilizing diuretics or blood pooling drugs do not sufficiently treat the condition. In cardiogenic pulmonary edema, treatment includes several pharmaceuticals that reduce pulmonary pressure. These include Morphine, Nitroglycerine, and Lasix. Both Morphine and Nitroglycerine pool blood in the capillaries, which effectively reduces pulmonary volume and pulmonary pressure. The use of Lasix causes diuresis and reduces overall blood volume, and eventually reduces pulmonary pressure. In all of these cases, reduced pulmonary pressure reduces the influx of serum into the alveoli. But, because chemically induced pulmonary edema is not caused by increased pulmonary pressure, these drugs are of limited value in correcting the condition.

It is recognized that providing treatment that reduces pulmonary pressure may be of help with chemically induced pulmonary edema simply by decreasing the pulmonary pressure. But, the leaking of serum into the alveoli is related to damage to the alveolar tissue; reducing pulmonary pressure will not correct the problem. Instead, the most efficient means of treating chemically induced pulmonary edema is through the use of increased alveolar pressure. This is accomplished by using CPAP. CPAP is the infusion of air or oxygen under pressure (using positive pressure) then allowing exhalation under normal means, but instead of allowing full exhalation, the use of a PEEP setting will maintain a pressure in the alveoli even at the end of the respiratory cycle when there is typically an equalization of inter thoracic pressure and atmospheric pressure. Utilizing CPAP does not force the additional fluid back into the bloodstream as commonly thought. Instead, what it does provide is pressure in the alveoli to reduce or stop the influx of more serum from the bloodstream. The fluid that has already made its way into the lungs is rapidly absorbed into the interstitial tissue. Pulmonary edema tends to disrupt the surfactant as well. The use of CPAP also ensures that the alveoli remain open even with the loss of surfactant.

After the exposure to a respiratory irritant with evidence of pulmonary edema (rales, decreased oximetry readings, and initial decreased CO_2 on exhalation), both pharmaceuticals and CPAP should be considered to maintain respiration and oxygenation.

Chemical Sensitivity

When reviewing chemical effects on the respiratory system, the sensitivity reaction cannot be ignored. Sensitivity to chemical exposure is a serious occurrence and one that can cause quick and deadly consequences. Sensitivity reactions can affect an individual with very low chemical doses (exposures) in the presence of others who are left virtually unaffected.

Sensitivity to a chemical is, in effect, an allergy or allergic reaction to a chemical or group of chemicals. It can take place after multiple exposures that do not create any dramatic effect but suddenly, the effects change and become much more serious. The one that comes to mind and seems to create issues in the healthcare environment is formaldehyde. Those healthcare laboratory workers who are exposed to low levels of formaldehyde over weeks, months, and years suddenly develop an intolerance to even very low exposures to the chemical. The intolerance may begin with some shortness of breath and minor wheezing but can rapidly progress to respiratory depression associated with bronchial occlusion, loss of vascular tone, and shock.

Types of Chemicals that Injure the Respiratory System

The two most common types of respiratory exposures are those involving asphyxiants and irritants. Other chemicals enter the body through the respiratory system but affect other organ systems. There may be little or no effect on the respiratory system because it is not the target organ system for that chemical.

Asphyxiants

Asphyxiants can be classified in two ways: those that displace oxygen (simple asphyxiants) and those that interfere with the transportation of oxygen within the blood or the uptake of oxygen by the cells (chemical asphyxiants).

Simple Asphyxiants

Simple asphyxiants cause a decrease in the amount of oxygen entering the external respiratory system. These asphyxiants displace oxygen, resulting in lower concentrations reaching the alveoli and ultimately supplying the cells. The end consequence is cellular hypoxia, acidosis, and eventually death. Examples of oxygen displacing asphyxiants are carbon dioxide, nitrogen, and halogenated extinguishing systems.

Chemical Asphyxiants

Some chemical asphyxiants enter through the respiratory system, while others can be absorbed through the skin or ingested. Once in the body, the chemical interferes with the transportation or usage of oxygen by the cells. For example, carbon monoxide, one type of chemical asphyxiant, combines with hemoglobin, disallowing it from carrying oxygen. This, in essence, starves the cells of needed oxygen, slowing cellular respiration and energy production. Therefore, carbon monoxide is an example of a chemical that interferes with the transportation of oxygen to the cells. Cyanide, on the other hand, combines with an enzyme found inside of the cells. This enzyme is necessary for the utilization of oxygen in the cells. In this case, there is plenty of oxygen circulating in the bloodstream, but the cell is unable to utilize it, and therefore, cellular respiration slows and becomes very inefficient. Cyanide is classified as a chemical asphyxiant because it interferes with the usage of oxygen by the cells.

If the patient is conscious and is breathing adequately, the treatment of exposure to asphyxiants is to always give 100% oxygen by non-rebreather (NRB) mask. If breathing is inadequate or the patient is unconscious, treatment should consist of intubation and PPV, utilizing a PEEP setting. Other, more specific treatments based on the offending chemical will be discussed in future chapters.

Irritants (Corrosives)

Irritants can cause several and sometimes severe symptoms involving the external respiratory system. Bronchial constriction due to tissue irritation, laryngospasms, tissue sloughing, bronchospasms related to sensitivity, and chemically induced pulmonary edema are the most common. Irritants can injure all levels of the external respiratory system. Chemicals that are water-soluble such as chlorine and ammonia, unless in very high concentrations, affect the upper areas of the respiratory system. Non-water-soluble chemical irritants gain access deeper into the respiratory system, causing damage to the fine bronchioles and alveoli, leading to chemically induced pulmonary edema.

Respiratory System Injury Recognition (Assessment) and Diagnostics

Today, the paramedic has several tools that can be used in assessing patients. Vital signs consisting of a blood pressure, pulse rate/rhythm, respiration rate/rhythm/lung sounds, and temperature are always the basic diagnostic tools. But there are now diagnostic tools using technology that may be critical for assessing chemically exposed patients. These diagnostic tools can also provide critical information after chemical exposure. Among those tools are pulse oximetry, capnography, and Rainbow, Masimo technology. Each will be discussed in detail and determine how the findings provided by this technology can assist in developing a differential diagnosis and eventually guide the emergency care provider to deliver the correct and appropriate treatment.

Pulse Oximetry

Oximetry reached the emergency response community in the 1990s. It is a very good diagnostic tool for determining the respiratory status of a patient injured with an irritant and asphyxiant. Before the pulse oximetry unit became portable and usable in the field, it was used in the hospital setting. In the 1980s, it became a tool used in the operating room to determine the oxygenation of patients under general anesthesia. It later moved into the emergency rooms, where it was used to assess the oxygenation of patients entering the hospital in acute respiratory compromise.

Pulse oximetry has been hailed as the fifth vital sign following blood pressure, temperature, pulse, and respiration. Although it is a very useful tool, it cannot be used properly without knowledge of what the readings signify and how inaccurate readings can mislead a diagnosis.

The pulse oximeter indicates, in a percentage rating, the amount of oxygen-bound hemoglobin delivered to the peripheral tissues by the circulating blood. The oximetry reading may indicate changes in circulating oxygen much before there is any recognized change in blood pressure, pulse, or level of consciousness. In this respect, the medical provider may be forewarned of impending changes in the patient's condition before any changes in the traditional vital signs.

The pulse oximeter unit works by determining the ratio between oxygen-rich hemoglobin (oxyhemoglobin) and oxygen-poor hemoglobin (reduced hemoglobin). This is done by evaluating a pulsating arterial capillary bed in an appendage of the body. The most common evaluation sites used are the finger, toe, bridge of the nose, or ear lobe.

The unit utilizes a probe that contains two sensors and a light source. The two-sensor unit is the most common type found for prehospital use. Other more sophisticated units have multiple sensors, but these are usually too sensitive for field use. Each sensor detects a slightly different color of light as it transilluminates the selected body part.

Oxyhemoglobin has a brighter and more transparent color than reduced hemoglobin which, in comparison, has a darker opaque color. Once sensed, the oximeter calculates a ratio based on brightness and transparency and displays a read-out that represents the percentage of hemoglobin saturated with oxygen. The probe is designed to detect only a pulsating capillary bed. The pulsation is what identifies the oxygenated arterial side of the capillary bed and filters out the desaturated vascular side.

Normal arterial oxygen saturation, measured by an oximetry unit, is between 94% and 99% (SaO_2 94–99%). Therefore, when and SaO_2 reading drops below 94%, the provider should suspect respiratory compromise. A reading of less than 91% SaO_2 oxygen should alert the care provider to consider the therapy of intubation and supportive respirations. The reading of 100% is also an irregular finding if the patient is not on oxygen supportive therapy. A patient complaining of shortness of breath and displaying an oximetry reading of 100% may be diagnostic. This will be explained in greater detail in the following chapters.

Although the pulse oximeter is recognized as a very useful tool, it has its shortcomings. It will only work on a patient with a well-circulating (pulsating) capillary bed. If the patient is cold or in shock, these area capillary beds may have a considerable decrease in circulation, making oximetry impossible. Warming the site will re-establish circulation in a cold patient and enable the oximetry to read the pulsations.

Specific Oximetry Considerations When Assessing HazMat Exposures

Carboxyhemoglobin There are changes in hemoglobin that can affect the ratio measured by the oximeter unit. For example, carboxyhemoglobin (HgCO), the combination of carbon monoxide and hemoglobin, is seen by the oximeter unit as oxyhemoglobin (HgO_2). HgCO and HgO_2 have similar properties; both are very transparent and bright red in color. Thus, a person acutely exposed to increased amounts of carbon monoxide will fool the oximetry unit and may register an unusually high SaO_2 reading, when in reality, the patient may be severely hypoxic. Capnography can be used to determine if the high oxygen saturation reading is inaccurate.

Methemoglobin (MetHg)   is a chemically changed hemoglobin molecule that is incapable of carrying oxygen. Methemoglobin is formed when a patient is exposed to organic nitrogen substances like fertilizer, nitrates, and nitrites like nitroglycerine or aniline dyes. Methemoglobin is chocolate brown in color and obscures the light as it passes through the capillary beds. In a normally circulated capillary bed, the oximetry reads opaque (poorly transparent) blood as deoxygenated blood, reflecting an inaccurately low reading. The pulse oximetry reading from a patient poisoned with a nitrogen-based compound will be very low due to the chocolate color and may not reflect the true oxygen saturation present. Depending on the concentration of the methemoglobin in the blood, the oximeter may not register at all. Therefore, the oximeter reading should not be relied on in patients suspected of this type of poisoning. Again, capnography may become the true indicator of a patient aerobic metabolism.

Keep in mind that a cardiac patient who has been suffering from chest pain for many hours and reports multiple uses of nitroglycerine might have a low pulse oximetry reading. Nitroglycerine can generate methemoglobin when used in excess. An emergency medical provider not understanding this concept may continue to medicate the patient by administering additional nitroglycerine and 100% oxygen and not see a significant change in either chest pain or pulse oximetry reading.

Cyanides and Sulfides   When a patient has been exposed to a cyanide-based compound or hydrogen sulfide, cellular respiration is slowed. These chemicals can stop anaerobic metabolism and push the cell into anaerobic metabolism. When the cells stop removing oxygen from the circulating blood, the results are full saturation. This is reflected as a 100% SaO_2 reading on the pulse oximeter. In this case, the reading is accurate as the blood is fully saturated, but the cells cannot utilize the circulating oxygen. Capnography reading should be disproportionally low as a result of the cell's lack of aerobic metabolism and, therefore, the lack of production of carbon dioxide.

Capnography

Capnography is somewhat new to emergency response but has also proven to be a useful tool. Capnography readings are a direct result of the body's ability to perform aerobic metabolism (see Figure 2.11). High and low patterns can help to determine what chemical is injuring the patient and how serious the exposure might be. In addition, it also can detect metabolic changes early in an event, even before symptoms appear. By evaluating the capnography tracing (capnogram), an emergency care provider can detect the early onset of bronchial constriction before the patient feels short of breath or suffers from labored breathing and is even noted before the pulse oximetry is affected. The change in carbon dioxide production by the cells is read on the capnometer almost immediately where changes in oximetry are somewhat delayed.

Capnography measures the amount of carbon dioxide (CO_2) in exhaled air. There are two components to capnography. The first is the capnometer that provides a numeric measurement of CO_2 production. The second is the capnogram which is the waveform that is drawn during capnography. Both the capnometer and the capnogram will be discussed in greater detail below.

Figure 2.11 "CO$_2$ is the smoke from the flames of metabolism." Ray Fowler, MD, Dallas, Texas.

Note: The capnogram is the waveform generated measuring the level of CO$_2$, while the plethysmograph or sometimes called the "pleth" is the waveform generated from pulse oximetry.

The capnometer measures the end-tidal CO$_2$ (ETCO$_2$ or PetCO$_2$), which is the level of carbon dioxide released at the end of expiration. Since CO$_2$ is produced during cellular metabolism, the reading of end-tidal CO$_2$ is a direct reflection of the efficiency of aerobic cellular metabolism. This number is reported in partial pressure as millimeters of mercury or mm/Hg. Normal (breathing and cellular metabolism at a normal healthy rate) end-tidal CO$_2$ is between 35 and 45 mm/Hg.

Capnography also captures this cycle of CO$_2$ production during the breathing process on a capnogram graph. The capnogram displays a continuous tracing of the level of CO$_2$ as the patient goes through the respiratory cycle. The end-tidal CO$_2$ is captured at the end of expiration which is displayed on the capnogram at the high point of the graph, as seen in Figure 2.12.

If a patient is hyperventilating with an increased respiratory rate, it stands to reason that the concentration of CO$_2$ in each exhalation would go down as the CO$_2$ is diluted with the increased volume of air. On the other hand, if a patient is hypoventilating, the rate of respiration would go down, and the concentration of CO$_2$ would go up. Generally, if a patient breathes faster (tachypnea), their pulse oximetry reading (saturation of oxygen) will also increase. If a patient's breathing slows down (bradypnea), the saturation of oxygen reading will also go down.

Understanding this concept, if the capnography reading goes down, but the oxygen saturation reading goes up, the emergency care provider must consider one of the chemical

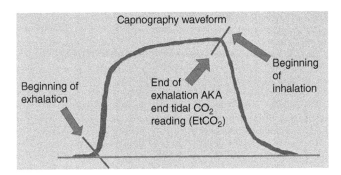

Figure 2.12 The waveform (capnogram) indicates the level of CO_2 found in exhaled air.

asphyxiants as the offending chemical. Carbon monoxide is a great example. The oxygen saturation on the pulse oximeter will read abnormally high, but the capnography reading may be well below 45 mm/Hg. Cyanide and hydrogen sulfide can generate the same type of readings (high SaO_2 and low CO_2). See the Diagnostic Chart in the references.

There are other diagnostic benefits to capnography in the guidance of determining correct treatment for a patient exposed to a hazardous material. For example, when the waveform begins to develop into a shark fin pattern (Figure 2.13), this is an early indication that the bronchioles are being obstructed and causing some lengthening of the expiratory process. This obstruction can be caused by bronchospasms, bronchial swelling, or increased mucous production. The shark fin pattern may begin developing well before the patient complains of shortness of breath and even before audible wheezing is noted. This change in pattern, especially postexposure to an irritant, will give the patient care provider an early warning about a serious condition that is developing and allow time to start aggressive treatment.

Another example of using patient assessment findings is to determine the possible diagnosis of either pulmonary edema or bronchial constriction. In the case of pulmonary edema, whether it is caused by congestive heart failure or chemically induced, the care provider may see an early decrease in CO_2 production (below 35 mm/Hg) well before a decrease in SaO_2 is

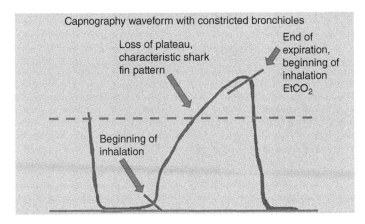

Figure 2.13 The shark fin pattern on the capnogram indicates constriction of the bronchioles from either swelling or bronchospasms.

noted. Eventually, an increase in respiratory rate may be seen related to the bloodstream holding more CO_2.

Lung sounds that are clear, and the respiratory rate is either normal or tachypneic, but the CO_2 levels are above 45 mm/Hg is an indication of increased metabolic rate and increase production of CO_2. This sign may provide a differential diagnosis of either a CNS stimulant exposure or a heat-related injury and possibly moving toward heatstroke.

Finally, one diagnostic use in the southern states involves the response to patients who are suffering from a loss of consciousness while conducting lawn maintenance in the summer. Most of the time these patients are thought to have a heat-related injury but are later found to have high levels of methemoglobin related to fertilizer exposure. Although the initial symptoms of both heat injury and exposure to nitrates are very similar, the use of capnography quickly determines which it might be. A patient suffering from heat-related injury will display a high $ETCO_2$, while a patient exposed to a fertilizer (high in nitrogen) will display a low $ETCO_2$ because of a decrease in cellular metabolism. Remember that a 1° increase in core temperature generates a 6% increase in metabolism and thus a 6% increase in carbon dioxide production.

Masimo™/Rainbow Technology

The newest diagnostic tool for emergency response is Rainbow, Masimo technology. This technology can provide information concerning toxic levels of carboxyhemoglobin and methemoglobin. These assessment tools can provide valuable information for the development of a differential diagnosis.

This technology is available on several different assessment tools. For emergency response, the most popular platforms are the integration of the Rainbow into the Life Pack 15 (and later models) and the RAD 57. On the LP 15 and RAD 57, the diagnostic tool utilizes a cable with a noninvasive finger sensor. This sensor is very sensitive to light, so the medical practitioner needs to make sure that light is not infused into the sensing area.

Unlike the pulse oximetry that evaluates both the color and transparency, the Rainbow evaluates seven wavelengths of light and light absorption. Using this technology, the Rainbow can accurately display the level of methemoglobin, carboxyhemoglobin, and oxyhemoglobin.

These levels are displayed in percentages. A normal methemoglobin level is about 1% but should not be a concern to health-care providers until it reaches above 5%. Carboxyhemoglobin is normal at 1–2%. For a smoker, this level can be as high as 10%. Over 10% for a smoker or non-smoker should be of concern. Both of these dyshemoglobins can affect the oxyhemoglobin levels.

Methemoglobin is formed after an exposure to nitrate or nitrate-containing material. Methemoglobin does not carry oxygen and is brown in color. When there is a measurable amount of methemoglobin in the blood, the actual color of the blood will change, causing the oximeter to inaccurately display a low level of oxyhemoglobin. The physiology of nitrite/nitrate poisoning will be discussed in greater detail in Chapter 4.

Carboxyhemoglobin is the combination of hemoglobin with carbon monoxide. Carboxyhemoglobin causes the blood to be bright red. Although the real amount of oxygen in the blood is severely reduced because of the carboxyhemoglobin, the oximetry will display an inaccurately high SaO_2 reading. The physiology of carbon monoxide poisoning will be discussed in greater detail in Chapter 4.

Cardiovascular Abnormalities Related to Exposure

Several cardiovascular conditions can result from exposure to toxic materials. Two types of cardiovascular dysfunctions are most commonly displayed as a result of poisoning; those that affect the venous circulation or vascular tone and those related to inadequate heart-pumping action. Serious cardiac dysrhythmias that either hinder cardiac output or over-stimulate cardiac output are the most commonly seen conduction malfunctions.

Description

Blood vessels are affected in many different ways. Thrombosis, excessive capillary leakage, bleeding from solid organs, vasodilation, and vasoconstriction have all resulted from exposure to toxins. The most common vascular response noted is persistent hypotension and shock. The provider must make an effort to determine if shock is caused by vascular space expansion or volume depletion. In both instances, there is a discrepancy between the capacity of the vascular space and the circulating blood volume. Determining the cause of hypotension will guide the provider in corrective treatment.

Vasogenic Shock

Vasogenic shock is defined as an increase in vascular space. Vasogenic shock can be attributed to one of two mechanisms.

The first is related to a defect in the responsiveness of vascular smooth muscle to neuro or chemical stimuli. Acute nitrite poisoning results in the relaxation of vascular smooth muscle and causes profuse vasodilation and hypotension. The symptoms witnessed are very similar to those seen with the overzealous administration of nitroglycerine.

The second can be caused by a decrease in the activity of vasomotor centers of the brainstem. Usually, due to a hypoxic event, the brain is unable to send neuro stimuli (signals) to the vascular muscles. This condition is usually the late result of an asphyxiant.

In the case of vasogenic shock, fluid challenges are of limited value. The true treatment would be the use of vasoconstrictors such as dopamine or norepinephrine. Hyper-oxygenation is also important to ensure a continued oxygen supply to the cells during the decreased circulation.

Hypovolemic Shock

Hypovolemic shock is defined as a critically small circulating blood volume. In chemical exposures, hypovolemic shock can be caused by abnormal permeability of blood vessels. This permeability causes leakage of blood or serum from the vessels into surrounding tissues, cavities, the lungs, and the gastrointestinal system. Dehydration may be another cause of low-volume shock.

Severe pulmonary edema can deplete the circulating blood volume causing a twofold problem. Initially, poor oxygenation will result from the decrease in oxygen crossing the alveolar–capillary membrane. Next, the decrease in circulating blood volume caused by the leakage of fluids into the lungs may result in the loss of up to two liters of fluid from the blood.

Increased pulse rate and orthostatic hypotension are early indicators of hypovolemic shock. Treatment of hypovolemic shock involves rapid volume replacement and oxygen therapy but care must be used in patients suffering from uncontrolled pulmonary edema. Treatment of pulmonary edema must occur concurrent with treatment for hypovolemia.

Heart Failure

Heart failure appears when pumping action does not keep pace with the circulatory requirement of the body. Heart failure, as a result of exposure to a toxin, is evidenced by several different syndromes.

Hypotensive shock (cardiogenic shock) is the result of a decrease in overall cardiac output. Definitive treatment usually involves increasing stroke volume by increasing contractility. The most common way of treating this condition is by digitalizing (using digoxin or digitalis) the patient, a treatment that should be done in the hospital setting under the direction of a physician. The prehospital treatment will be supportive. The use of dopamine is suggested because it will increase blood pressure and increase cardiac output by increasing preload.

Cardiogenic pulmonary edema is the result of an unbalanced cardiac output where the left heart fails to keep pace with the right heart. The result is a backup of blood trying to enter the left side. This backup causes an increase in pressure in the pulmonary circuit resulting in fluid being forced into the alveolar space.

The treatment is the same as if the condition were of nontoxic origin. Drugs used to decrease the preload and volume are useful. Lasix creates diuresis, morphine increases capillary pooling, and nitroglycerine increases vascular space and all work to decrease the pulmonary pressure. The definitive care for chemically induced pulmonary edema is the use of PPV with a PEEP setting of 10-12 cm/H_2O. This reduces the influx of blood serum into the alveolar space.

Chemical agents may also impair cardiac output by producing many types of cardiac dysrhythmias. An EKG will identify the rhythms, but there is no way initially to determine if disorders are of toxic or nontoxic origin. Tachycardic rhythms such as supraventricular tachycardia can be one of the most dangerous. If not treated, it may lead to myocardial infarction and cardiac arrest. Chemicals like solvents or one of the halogenated hydrocarbons are prone to cause cardiac sensitization stimulating ectopic beats and tachycardia.

NOTE: A tropical flower, Angel Trumpets, used by some drug abusers to make hallucinogenic tea, causes runaway heart rates of greater than 200 bpm. Angel Trumpets contain a narcotic Atropa Belladonna that causes runaway tachycardia. The foliage and flowers are extremely toxic. These toxins include atropine, scopolamine, and hyoscyamine, which cause bizarre delirium, hallucinations, and tachycardia. Ecstasy, a hallucinogenic amphetamine, has also reportedly caused supraventricular tachycardia. Both of these toxic responses are difficult to treat using the standard means to reduce heart rates.

Many drugs are available for treating these conditions. Verapamil, Brevibloc, and Adenosine are some of the most commonly used in the prehospital setting. Other treatment is supportive and geared to maintaining rate, rhythm, and oxygenation. In any case, tachycardia that is stimulated from chemical exposure is difficult to break and often continue until the chemical is detoxified by the body.

Neurological Abnormalities Related to Exposure

Central Nervous System (CNS) Exposure

CNS Depression

CNS depression is a physiological depression of the CNS that can be seen by a decreased rate of respiration, slow pulse rate, hypothermia, and an altered level of consciousness to the point of coma and death. Furthermore, it can cause confusion, dizziness, affect intellectual performance and impair judgment. CNS depression is the physical inhibition of brain activity.

Some chemicals and drugs can cause a depressed CNS. Probably an excellent example of a CNS depressant is alcohol. After seeing the signs and symptoms of a person who is intoxicated from drinking alcohol, it is easy to see the overall effects of a chemical causing CNS depression.

Similar CNS depression can be seen with those using marijuana or huffing paint/glue. Both the chemical in marijuana (THC) and the solvents found in spray paint or xylene found in model airplane glue can give similar symptoms as alcohol does.

In industry, a person exposed to the chemicals phenol (also called carbolic acid) or acetone will display drunk-like symptoms, including a staggered gate, slurred speech, drowsiness, lethargy, slow/shallow breathing, and a lower core temperature. Supportive care is all that is needed for these exposures. There is no identified antidote and no miracle cure except for providing support by maintaining respiration, blood pressure, and core temperature.

Assessing a patient who is suspected of being exposed to a CNS depressant will have:

1) Pupil size – constricted
2) Reaction to light – difficult to discern
3) Pulse rate – slow
4) Blood pressure – low
5) Temperature – low
6) Capnography – low/normal or below 35 mm/Hg

CNS Stimulation

There are some drugs and chemicals that can stimulate the CNS. These chemicals cause rapid pulse and breathing, hyperactive activity, constrict blood vessels, dilation of pupils, generate hypertension, and cause manic activities.

In the pharmaceutical world, these stimulants are used to treat narcolepsy and hyperactivity disorders. They are also used for appetite suppression in cases of severe obesity.

For recreational purposes, cocaine and methamphetamines are popular drugs. Both of these drugs stimulate the user giving them the feeling that they are thinking faster, are clearer in their thoughts, and have unending energy. Over the past couple of years, pure caffeine has become popular. Caffeine is not regulated and can be legally purchased. People add caffeine to their food and drinks to get the stimulation for recreational purposes.

Assessing a patient who is suspected of being intoxicated with a CNS stimulant will have:

1) Pupil size – dilated
2) Reaction to light – slow
3) Pulse rate – up
4) Blood pressure – up
5) Temperature – up
6) Capnography – high normal or above 45 mm/Hg

Parasympathetic Nervous System

Parasympathetic Stimulation

The most common type of parasympathetic stimulation is seen after exposure to an organophosphate pesticide or a carbamate pesticide. Both of these chemicals stimulate the parasympathetic pathways by inhibiting the enzyme responsible for the removal of the neurotransmitter from the synapse. In this case, an organophosphate or carbamate bind with acetylcholinesterase rendering it inactive. This allows the neurotransmitter acetylcholine to enter the synapse and remain in play, causing the spontaneous release of nerve stimulation down the pathway. The overstimulation causes an overproduction of sweat, saliva, urine, tears, and other uncontrolled autonomic nerve reactions.

Nerve agents that were developed for wartime activities are an extremely strong form of organophosphates. The intent was to stimulate the parasympathetic nervous system to the point that a soldier could no longer fight. Without treatment, the overstimulation would go on for days and possibly cause death. In any case, it would remove the soldier from the battleground.

This is covered in detail in the toxidromes section on organophosphates.

Parasympathetic Depression

Parasympathetic depression can also take place from an exposure. One of the biological agents developed for wartime use is botulinum neurotoxin. Botulinum works the opposite of an organophosphate. Botulinum suppresses the release of acetylcholine into the synaptic junction. The result is a loss of muscular tone, including those responsible for breathing. Botulinum toxin is one of the most powerful biological toxins known.

Other chemicals that can depress the parasympathetic nervous system include atropine and scopolamine. Both of these chemicals are found naturally in tropical flowering plants called "Angel Trumpets" and "Deadly Nightshade" (Datura stramonium and Atropa belladonna). These plants have been used for years as a type of recreational drug. The flowers and leaves are boiled and the resulting liquid is high in scopolamine and atropine. Drug users call this mixture "Angel Trumpet Tea" and is consumed for it euphoric effect.

Integumentary System (Skin)

Usually, a person's first contact with a chemical is through a dermal exposure. Therefore, it stands to reason that dermal exposures make up the majority of work-related injuries associated with hazardous materials. It is also important to note that many injuries also occur away from work in the home or other public places. Just assessing some of the common chemicals used at home include items such as oven cleaners, drain clog removers, paint solvents and paint removal agents, disinfectants, pesticides, and the list goes on and on. Medical responders need to be familiar with the effects common to these common household chemicals.

Skin Anatomy and Physiology

By weight, the skin is the largest organ of the body. On an average adult human, the surface area of the skin is around $3000\,in.^{2}$. This is about the same surface area as a tournament pool table. The thickness of the skin varies from about 0.02 to 0.12 in. thick and makes up about 10% of the total body weight. When exposure takes place, those areas of skin that are inherently thin will absorb chemicals faster than those areas that are thicker. Generally speaking, skin is thicker on the dorsal (back) side of the body and thinner on the ventral (front) side. More specifically, the thickest skin on the body is found on the plantar surface of the feet, palms of the hands, knees, and elbows, respectively. Skin is the thinnest on the face (eyelids in particular) and in the groin area (scrotum and labia). These thin areas can absorb chemicals much faster, causing systemic exposure if the chemical is not removed quickly after exposure.

Structure

Understanding the structure of the skin and its components will assist in an overall assessment of the exposure and injury. With this basic understanding of the structure, an emergency responder can sensibly estimate the amount of injury and provide appropriate treatment for a victim exposed to a hazardous chemical. In addition, knowledge of the structure of the skin will also guide responders in their decontamination efforts.

The skin consists of two principal parts, each made up of sublayers (Figure 2.14). The epidermis makes up the top layer, and in comparison, is the thinner layer of the skin. It is cemented to the thicker part called the dermis. The epidermis varies in thickness depending on the protection or durability needed for the differing areas.

The epidermis generally consists of four layers of tissue. Where friction is most likely, such as on the soles of the feet or palms of the hands, the epidermis is five layers deep.

The top layer of the epidermis is the stratum corneum. This layer has also been called the horny layer because of its appearance under a microscope, where the dead skin cells look a lot like the horns of cattle. The stratum corneum provides the first line of defense against chemical exposures because of the chemical resilience offered by it. The stratum corneum is made up of 25–30 rows of flat dead cells. The dead cells of this layer are the remains of the cells from lower levels of the epidermis that rise to the surface as they die. These cells are brushed, scraped, or flake off continuously. Since they are already dead, these cells are not injured by direct contact with a chemical and act to either repel a chemical or absorb it, and these cells are easily removed during most types of decontamination efforts. Because this layer provides so much protection from chemical absorption into live tissue, it is important that rigorous brushing of the skin not take place before decontamination. Brisk rubbing of the skin disrupts this layer and allows chemicals an opportunity to easily invade the underlying living cells.

The other three layers, stratum granulosum, stratum spinosum, and the stratum basale, are all living cells. Where friction is expected to be the greatest, a fifth layer, called the stratum lucidum, exists. This layer is like the stratum corneum in the fact that it is made of flat dead cells that are expected to take the brunt of external injury or chemical exposure.

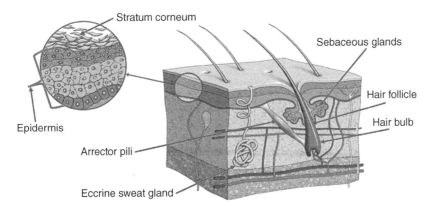

Figure 2.14 Skin represents the largest organ of the body by weight. Skin exposures are the most commonly reported exposure in the workplace.

Note: Men and women having facials performed, including chemical peels, are exposed to an intensive cleansing of the face to bring out brighter, smoother, younger-looking skin. This cleansing is nothing more than a controlled chemical burn, many times done with phenol (carbolic acid), to remove the rough dead skin associated with the stratum corneum. What is left of the stratum corneum is then moisturized with a lipid-based moisturizer, essentially increasing permeability greater than ten times. In these cases, the initial protective feature of the skin is removed, effectively creating an opportunistic means for chemical invasion through the surface.

The second principal section of the skin is the dermis. The dermis, which covers a layer of subcutaneous tissue, is anchored in place with fibers from the dermis extending into the subcutaneous layer. It is thicker than the epidermis and is thickest on the palms and soles. It is thinnest on the eyelids, scrotum, and penis. It is important to recognize these thin areas as they are easily missed during decontamination efforts. Embedded in the dermal layer are blood vessels, nerves and nerve endings, glands, and hair follicles. This layer is made up of connective tissue containing collagenous and elastic fibers that allow the skin to stretch without tearing. Adipose tissue (fat cells) is found dispersed throughout the dermis.

Function

Skin provides the external covering for the body. It protects the underlying structures from the invasion of parasites and bacteria, assists in temperature regulation, controls hydration/dehydration, excretes electrolytes and other organic materials, is a reservoir for food, is a source for vitamin D, senses cutaneous stimulation, and protects the body from external injuries. It also provides surprisingly good protection from the invasion of chemicals.

Types of Chemical Injuries to the Skin

The skin can be injured in many different ways. It can be exposed to extremely hot or extremely cold substances destroying tissue. It can suffer from mechanical injuries such as abrasions, lacerations, and contusion. These types of skin injuries disrupt the protective features of the skin, allowing a chemical to gain easy access for absorption. Traumatic injuries are not the intent of this text. Chemical exposures will be addressed here, along with the appropriate treatment of each type of chemical injury.

Chemical-Related Irritation

It makes sense to first talk about the least serious type of skin injury-related chemical exposure. Irritation of the skin is caused by exposure to chemicals that are not extremely toxic to the skin or are in lower concentrations of a mixture. Typically, these include diluted forms of acids, alkalis, and solvents. This type of reaction is commonly called contact dermatitis. It causes inflammation of the skin in the area of contact. If the area is decontaminated within a reasonable amount of time, the injury is limited to localized redness, minor swelling, and infrequently minor blistering. The symptoms may develop immediately or be delayed and can last from hours to days. In most cases, the injury is self-limiting and heals without adverse long-term effects or scarring.

An example of this type of injury is experienced by many people while operating a pump at a gas station is the splattering of gasoline onto the backside of a hand. Gasoline, being a solvent, causes dehydration of the skin cells, stimulating an injury that is usually temporary. Within a few minutes, a burning or stinging sensation develops and continues until the gasoline is washed or wiped from the skin. In some cases, the burning is the only symptom: other times, contact dermatitis appears in the form of reddened, raised spots. In either case, the injury subsides once the product is removed from the skin.

Chemical Skin Burns

Chemical burns represent a more serious injury because the exposure causes the destruction of cells and has a much longer healing time than an irritation does. Many times, burns leave permanent scars and lasting evidence that an exposure took place. Burns are caused by chemicals found in a higher concentration. Most commonly, acids and alkalis are the chemicals assaulting the skin. They have significantly different physiologic effects on the skin and underlying tissues (see Figure 2.15). Because acids are one of the most produced chemicals in the United States, therefore, injuries related to acid exposures make up a high percentage of hazardous material injuries.

Acid (low pH) burns are one of the most common chemical injuries to the skin. Acid burns can be devastating, but most are self-limiting due to the effect acids have on the proteins within the skin. Proteins exposed to acids are precipitated, forming a thick, mostly impenetrable layer that protects underlying structures. The injury caused by the destruction of cells and precipitation of proteins is called *Coagulation Necrosis*. The coagulum formed within the skin as a result of an acid exposure is very similar to the effect of dropping a raw egg on a hot grill. A hot grill turns the clear protein of a raw egg into a white rubbery coagulum. This same type of coagulum forms in the skin as a result of acid exposure. This coagulum is tough and, for the most part, impenetrable by the chemical intruder. Stronger acids, such as those with a pH of less than two, still can penetrate further into the skin and cause extensive soft tissue damage. These acids are an exception to the rule, and finding a patient exposed to them would be a rare occurrence.

One thing that should be noted is the fact that acids cause almost immediate pain at the exposure site. This pain usually stimulates a victim to immediately self-decontaminate by either washing or wiping the skin to remove the chemical. This further reduces the overall damage that could be experienced by a person exposed to an acid.

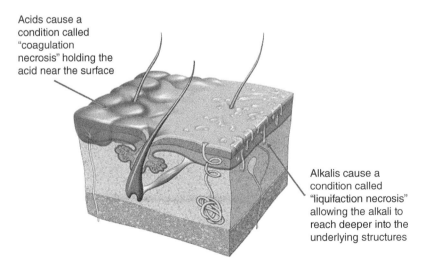

Acids cause a condition called "coagulation necrosis" holding the acid near the surface

Alkalis cause a condition called "liquifaction necrosis" allowing the alkali to reach deeper into the underlying structures

Figure 2.15   Skin acid/alkali injuries.

Treatment of acid injuries involves removing the chemical from the surface of the skin. Most of the time, water is the physiologic decontaminating substance, although certain acids require special decontaminating agents. For example, phenol (carbolic acid), an organic acid, is only partially soluble in water. Typically, non-water-soluble agents (fat-soluble or lipid-soluble) are removed using soap and water, but with phenol, this has only been partially successful. It has been found that decontamination only with soap water may not completely remove phenol but, by utilizing another nontoxic lipid-soluble agent, such as olive oil, mineral oil, or vegetable oil will bond with the phenol which can then be efficiently decontaminated with soap and water. Care must be taken in selecting these specialized decontaminating agents. Some chemicals, when mixed with fat-soluble agents, may increase the absorption rate through the skin.

Alkali (high pH) burns cause a change in the fatty substances in the skin. An alkali causes a liquefaction of fatty substances, allowing the chemical to gain access deep into the tissue. The liquefaction process caused by the alkali is called *Liquefactive Necrosis*. The chemical term used to describe this process is *saponification* (the conversion of fat by a strong alkali into soap). This process was used when making lye soap. Lye (sodium hydroxide) was added to animal fat, forming a slick, wet paste. The paste was then pressed into bars and left to dry.

Generally speaking, an alkali injury as compared to an injury from an acid of the same strength is 20 times worse. This is related to a couple of characteristics of the burns. First, acid injuries result in the formation of a coagulum that limits the penetration of the acid. Second, an acid injury causes almost immediate pain, causing the victim to respond with efforts to remove the chemical. Because the chemical has been kept to the surface by the coagulum, decontamination is very effective. On the other hand, alkalis liquefy the skin and allow the chemical to absorb deeper into the tissue. With alkalis, the pain response is delayed, and when it occurs, the alkali is well below the surface and does not respond to efforts to remove it.

Although acid injuries may produce minor blistering as a result of the chemical reactions between the tissue and the acid, these are generally rare and usually seen on the fringe of the injury. On the other hand, alkalis tend to cause significant blistering related to the depth of the injury. Because alkalis liquefy the tissue allowing deeper access into the skin and cause cellular destruction below the surface, the excessive fluids (both extracellular and intracellular) gather below the surface of the skin causing fluid-filled blisters. Typically, if large blisters are seen, the offending chemical is an alkali.

NOTE: If you have ever washed your hands with a degreaser or with household bleach (both are alkalis), you have experienced liquefactive necrosis. Once the hands are washed and being rinsed, it seems that the chemical (bleach or degreaser) is still present because your hands feel very slick. The slick feeling is because some of the fats in your hand have gone through saponification, essentially, making lye soap. The slick feeling does not go away until the hands are left to dry. Once dry, the damage to the tissue can be seen as the skin is usually very dry and cracked.

Alkaline Metals are metals that, in their pure form, will react violently with water. The elements sodium, lithium, magnesium, and calcium are the most commonly found alkali metals. Once in contact with the skin, they must not be washed away with water. The reaction with water forms hydrogen gas, heat, corrosive alkali material, and eventually ignition. The excess chemical should be brushed away, and then the affected area covered with cooking oil to prevent a reaction with air. The metal itself is a strong, caustic irritant to tissue and, when exposed to water, forms a strong alkali (hence the name alkali metal).

It is used for the production of alkali liquids and solids in nonglare highway lighting and as a heat transfer agent for solar-powered electric generators. Lithium is found in batteries and if damaged and exposed to water, the reaction can be violent including high-temperature fires.

Other chemicals can cause injuries to the skin, but these chemicals account for only a small percentage of the overall injuries related to exposure. Phosphorous (red, white, and black) are examples of these unusual chemicals. If the medical responder is unfamiliar with their properties, the injury can be made much worse.

Phosphorus is a nonmetallic element that has a variety of uses in industry. It is produced in white (yellow), red, and black forms. There is presently no industrial use for black phosphorus, so you would not expect to find it in industry. Phosphorus was also used in the production of methamphetamine (crank) in clandestine drug labs but there are currently other methods of methamphetamine production that are much safer and do not require the use of phosphorus. Phosphorus is also used in munitions, pesticides, herbicides, and fungicides. All of these may provide a source for this and other toxic injury types.

White phosphorus, also referred to as yellow phosphorus, is a crystalline wax-like, transparent solid. It may be used in rodenticides, analytical chemistry, and munitions. White phosphorus is dangerous when in contact with the skin because it is known to self-ignite at temperatures greater than 86 °F (30 °C). If the skin is contaminated, the excess chemical should be brushed away, followed by submersion in cool water to prevent ignition.

- During World War II, munitions specialists handled white phosphorus during loading procedures. White phosphorus was called "willie p" by the specialists who handled it often. These specialists suffered from the chronic effects of exposure to the chemical, which included a breakdown of the bone materials in the jaw. In addition, the jaw bone would glow a greenish-white color in the dark. Exposure to phosphorous also causes serious brain damage. The condition was termed *phossy jaw*.

Red Phosphorus is not as serious of a danger to the skin but will spontaneously ignite if exposed to oxidizing materials. It reacts with oxygen and water vapor to form poisonous phosphine gas. Red phosphorus is used to produce semiconductors, incendiaries, pyrotechnics, and safety matches. Under certain conditions, heating red phosphorus may form white phosphorus, which can self-ignite at temperatures greater than 86 °F.

Ocular Exposure and Injury

Every chemical that can injure the skin will injure the eye. Most chemicals that cause injury to the skin will cause a more severe injury to eye tissue. Of surface chemical injuries, eye exposures pose one the greatest emergencies. Simply put, an emergency is an occurrence that, with quick appropriate action, can significantly lessen a negative outcome. Chemicals in the eyes can cause irreversible injury unless the action of the emergency responder is quick, appropriate, and deliberate to remove the chemical and reduce the exposure time.

The eye is a very complex, very delicate structure that allows a person to process visual stimuli into thought-projected images. Rapid, appropriate action by the rescuer can make the difference between the loss of sight or complete recovery of a victim injured by chemical

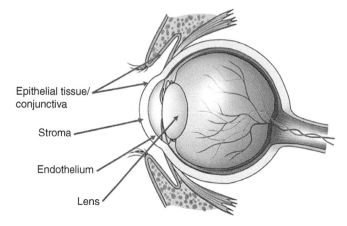

Epithelial tissue/
conjunctiva

Stroma

Endothelium

Lens

Figure 2.16 The eye is 1 in. in diameter with only one-sixth exposed to the outside environment. Epithelial tissues isolate the external surface of the eye to the outside.

exposure. This section will overview the structures of the eye (see Figure 2.16) and how different chemical exposures affect these structures and identify means for emergency treatment of a chemically exposed eye.

Eye Anatomy and Physiology

The Globe

The globe of the eye is about 1 in. in diameter. Of the total surface area, only the anterior one-sixth of the globe is exposed to the outside environment. The remainder is well protected within the socket or orbit in which it fits. When chemical exposures occur in the eye, only the anterior one-sixth of the globe and its underlying structures are usually affected. Understanding the anatomy of the surface of the eye and its underlying structural components is important in the process of conducting a good assessment of the injury and in determining the appropriate treatment.

The Ocular Surface

The ocular surface is made up of several layers consisting of the conjunctiva or *epithelial tissue, cornea, sclera,* and *tear film*. The ocular surface provides the first line of defense against chemical exposure. Each level that is affected by the chemical represents a differing degree of injury. The means of grading the injuries are identified in the *Assessment of the Eye Injuries* portion of this chapter.

The tear film is a clear fluid layer that covers the open surface of the globe. Its purpose is to provide lubrication for the lid to slide easily over the globe. It also traps invasive particles so that the lid can sweep them into the lower canthus for movement out of the eye. In addition, the tear film allows the surface tissues to stay moist and transparent to allow light movement into the inner structure of the eye.

The tear film is made up of *mucin* covered with saline water and a lipid layer. Mucin is a gel-like secretion that lubricates the eye and participates in the formation of a chemical barrier. Mucin is excreted by goblet cells located in the epithelium (surface tissue layer) of the conjunctiva. Combined with the saline solution excreted by the lacrimal glands, mucin decreases the surface tension of the aqueous tear allowing the tear film to coat the globe completely. The tear film is covered with a lipid layer to decrease evaporation between blinks.

The outer layer of the eye and surrounding tissue consists of the *conjunctiva (epithelial layer)*, which covers the interior segments of the upper and lower lids, then just inside of the socket loops around and covers the sclera and clear layers of tissue over the cornea.

The epithelium is the top layer of tissue and is composed of lipoproteins that readily allow the passage of lipid-soluble chemicals. Once the epithelium is injured, chemicals gain easy access to the second layer, known as the *stroma*.

The epithelium not only covers the anterior surface of the exposed tissue of the eye. The epithelium also loops around in the socket and covers the under surface of the upper eyelid and the lower eyelid. This essentially creates an outer envelope protecting the inner surfaces of the globe from foreign materials entering the tear film. If a chemical causes damage and disrupts the epithelial tissue, the chemical can gain access to the tissue located toward the back of the globe, or even worse, the optic nerve creating permanent damage to the eye globe.

Stroma is the middle layer of tissue and makes up 90% of the corneal thickness. It is permeable to water-soluble and polar solvent chemicals. It is also the tough membrane that provides rigidity and puncture resistance. Once a chemical injury has reached the stroma, scarring, opacification, and irregularities can result.

The endothelium is the third level and makes up the basement membrane of tissue of the cornea. It consists of a single layer of cells that cannot regenerate in adults. If the endothelium is affected by chemicals, the result is usually a loss of visual acuity, loss of sight, or a total loss of the globe.

When chemical exposure occurs, the epithelium protects the underlying tissue by taking the brunt of the injury then sloughing off, carrying with it much of the chemical contamination. After the injury, new epithelial tissue is generated to replace damaged sloughed tissue. During the healing process, a patient is usually treated with an antibiotic ointment or solution to prevent infection of the exposed surfaces.

The sclera is a tough, vascularized, white, fibrous membrane that covers the globe and extends from the cornea to the optic nerve. It is made of fibrous material that protects and gives shape to the globe. The toughness comes from the tissue fibers that crisscross over the eye surface materials. Throughout most of the body, tissue fibers run parallel to each other, which allows tissue to tear. The eye will usually not tear and is extremely difficult to puncture because of the way that the surface tissue lays in a multidirectional pattern.

The sclera is referred to as the white of the eye, although upon close examination, the color is off-white and is interlaced with numerous blood vessels. If, during the evaluation of an injured eye, it is noted that the sclera is truly white or porcelain in nature (*porcelainized*), then there is reason to suspect that circulation to that area has been compromised. Because the eye cannot form scar tissue to replace ischemic areas, the condition will almost certainly lead to necrosis, perforation, and loss of the globe.

All of the tissues of the eye require a supply of nutrients and oxygen to sustain life. This supply comes from a circulatory system that provides an uninterrupted flow of blood for these tissues. The system is called *perilimbal circulation* and is evidenced by the vessels seen in the close exam within the sclera.

Assessment of Eye Injury After Exposure

The eye has several assessment points that need to be evaluated after chemical exposure. All are related to the eye structures that are affected by the exposure and the depth of the injury. Usually, the crucial and more serious signs are related to opacification of the stroma and endothelium and the loss of circulation. These indicators can be examined in the field and relayed to the hospital before or during transport. The key to a rapid assessment is knowing what to look for and realizing what is seen. Opacification or sloughing of the epithelial layer, generally from an acid, is not necessarily a critical sign. Many times, the epithelial tissue is damaged, protecting the underlying structures. Then the tissue sloughs off, exposing a virtually undamaged cornea stroma beneath it. For simplicity reasons, assessment of burned eyes can be broken down into mild, moderate, and severe injuries.

Normal Eye Assessment

An eye assessment includes looking closely at the structure of the eye (Figure 2.17). The first part of a rapid assessment will be to determine if the eye is wet. A dry eye globe after a chemical exposure can lead you down several paths. If the exposure was to a polar solvent, such as alcohol, the immediate drying of the eye would be expected. After rapid irrigation of the eye, tears should reappear, and the eye would return to the pre-exposure wetness. A dry eye can also indicate injury or swelling to the tear glands located in the upper eyelid that would prevent tears from entering the eye.

The next assessment point would be to evaluate the surface tissue over the eye. When the surface tissue or epithelium is exposed to acid, it will develop coagulation necrosis, turn opaque and white, and normally peel away or slough off of the eye globe. When alkalis are involved, liquefactive necrosis takes place in the epithelium, causing it to slough off. In the case of an alkali injury, the necrotic epithelium appears more like a jelly and is not opaque in color. Because of the ability of the alkali to penetrate tissue quickly, discoloration of the stroma is often seen following the loss of the epithelium.

There are times when this injured and discolored epithelial layer will remain intact for some time and may mislead a medical responder into believing that the stroma has turned opaque from the injury. Once irrigation of the eye is started, the injured tissue usually begins sloughing, and the underlying cornea can be evaluated.

Further assessment includes the evaluation of the cornea. The rapid evaluation of the eye should include a close assessment of the clearness of the cornea. This can be evaluated by looking closely at the iris using a penlight. The iris contains lines called iris detail. These can

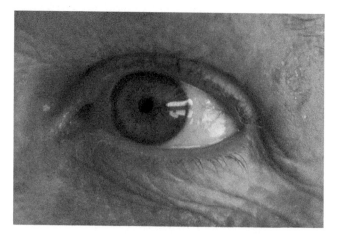

Figure 2.17 A healthy eye is wet, has a clear iris with easily seen iris detail, a well-defined pupil, and visible blood vessels in the sclera.

be seen, even with very dark-colored eyes, using a penlight. The absence of these details generally means that the cornea (stroma) has become somewhat opaque from the injury. Another indicator of the opaqueness of the cornea can be seen over the pupil. A white and cloudy pupil can indicate that opaqueness is taking place. Discoloration of the cornea can occur as a result of an acid that has penetrated the epithelial tissue or, most often, by an alkali that is being absorbed into the corneal structure. Corneal discoloration can range from slight to complete opaqueness. Remember that the stroma is 90% of the thickness of the tissue making up the cornea. This tissue does regenerate at a slow rate, and the older the patient, the less effective regeneration is. If the chemical reaches the endothelium and discolors this tissue, the injury is usually permanent, and the discoloration is there for the life of the tissue.

A normal healthy eye generally has a pupil that is almost perfectly round. It is important to note that abnormally shaped pupils have two causes for this condition. The first is a birth defect, and the second is a past or current trauma. An abnormally shaped pupil (not constricted or dilated) is not an effect seen from chemical exposure. If this condition is found during a rapid assessment, you must ask the patient if this is a normal condition for that patient. If this is an acute occurrence, then you must assume that the patient has suffered from not only a chemical exposure but also a traumatic injury.

The final assessment point would be to determine the blood circulation of the eye. If the surrounding tissue (the sclera) is pink or red, this is one indication of good circulation. When the eye is subjected to an irritant, the blood vessels dilate to provide better circulation and dilution of the irritant. These blood vessels should be easily seen on exam. If they are absent and the sclera is bright white without the evidence of blood vessels, this represents a negative sign. A sclera that is very white and does not show any signs of blood vessels is called *porcelainizing*. This would be an indication of the loss of blood flow to the eye and ultimately the destruction of eye tissues.

Note: A healthy eye is wet, well circulated, and clear.

Chemical Eye Burns

Mild eye injury is distinguishable by the several signs noted on rapid assessment. They are characterized by sloughing of epithelial tissue over the cornea and surrounding tissue. The sclera and associated tissue remain well circulated, evidenced by intact blood vessels in the sclera and reddened or pink tissue surrounding the eye. The iris is clear, and the iris detail is easily seen.

Moderate eye injury is characterized by the signs and symptoms noted on the mild burn but include haziness of the cornea, indicating that the chemical has reached the stroma. Circulation of the eye and surrounding tissue remains intact. Depending on the chemical and the amount of time exposed, the prognosis for saving the eye is good. There will probably be some loss of visual acuity but not total blindness. This injury is usually associated with acids, mild alkalis, or strong solvents.

Severe eye injury is usually found when there is exposure to alkalis or very strong acids. The cornea and sclera usually have a total epithelial loss. The cornea appears very hazy or opaque, masking the iris detail. The sclera may be white and porcelainized, indicating ischemia to that area. The prognosis of a severely burned eye is very poor. The ischemia, as mentioned earlier, will lead to necrosis, perforation, and loss of the globe. For example, concrete dust causes severe burns that have a slow onset but can result in complete loss of vision and globe.

Surface Toxins

Identification of the toxin is not immediately imperative for initial treatment to begin. Eventually, identification is important for definitive care. Straightforward first aid such as irrigation is relatively simple and safe. Irrigation should begin immediately upon determining scene safety. Questions about the chemical offender should be asked after initial irrigation has started. Irrigation can be accomplished utilizing any nonirritating water-based solutions. It can include tap water, hydrant water, lactated ringers, or preferably normal saline or sterile water and should continue until the patient reaches the hospital. Many references state that irrigation of the eye after acid exposure can be stopped once the pH of the tears reaches 7. A better practice would be to irrigate the exposed eyes, regardless of the chemical, until the patient reaches the hospital and a physician evaluates the injury. It cannot be stressed enough that immediate irrigation can make the difference between maintaining sight or losing the globe.

Exposures of the eyes to toxic substances may bring on many diverse symptoms such as altered color perception, decrease in visual acuity, photophobia, excessive tearing, and blindness. Destruction of the eye globe may result from an overwhelming exposure to a toxic material. Regardless of level of exposure or the chemical involved, the patient should receive immediate irrigation and transportation to the hospital.

Corrosives represent the most common type of eye exposure found in hazardous materials incidents. Acids and alkalis are not only found in industry but are also common chemicals found around the home. As mentioned earlier in the chapter, acids are one of the most commonly produced chemicals in the United States. As would be expected, acids comprise the most common corrosive injury. Alkalis cause the most devastating injuries to the eyes.

Alkali exposures are particularly devastating to the corneal structures of the eye. The process in which alkalis produce damage is through a liquefaction of the fatty substances within the tissue of the eyes. The liquefaction process allows the alkali to penetrate deeper into the underlying structures, causing a much more devastating injury as compared to acidic chemicals. As mentioned under the skin exposure section of this chapter, alkali injuries are 20 times worse. This is no different for the tissues of the eyes. Alkaline chemicals with a pH of greater than 11.5 cause particularly divesting results. The more severe the penetration and damage accrued by the alkali, the less pain will be experienced by the victim. Much like third-degree thermal burns, deep alkali burns destroy the corneal nerves and desensitizes the tissue as it damages it. Penetration is also related to the toxicity of the chemical involved. For example, ammonium hydroxide penetrates the fastest, followed by sodium hydroxide, potassium hydroxide, and calcium hydroxide, respectively.

Damage to the circulatory system or underlying structures of the eye may lead to necrosis of tissue and loss of the globe. Opacification of the cornea and related underlying structures may lead to blindness without loss of the globe. In either case, alkalis can evoke serious injury to the eye and must be treated aggressively by the first responder as soon as they arrive with the patient, and it is safe to provide care.

Immediate treatment involves irrigation of the affected eyes. This irrigation should continue for a prolonged period. It is suggested that once irrigation is established in the field, it should be continued throughout transportation to the hospital until stopped by an ophthalmologist or emergency department physician. Alkalis continue to burn even after the surface chemical is washed away because of their ability to penetrate the upper layers of the tissue through liquefaction of the fats (saponification). Unlike acids, when alkalis are the offender, stabilization of the surface pH is not an indicator for irrigation to stop.

Using chemicals to neutralize acids or alkalis should never be considered on the surface of the eye due to the generation of heat during the neutralizing reaction, furthering the injury. The only treatment for this type of injury in the field should be immediate irrigation and transportation. Once irrigation has started, and the eye is washed of excessive chemicals, an irrigation lens, such as the Morgan Lens, can be placed to make the irrigation process more efficient. The placement of a therapeutic lens is discussed in the section on Specialized Eye Equipment.

Acid exposures to the eye are usually not as devastating as alkali injuries. The epithelial tissue reacts with acids forming a tough coagulum. Acids act on the proteins found within the tissue causing a process called coagulation necrosis. This is the same process that takes place when skin is exposed to acid materials. The tissue becomes thickened and rubbery slowing the absorption of acids below the surface. The process acts as a natural protective barrier against acid penetration. Highly concentrated acids, those with a pH of less than 2, can overwhelm the protective barrier and cause deeper, more severe injury, similar to those caused by alkalis.

Again, as with all chemical exposures to the eyes, treatment consists of rapid irrigation of the eye by the first responder. Because the physiology involved in an acid burn is different than that evoked by the alkali, irrigation may be (but normally not recommended) halted in the field after 30–60 minutes. To determine if enough irrigation has been done, test the pH of the fluid found in the pocket of the lower lid. pH reading can be done utilizing litmus paper or a urine dipstick by placing the reactive surface in the tear found in the pocket formed by the lower lid of the affected eye. Although, many medical reference/treatment books suggest that a medical provider can stop irrigation if the pH is found to be at 7 but, in most cases, it is more feasible to irrigate until the patient reaches the hospital, then allow the emergency department physician to decide to stop irrigation.

Using an irrigation lens, such as the Morgan Lens, will allow the medical provider to irrigate the victim's eyes for longer periods without having both hands involved in the operation. The lens can be placed after rapid irrigation has already started, then irrigation can continue throughout transport to the hospital.

Solvent Exposures such as gasoline, alcohol, toluene, and acetone are commonly used in industry. These chemicals are fat-soluble, meaning they mix with and disrupt fats found in the tissues of the eyes. Unlike alkalis that cause death to the cells by liquefying the fats, solvents disrupt the fats causing normally transparent tissue to become opaque and dehydrated. The effect of the injury includes epithelial sloughing and pain. The symptoms generally persist for several days, and most heal without lasting damage. In industry, most solvents are heated to accentuate their effect. Workers injured with heated solvents are suffering not only from the chemical effects of the solvent but also from the thermal effects.

Treatment in the field is, once again, continuous ocular irrigation and rapid transportation to the hospital for definitive care.

Surfactants and Detergents are used in industry to promote wetting, disperse and dissolve fatty substances, and decrease foaming. Short-term exposures to such substances that are rapidly rinsed from the eye will only cause epithelial sloughing and pain but usually not a long-term injury. Everyday soaps and shampoos are an example of this class of chemicals. Most people have experienced the discomfort of soap in the eyes. Much stronger versions are used in the industry, and the response to exposure is also much worse. When an industrial surfactant or detergent is splashed in the eye, severe discomfort, epithelial cellular injury, and nerve irritation result, rapid irrigation will minimize injury, and if done within a reasonable period, recovery is usually complete.

Lacrimation Agent Exposure stimulates corneal nerve endings, causing reflex lacrimation and stinging pain. This effect is usually self-limiting and many times can be dealt with by the use of short-term irrigation followed by a topical anesthetic. In higher doses, corneal and conjunctival inflammation may result but this injury is most often self-limiting, and does not require follow-up care.

Tear gas is the most common lacrimator associated with ocular injuries. The police departments today use a spray that contains capsicum, a chemical found in hot peppers. Capsicum (pepper spray) is a strong lacrimator usually with no lasting effects. There have been reports of severe reactions to capsicum, but the reports are rare, and to this date, none have resulted in permanent damage. If treating a patient for capsicum in the eyes, usually a topical anesthetic, like Tetracaine®, will take care of the burning and lacrimation. By the time the anesthetic effect wears off, the chemical is no longer effective in the eye, and the symptoms do not reappear. More serious injury to the eye is usually the result of mechanical damage received from the propellant being sprayed on the eyes rather than an injury resulting from the chemical. This is covered in much greater detail in the Toxidrome section of this text, Lacrimatory Agent Exposure.

Metals exposure, like an exposure to a metallic salt, can cause damage to the ocular surface. The injury can be as mild as irritation or as severe as tissue necrosis. Many of these metals bind with the proteins and form metallic complexes, which may result in the formation of permanent granular deposits within the ocular tissue.

Solubility within the cornea determines the toxicity of the metallic salt. Organic mercury has the highest solubility, followed by tin, silver, copper, zinc, and lead. Iron, on the other hand, causes little damage to the ocular tissue but, as a foreign body, causes staining of surrounding tissue, referred to as a "rust ring."

Treatment of Eye Exposure

Understanding that toxic exposure of the eye continues until the toxin is removed, time is the most important factor. The sooner the toxin is removed or diluted, the better the outcome from the exposure. Irrigation must begin immediately and continue until the injury stops. The most accessible bland solution, preferably water or saline, should be used. NEVER attempt to neutralize an acid or alkaline irritant in the eye. After a chemical insult to the eye, a spasm of the upper and lower lid occurs, causing the eye to involuntarily close tight. This forced closing of the eyelids is called *blepharospasm*. This occurs as a natural protective feature of the eye. The victim should be provided with all available assistance to open the eye and provide irrigation. Many different means for irrigation are available to emergency responders, but none are efficient unless the eye can open. Digital opening of the eye must be done to ensure proper irrigation.

The eyelids provide a watertight and airtight seal when tightly closed. Irrigation without opening the lids is useless, so care must be taken by the medical provider to open the lids without placing excessive pressure on the globe. As long as the blepharospasm continues, the patient will be unable to hold his own eyes open; therefore, an attendant must continue to assist with the eye-opening during irrigation.

An old trick to irrigating the eyes involves the use of an oxygen administrating nasal cannula to apply the irrigation solution. A nasal cannula placed on the bridge of the nose and connected to an IV solution provides a means for applying irrigating solution evenly to both eyes. This apparatus frees a caregiver's hands to open the lids and ensure that the irrigation solution reaches the globe surface. Once the chemical is diluted and pain subsides, the blepharospasm will subside, and the patient will be able to assist in holding the eyes open for irrigation.

Utilizing an anesthetic solution such as Ponticaine®, or Tetracaine, will minimize or relieve the blepharospasm. The use of an anesthetic alone allows irrigation to be accomplished more easily. It also minimizes the pain suffered by the patient and reduces anxiety. If Morgan lenses are available, the anesthetic agents are necessary before placing the lens in the eye. As a point of interest, most of the optical anesthetic solutions are heat sensitive, so if they are carried in an emergency vehicle for field use, they must be kept in a cool environment.

Specialized Eye Equipment

The Morgan Lens

The Morgan therapeutic lens was developed to ease the process of irrigation when an eye is exposed to a chemical. It provides the eye with a constant flow of physiologic saline evenly distributed over the exposed globe. If properly used, it is placed with no additional trauma. This is not true of other irrigation techniques.

Once an irritating substance reaches the eye and an injury starts, the lids are forced closed due to the protective response of blepharospasm. This reaction disallows further exposure of the eye to the chemical. The medical provider arriving on the scene is faced with two problems involving irrigation of the eye. First, holding the eye open in an attempt to irrigate the globe properly is extremely difficult or, at times, impossible without specialized equipment. This is not because the patient is uncooperative but because the patient is unable to open the eye due to the spasm. Second, there is no efficient way one person can pour the solution into the eyes once the lids are opened digitally.

The Morgan Lens is designed as a contact-styled lens. It is to be used in conjunction with an anesthetic solution such as Ponticaine or Alcaine®. Once the anesthesia has taken place, the lens is moistened with a steady flow of solution.

The solution of choice is normal saline (0.9% sodium chloride) and is attached via an IV administration set to the lens. The technique is to first place the lens in the lower lid while instructing the patient to look up. Then, place the lens under the upper lid against the globe while instructing the patient to look down. The drip is adjusted to provide a continuous flow over the cornea and under lids (see Figure 2.18). Just like intravenous lines established in the field, great care should be used to attach the tubing to the patient.

So often, when moving patients, the solution bag goes in one direction while the patient goes in another. Since the Morgan Lens is shaped much like a suction cup, it appears that mechanical damage would probably result from a sudden, violent removal of the lens from the eye. Securing the lens can be accomplished by looping the tubing on the forehead and taping it there. Removal of the lens is done by following the procedures for placement in reverse. When attaching the tubing, make sure that there is sufficient slack in the tubing on the lens side to allow for periodic removal. If irrigation is continued for long periods (in excess of 15–20 minutes), then the lens should be removed and the eye re-medicated with the anesthetic agent.

The inside cup of the lens is manufactured smooth, so no mechanical injury can be caused over the injured eye. The Morgan Lens and other contact-style lenses are comfortable to the patient and provide better irrigation to an injured eye than a medical provider struggling to hold open an injured eye while providing a steady flow of solution over the globe. If care is taken with installation and removal, little or no trauma should result from the irrigation techniques.

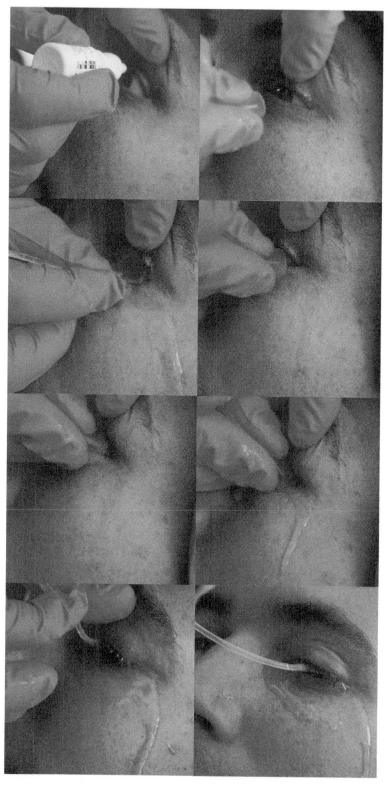

Figure 2.18 Morgan lenses are easy to use and are the only way to provide truly hands-free irrigation.

The lenses are not recommended when foreign particles or contact lenses are present. Both need to be removed before the installation of the lens. If a contact lens cannot be removed, due to the corrosive nature of the chemical exposure, then traditional irrigation should continue without the use of the Morgan Lens.

Nasal Cannula for Eye Irrigation

Using a nasal cannula over the bridge of the nose is an optional technique available for those providers who do not have access to irrigation lenses. By placing the nasal cannula and supplying it with a steady flow of irrigation solution via an IV administration set, a provider can wash the eyes from the medial to the lateral canthus, a practice preferred for this type of irrigation. Interestingly, most nasal cannulas are compatible with the Morgan Lens connector that slide on tightly over the prongs of a nasal cannula, forming a type of manifold to irrigate both eyes using one IV solution bag. This serves to solve the problem of fastening the lens in place to avoid accidental removal. The nasal cannula is attached to the patient in the same manner to provide oxygen, but instead of the prongs being inserted in the nose, they are placed across the bridge of the nose at the eyes. The lens's tubing can then be attached to the prongs.

Providing initial irrigation in the most rapid means is of vital importance. Once initial irrigation is established, then a provider can take the time to set up a more elaborate means of irrigation to the eyes. In the unlikely event that the runoff solution is in a high enough concentration to further injure the patient or the healthcare provider, consideration should be taken to limit the spread of the irrigation solution once it has left the eye. Control and/or contain runoff to reduce the chance of secondary contamination of both the provider and the victim should be considered.

Gastrointestinal Exposure to Toxic Materials

When any material is swallowed, it moves into the gastrointestinal system, where it is eventually absorbed. The gastrointestinal system is made to absorb and, to a large extent, does not care whether it is a nutrient or a chemical. The absorption of substances starts almost immediately. Once the toxic substance reaches the stomach and small intestine, the highest absorption occurs.

Absorbing Chemicals and Nutrients

Most of the chemicals and nutrients that are ingested pass through the lining of the small intestine into the bloodstream. The lining of the small intestine is covered by microvilli which are tiny finger-like protrusions that make the lining of the intestine look like velvet. Each microvillus contains blood capillaries that absorb the ingested products into the bloodstream via the hepatic portal vein. This blood contains dissolved nutrients as well as any ingested toxic materials. Before these toxic materials can harm other tissue, it enters the liver for detoxification.

Liver

The liver's job is complex, with many functions. It makes glucose from carbohydrates and fats. It stores glucose for times when the body needs it. But most important for this text is the liver's ability to metabolize (or break down) harmful substances such as ammonia and other toxins. The liver is very vascular and holds about one pint or about 13% of the total blood volume.

Detoxification takes place in the liver through several different pathways. These are referred to as Phase I and Phase II detoxification pathways. Almost every chemical is broken down by enzymes inside the liver cells. Many of the toxic chemicals that enter the body are fat-soluble and are not dissolved in water-based materials. This makes them more difficult for the liver to detoxify and for the body to excrete. If the liver is unable to immediately detoxify these chemicals, they make their way back into the bloodstream. Once in the bloodstream, fat-soluble substances are quickly absorbed into fat tissues and cell membranes and may be stored for years. They can be released during exercise or dieting. During the release of these toxins, a patient may suffer from headaches, stomach pain, nausea, fatigue, and dizziness.

NOTE: This was an issue when Lysergic acid diethylamide, better known as LSD, was popular during the 1960s. First, this drug was fat-soluble. It was taken because it caused vivid hallucinogenic effects (these hallucinations were called a "trip"). Eventually, the drug was absorbed into fat tissues, and the effect would go away. When servicemen were drafted and eventually sent to Viet Nam, they would be sent out on patrol in the Vietnam countryside. This expended a great deal of energy under hot and humid conditions that caused the rapid breakdown of fat tissue, releasing the LSD back into their system, causing a second hallucinogenic effect. Needless to say, this was not a welcome occurrence while fighting a war.

Phase I and II Detoxification

Phase I detoxification chemicals, especially those that are fat-soluble, are taken through a process of oxidation, reduction, and/or hydrolysis in an attempt to make them water-soluble. This is accomplished using a group of different enzymes collectively called the cytochrome P450 system. Many substances can inhibit the cytochrome P450 system. For example, eight ounces of grapefruit juice contains enough flavonoid naringenin to decrease cytochrome P450 activity by up to 30%.

If Phase I is successful in making the toxin water-soluble, the chemical compound is then released into the blood stream for filtration by the kidneys and eliminated in the urine. If the compound is still not water-soluble (polar), it is sent into Phase II of liver detoxification (see Figure 2.19).

During Phase II, detoxification conjugation (bonding of a chemical with another substance such as an enzyme) takes place involving a series of enzymes known as transferases. This prevents the toxins from doing further harm to healthy tissues in the body and increases

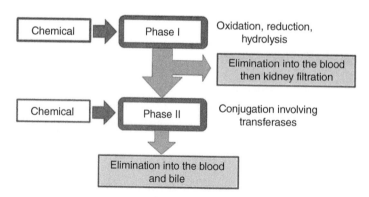

Figure 2.19 Simplified example of the liver's detoxification process. Other processes also contribute to detoxification.

polarity making the compound more water-soluble. Once these are processed, they can be effectively eliminated through the kidneys into urine or through bile into the G.I system and eliminated in the feces.

It is important to note that chemicals can enter the liver and go directly into Phase II, but these situations are not common. For a more detailed description of Phase I and Phase II detoxification, see Phase I and Phase II Reactions in Chapter 5.

Phase I
- **Simply Put:**
  - Using the chemical reactions below attempts to convert non-water-soluble toxins into water-soluble compounds for elimination. These reactions are:
    o Oxidation – the process of removing electrons
    o Reduction – the process of adding electrons
    o Hydrolysis – the process of breaking a substance apart
  - If not successful in making the chemical water-soluble (polar), then the chemical is sent to Phase II.
  - If the substance comes in as water-soluble, then these processes make it more soluble and pass through the system.

Phase II
- Phase II involves a complex process utilizing enzymes to make lipid-soluble toxins water-soluble, so they can be excreted through bile or urine.
- In addition, the liver cells add another substance (e.g. cysteine, glycine, or a sulphur molecule) to a toxic chemical to render it less harmful.

NOTE: During Phase I detoxification, the compounds produced in this phase can become more toxic than the initial chemical.

For example, dichloromethane (also called DCM or methylene chloride, CH_2Cl_2) is an organic compound found in spot cleaners and paint strippers. During Phase II detoxification, dichloromethane is partially changed into carbon monoxide (a water-soluble material) and released into the bloodstream for kidney filtration. But what occurs is carbon monoxide enters the bloodstream and binds with hemoglobin causing carboxyhemoglobinemia, which is a potentially fatal condition.

Environmental Exposures

The Hot Environment – Hydration and Hyperthermia

Hydration plays such an important role in the health and well-being of emergency responders. Especially in hot environments, the loss of critical fluid volume leads to heart attacks, strokes, hyperthermic injuries, and even decreases the body's ability to detoxify chemicals. The emergency worker, whether representing the fire service or emergency medical services, needs to recognize the importance of hydration while working in an abnormally hot environment. This section will highlight the heat-related injuries that may occur when working in encapsulating chemical protective equipment as well as other emergency personal protective equipment such as firefighter turn-out gear.

The idea that workers can survive and produce more if kept properly hydrated in hot environments has been studied for many years. During WW II, Field Marshal Rommel realized he was dealing with extremely hot conditions with his German troops in the Sahara Desert.

Rommel wanted to find what would allow his troops to function proficiently in this atmosphere. He devised an experiment that would indicate the best way to keep his troops hydrated. The experiment divided a segment of his troops into three teams, Team A, Team B, and Team C.

Team A was sent out to march as far as they could without consuming water. They were able to reach about 10 mi before becoming exhausted and unable to continue.

Team B was told to march as far as they could but were given as much water as they desired. Team B was able to march approximately 16 mi. This represented a significant improvement above what Team A accomplished without water.

Team C also marched as far as they could but were forced to periodically drink even beyond their level of thirst. Team C was able to march 26 mi. This was a total of a 260% increase in distance over Team A.

Although this study was not conducted using exact scientific specifications, it brings to light the importance of hydration and accentuates the improved level of performance gained through good hydration. It also gives some insight into forced hydrations or drinking beyond the desire of thirst.

Physiology

The human body maintains a complex balance between heat loss and heat production. Heat is produced as a by-product of metabolism. Simply put, heat is generated from chemical reactions necessary for cellular work in the production of energy, sustaining life itself. The heat produced from cellular metabolism is lost or maintained by the process of radiation, conduction, evaporation, and conduction. These processes are necessary for temperature homeostasis within the body.

Some important factors determine the rate of metabolism and, therefore, the rate of heat production. First, the Basal Metabolic Rate (BMR), which is the cellular production of energy and an increase of temperature caused by the cell's energy production. The BMR refers to the heat production of an individual while at rest but not asleep. The average BMR for an adult at rest is 70 calories/h.

Under stress, whether the stress is due to eating a large meal, emotions, or hard work, the metabolic rate can increase dramatically, adding further to heat production within the body. In addition to these natural factors, some chemicals can increase metabolism and heat production. Any CNS stimulant will raise metabolism and cause an increase in core temperature. Some of the most common are cocaine and methamphetamines. Both cause a great increase in metabolism and core temperature. Not only is the core temperature raised in these circumstances, pulse and blood pressure also increase and the increase of metabolic rate is also evidenced on the end tidal CO_2 (Capnography) readings.

External stressors are also a factor. High or low external temperatures also affect metabolic rates. Warmer ambient temperatures initially increase metabolic rates, but once the body core temperature increases due to ineffective cooling, the metabolic rate decreases dramatically. This robs the body of energy, causing apathy.

Fluid is lost in a variety of ways on a second-by-second basis. The movement and loss of fluid are made to maintain homeostasis both for the osmotic balance of fluids within the body and for the maintenance of temperature. The gastrointestinal system loses about 100 ml/day, the kidneys 30–60 ml/h (700–1400 ml/day), the respiratory system 15 ml/h (360 ml/d), but the major loss of fluid in the body occurs from the skin in the form of sweat. This loss can be as high as 2–4 l an hour, even during rest in extreme heat.

During exercise, respiratory loss can rise 10 times, and sweating increases significantly. Two hours of strenuous exercise, such as running, firefighting, or hard work in an encapsulating suit, cause the plasma volume to decrease 12–14% just within the first 10 minutes. For the next 110 minutes, the loss of plasma volume is only 2–4%. The loss from sweating can be offset somewhat by water transference within the body. The redistribution of water helps maintain fluid volume within the bloodstream.

For example, a by-product of cellular respiration (glucose + oxygen = energy [ATP] + heat + carbon dioxide + water + small amounts of carbon monoxide) is water and can produce up to 1 l. During exercise, the body produces energy by metabolizing glucose. Once the blood glucose is depleted, the body draws energy from glycogen stores in the liver. The use of each gram of glycogen releases 3–4 g of water. Complete depletion of glycogen stores releases up to 2 l of water. Therefore, while exercising, the body can partially compensate by a redistribution of up to 3 l of water into the blood volume. The compensatory mechanism is only somewhat helpful and works best in physically fit aerobically trained individuals.

When the glycogen stores have depleted, the loss of energy is great, and the production of energy comes to a screeching stop. Marathon runners call this the wall. When they hit the wall, they can no longer continue to run, and the energy needed to complete the marathon is gone. Training for a marathon involves running for longer and longer periods which conditions a body to preserve and use the glycogen more efficiently so it will last for the entire 26 mi. Firefighters need the same conditioning to complete longer firefighting events where the high use of energy is required. When a firefighter "hits the wall," they are both dehydrated and have used the glycogen stores to depletion and are physically unable to continue to work at a high level.

When it is anticipated that a responder will be working in a hot and humid atmosphere, it is important to pre-hydrate those individuals. Because the absorption of water is a slow process, hydration must be started early. If not, the fight against dehydration becomes a losing battle.

Absorption of Water

It is important to understand the movement of water from consumption through the mouth into the intestines where it is eventually absorbed (as demonstrated in Figure 2.20). The cardiac (esophageal) sphincter located at the top of the stomach prevents stomach contents,

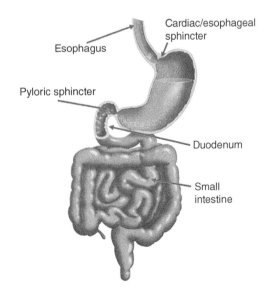

Figure 2.20 Rehydration fluid is held in the stomach by the pyloric valve until sugar is absorbed. Water is absorbed in the small intestine.

Cardiac/esophageal sphincter

Esophagus

Pyloric sphincter

Duodenum

Small intestine

including food, enzymes, acid, and liquid, from regurgitating back into the esophagus after it reaches the stomach.

The pyloric sphincter (valve) is located at the base of the stomach and is the regulating mechanism allowing the passage of nutrients and fluid into the intestines (this process is referred to as gastric dumping). Because the absorption of water into the bloodstream takes place largely in the small intestines, the pyloric valve becomes the deciding factor as to how rapidly consumed fluid gets into the bloodstream.

The pyloric sphincter at its completely open position only allows approximately 800 ml/h to pass through it. Other factors slow the movement through the valve as well. Gastric emptying is significantly slowed when a fluid or food material with a sugar concentration of greater than 5% is consumed. Most "power" drinks have a sugar (sucrose, glucose, and fructose) concentration much higher than 5%. Many have concentrations up to 10%. Although there is no nutritional benefit to plain water, it is absorbed 35–39% faster than many power drinks. Since the glucose stores can be depleted along with important electrolytes, it may be necessary to consider a fluid replacement that can address both problems.

As stated earlier, if hydration is not started early, then we will be fighting a losing battle. Remember, the greatest loss of plasma (12–14%) occurs in the first 10 minutes during two hours of exercise. Also, gastric emptying can only occur at a maximum of 800 ml/h. That means that only a maximum of 800 ml can reach the small intestine, where absorption takes place. Since a person working in a hot environment, like an encapsulating suit or firefighter turn-out gear, can lose between 2000 and 4000 ml/h, it is easy to see that pre-hydration and forced hydration is important. The importance is further enhanced when the ambient air temperature can reach 100 °F, as it so often does in the South, Midwest, and West Coast.

Before placing an entry team member (one who enters a hazardous atmosphere) in an encapsulating suit in these conditions, it is suggested to pre-weight and pre-hydrate them. This accomplishes two goals. First, it gives the entry team members a head-start on hydration. Second, it gives the medical support personnel a guide to rehydrating the team member post entry.

A loss of 5% body weight between weigh-ins indicates a worker that may be close to a serious dehydration condition. For example, a 200-lb worker losing 10 lb (5%) post entry has lost greater than 3.5 l of fluid. For a worker to recuperate from the fluid loss of this magnitude, he would need approximately 4.5 hours just for his body to reabsorb that amount of fluid. Furthermore, this amount of fluid loss could indicate that this worker may be teetering on heat exhaustion or stroke.

Dehydration also causes more systemic injury. If the entry team member happens to be exposed to a chemical, an intact and well-functioning circulating system is needed to detoxify the body. If a person becomes dehydrated, the first function of the body is to shut down filtration systems to preserve fluids. The kidneys stop filtering the blood allowing chemicals to stay in the body longer and cause more long-lasting injuries (as seen in Figure 2.21).

In addition, dehydrated blood also becomes more viscous. The increased viscosity of blood caused it to move slower through constricted vessels. This becomes critical in responders who have some pre-existing heart disease and as a result of the increase in blood viscosity are now unable to provide sufficient supplies of oxygenated hemoglobin to the capillary beds. This can lead to a hypoxic heart and sudden cardiac arrest. Typically, this occurs with firefighters who are older than 45 years old and die suddenly after the extinguishment of the fire. They die in the truck on the way home or once back at the station. Their death is often related to a heart attack when in reality, dehydration and high blood viscosity were the biggest cause in conjunction with some pre-existing heart disease.

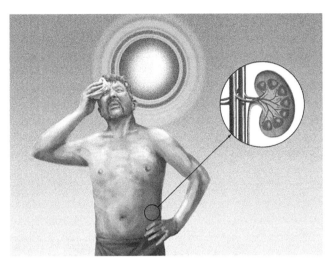

Figure 2.21 A chemically exposed person who is also dehydrated is at greater risk. Filtration by the kidneys stops to conserve water allowing toxins to stay in the body for much longer periods.

Currently, most emergency response departments require any emergency responder who was treated with an IV to either be seen in an emergency room or go home for the rest of the shift. Then, to make matters worse, requires the emergency responder to gain a "return to work" note from a physician. Policies allowing the infusion of 1000–2000 ml of Saline then returning to work allows firefighters who know that they are dehydrated to seek help before a more serious condition occurs.

The ability of EMS agencies to rehydrate using IV solutions either without a medical director's order or under a standing order is prudent to provide needed care and allows life-saving efforts in dehydration illnesses. In addition, a policy that allows paramedics to treat (with an IV), then release to return to work, allows a firefighter to seek help without retribution.

Note: It is interesting that IV lounges are popping up in many communities. These lounges allow a physician assistant or nurse practitioner the ability to start an IV and infuse various electrolytes and vitamins into the general public, some who were intoxicated the night before, and be released to drive home but, a well-trained paramedic cannot infuse normal saline into a dehydrated firefighter without a direct physician order. This stringent policy needs to be rethought to allow the hydration of on-scene workers who are sweating profusely and dehydrated to receive IV fluids for rehydration purposes and to reduce the occurrences of heat-related injury, possible myocardial infarctions, and other circulation-related injuries.

In extreme environments, pre-hydration is of utmost importance. During long incidents where workers are exposed to high temperatures for extended periods or are working hard in extreme temperatures, it is important to monitor fluid loss and intake, forcing hydration if needed to maintain fluid balance and increase the energy output of workers.

If emergency workers are only allowed to drink when thirsty, the battle may be lost, and heat exhaustion or more tragic heatstroke may be the outcome. An educated observance by medical personnel on the scene may be the deciding factor to whether the battle against dehydration is lost or won.

Acclimation

Acclimation is an important aspect to consider when examining the likelihood of suffering a heat-related injury. The human body can make subtle changes to adapt to environmental conditions, even a hot environment. If emergency workers perform daily in a hot environment, their bodies adapt by shifting and storing a higher quantity of water for its expected use. Another part of adaptation is the ability to sweat. Sweating involves a muscular contraction to open and close the gland. A person who exercises these glands regularly develops a stronger, more active muscle and, as a result, sweats much easier than a person who is not often exposed to hot environments. Studies have indicated that persons acclimated to hot environments have a greater plasma volume, allowing them to lose more volume before becoming critical. These studies have also indicated that the acclimated person loses fewer electrolytes during the sweating process.

Every few years, heat waves sweep across the northern states, causing deaths to hundreds of persons living there. Yet, in the South, temperatures range in the upper 90s–100s with humidity factors around 90% without a great loss of life. Much of this is due to the acclimation of the persons living in warmer environments. The opposite is true when the cold waves sweep through the South. Understanding acclimation gives medical personnel insight into predicting effects suffered by emergency workers as well as citizens in their districts.

Metabolic Thermoregulation

As discussed earlier, the body uses several methods for maintaining the balance of temperature within the body. Conduction, convection, radiation, and evaporation are all methods the body uses to lose or, in some cases, gain heat. Emergency responders dressed in either firefighting or chemical protective ensembles lose the ability to cool their bodies as metabolism creates heat (see Figure 2.22).

Conduction is of limited use on the emergency scene. Conduction is defined as the heat transferred to a substance or object in contact with the body. It accounts for approximately 3% of heat loss. Unless the emergency worker is submerged in cool water, conduction plays a limited role in maintaining thermal balance.

Figure 2.22 During physical activity, heat is rapidly rising in the body. Heat is lost through conduction, radiation, convection, and evaporation. None of these are effective while wearing PPE.

Radiation is the loss of heat in the form of infrared heat rays. This form of heat loss can account for approximately 60% of normal heat loss. When the temperature of the body is greater than the surroundings, a greater quantity of heat is radiated from the body to cooler objects in the surroundings, such as floors, walls, and ceilings. Conversely, heat is gained by the body when the surroundings are at a higher temperature. The efficiency of radiation is demonstrated to firefighters when the heat from an actively burning fire is apparent from hundreds of feet away.

Convection is the circulation of cooler air around a person's body and accounts for 10–15% of our heat loss. The more air movement provided around the body, the higher the heat exchange. This movement of air is termed convection.

Evaporation accounts for 20–30% of the body's heat loss that is generated when sweat is produced and vaporized, absorbing heat. The rate of heat lost to water through evaporation is many times greater than the heat lost to air at the same given temperature.

The loss can be further described as a loss of 0.58 calories of heat for every gram of water that evaporates. This loss causes continued heat loss at a rate of 12–16 calories/h. Under normal basal rates, 150 ml of water will remove all heat produced per hour.

When the body becomes overheated, large quantities of sweat are excreted to the surface of the body to cause evaporative cooling of the skin. This occurrence takes place when high core temperatures stimulate the preoptic area of the hypothalamus, which sends excitation signals via the autonomic nervous system to the sweat glands. Cooling of the skin directly affects the core temperature of the body. This cooling is done through the conduction of heat from the blood supply found at the surface into the skin.

When the core temperature elevates, profuse dilation takes place in the capillary beds of the skin. During cool temperature conditions, the blood flow to these capillary beds may, in theory, be virtually zero. During high temperatures, when core temperature rises, the capillary beds may profuse up to 30% of the cardiac output. It is easy to understand why the skin is termed the "radiator of the body."

Cooling vests have been popular items used by hazmat teams to keep their entry team personnel from overheating. These vests are filled with cold water and worn over the thorax in an attempt to cool the core temperature. Many theories developed from the use of such vests, including one stating that these vests may contribute to heat emergencies. The theory based its findings on the principle state previously. Cooling of the core takes place on the fact that dilation of the surface vessels allows 30% of the cardiac output to rise to the surface where heat is given off. When a cool vest is worn, the surface vessels constrict, holding in the heat at the core even if the surface is cool. It is because of theories such as these that cooling vests are not often used by emergency responders, and when they are, extra care needs to be taken to ensure that proper rehab takes place after entry.

Heat generation in the body at rest is a direct result of the BMR. Unfortunately, when the core temperature rises, so does the metabolic rate, which results in more heat production. Under extreme conditions, the body can excrete up to 4 l of water per hour. This converts to 200 calories of heat loss or 28 times the normal BMR of heat production. The increase in heat production can all too easily get out of hand. The limits of temperature the human body can withstand are directly dependent on the humidity. If the air is dry and there is sufficient airflow to allow evaporation, the body can withstand temperatures up to 150 °F for an extended time (up to two hours). On the other hand, if the air is 100% humidified, the maximum temperature withstood for a limited time is 94 °F. The maximum temperature safely withstood drops to 85–90 °F if the person is involved in work.

Determining Severity of Heat

There are several methods of determining the severity of heat an emergency worker will be exposed to. Three of the most common indexes are the wet-bulb/globe temperature index (WBGT), the temperature–humidity index (THI), and the humiture (H), all are useful to a safety officer who is concerned with the hostility of the environment his workers are exposed to.

The WBGT takes into account the air temperature, humidity, and radiant heat. The Marine Corps uses this index to determine if the weather is too hostile for training purposes. It is a bit more accurate in predicting the actual severity of the conditions an emergency worker is exposed to. The THI, on the other hand, takes into account the humidity, dry-bulb temperature, and the dew point to calculate a realistic temperature, much like the wind chill factor. The humiture utilizes the dry-bulb temperature and dew point to calculate a realistic temperature. The humiture is much easier to assess with only limited information. All of the heat indexes are useful in assessing how long a worker can be expected to function in the environment and how much recuperation time may be needed. During the summer months, all EMS units should have a chart readily available for use in the field or have an app on their cell phone that would help to determine the severity of the heat and humidity during training or a real incident.

Wet Bulb/Globe Temperature Index Considering air temperature, humidity, and radiant heat, the WBGT is calculated as follows:

$$\text{WBGT} = 0.2T_g = 0.1T + 0.7T_w$$

where T_g = globe temperature, T = ambient air in °C, and T_w = wet-bulb temperature.

Effective Temperature. The temperature of saturated air given the same thermal sensation as any given combination of temperature and humidity, effective temperature (ET) is a direct ratio between ambient temperature and humidity.

Temperature–Humidity Index. Formerly called the discomfort index, the THI is calculated as follows:

$$\text{THI} = 0.4\left(T^* + Tw\right) + 15$$

where T^* = ambient air in °F and Tw = wet-bulb temperature.

Humiture. Using dry-bulb temperature and dew point, humiture (H) is figured as follows:

$$\text{H} = T^* + \left(p - 21\right)$$

where T^* = ambient air in °F and p = the vapor pressure of ambient air in millibars.

Effects of Heat in an Encapsulated Suit

Temperature, humidity, and solar load (full sun without the benefit of clouds or any shade) can greatly increase heat stress on all response personnel. The hostile environment presented by these factors will mostly affect those working in Level A and Level B ensembles; however, it can, to a lesser or greater extent, affect anyone on the scene of an incident.

To appreciate the temperature that will be sustained within a suit, the adjusted temperatures as they relate to the solar loading must be examined. Although it is uncertain if this model truly pertains to a Level A encapsulated body, it will give some insight into what an entry team member might be experiencing. The following calculations have been used when determining solar loading and also in biological monitoring. It is felt that the same comparison can be made when the wearer of an encapsulating suit, although it has not been proved to have a direct correlation.

Within an encapsulated suit, a team member is affected not only by the effects of solar loading but also by the effects of heat generated inside of the suit.

The average temperature can be equated to 70% of the internal body temperature plus 30% of the surface temperature of the suit, or:

$$\text{Average Temperature} = 0.7(\text{internal temp.}) + 0.3(\text{surface temp.})$$

Internal temperature is denoted as the core body temperature, which is usually 0.6–1° higher than the oral temperature of 98.6 °F. So, let us work through a practical scenario. As the individual enters the suit in a relaxed state (normal body temperature without any environmental stress) and within a shaded area of approximately 85 °F with an 80% humidity, the suit temperature would be roughly:

$$\text{Average Temperature} = 0.7(\text{internal temp.}) + 0.3(\text{surface temp.})$$
$$= 0.7(99.6) + 0.3(85)$$
$$= 95.2 \text{ °F adjusted to humidity} = 97 \text{ °F}$$

Now let us assume that the core temperature remains the same for the first five minutes; however, once in the sun, the external suit temperature rises, giving an internal temperature as follows:

$$\text{Average Temperature} = 0.7(\text{internal temp.}) + 0.3(\text{surface temp.})$$
$$= 0.7(99.6) + 0.3(90)$$
$$= 96.72 \text{ °F adjusted to humidity} = 136 \text{ °F}$$

Another five minutes go by with the external suit temperature increasing to 100 °F and internal humidity increasing from the ambient 80–90%:

$$\text{Average Temperature} = 0.7(\text{internal temp.}) + 0.3(\text{surface temp.})$$
$$= 0.7(99.6) + 0.3(100)$$
$$= 99.7 \text{ °F adjusted to humidity} = 156 \text{ °F}$$

A few minutes later, the core temperature starts to rise, and the external suit temperature, due to intense sunlight, has risen to 110 °F. The internal temperature could rise as follows:

$$\text{Average Temperature} = 0.7(\text{internal temp.}) + 0.3(\text{surface temp.})$$
$$= 0.7(99.6) + 0.3(110)$$
$$= 102.79 \text{ °F adjusted to humidity} = 191 \text{ °F}$$

Another 5 minutes go by, with a total of 20 minutes within the suit. Humidity has now reached 100% with internal temperatures around 102 °F and a solar load of 115 °F:

$$\text{Average Temperature} = 0.7(\text{internal temp.}) + 0.3(\text{surface temp.})$$
$$= 0.7(102.79) + 0.3(115)$$
$$= 106.452 \text{ °F adjusted to humidity} = 205 \text{ °F}$$

Remember that this is just a theory based on hypothetical temperature rises utilizing a heat stress formula incorporating humidity factors. The numbers show a dramatic increase in temperature once the suit is closed and work is performed associated with the solar load. The normal core temperature for healthy humans is between 98° and 99 °F. However, humans can withstand variations in this temperature that can span between 95 °F up to 104 °F without sustaining an injury once the core temperature reaches around 107 °F. Body temperature thermoregulation fails, and heatstroke becomes imminent. This example highlights the importance of having a medical officer who is knowledgeable on the subject of heat stress.

Factors Contributing to Heat Emergencies/Injuries

Heat Exhaustion   Of all the heat-related injuries, heat exhaustion is one of the most commonly seen but the least serious heat emergencies. It is caused by the loss of fluid and sodium as a result of excessive sweating. The dehydration and sodium loss accounts for the symptoms generally seen by medical responders that include positive orthostatic vital signs with accompanying dizziness or syncope, headache, nausea and vomiting, diarrhea, and decreased urinary output. This is combined with the history of working in a hot environment and possibly poor fluid intake. Treatment includes moving the patient to a cool environment and rehydrating them with an IV of 0.9% sodium chloride or lactated ringers solution; if heat exhaustion is left untreated and allowed to progress, heatstroke my result.

Heat Cramps   Heat cramps result from a disproportionate loss of fluid and electrolytes from the body. In a hot environment, sweat, which is high in sodium, is suddenly lost. This loss results in the skeletal muscles suffering from a deficit of sodium, causing intermittent cramping. The cramping usually occurs in the abdomen, legs, arms, and fingers. Treatment includes moving the patient into a cool environment and giving fluids. It is not usually necessary to replace sodium because the sodium has just been disproportionately lost and will be replaced on transported from other areas of the body.

Heat Stroke   The most serious and likely life-threatening heat injury is heatstroke. Heatstroke is the complete failure of the thermoregulatory mechanism and can be observed by cerebral dysfunction. Symptoms include tachycardia followed by bradycardia, hypotension, rapid shallow respirations, seizures, and decreased level of consciousness into a coma. Heatstroke becomes imminent when core temperature approaches 42 °C or 108 °F. Mortality results from cerebral, cardiovascular, hepatic, or renal failure. The estimated mortality from heatstroke is between 50% and 80%. The wide variation is due to the chronic injuries sustained from the initial insult that causes death at a later date.

As the core temperature rises 1 °F, the metabolic rate rises approximately 6%, further adding to the core temperature, and once the core temperature rises to 110 °F, the BMR doubles. Furthermore, when the hypothalamus becomes overheated, its heat-regulating ability becomes greatly depressed, and sweating and capillary dilation diminish. This vicious cycle, unless abruptly interrupted, will eventually cause death.

The patient is often found unconscious. The onset of symptoms usually happens quickly and without warning. There is sudden collapse with a rapid loss of consciousness. Normally, the skin is hot and dry, although there have been some reported cases of continued sweating. CNS damage and cerebral edema due to hyperthermia are evidenced by seizures, confusion, delirium, and disorientation. If not rapidly treated, the patient loses bowel and bladder function and develops a deep coma.

Dehydration has a direct effect on the thermostatic setting of the hypothalamus. Those that become dehydrated have a higher core temperature. Therefore, it becomes much easier for those who are dehydrated to reach that point of no return due to the increased rate of metabolism and decreased rate of sweating. This further explains the need to maintain an adequate intake of water in hot surroundings. How would the temperature of your automobile be affected if the radiator was only half full?

When medical personnel deal with hazardous materials team members who become exposed to hostile temperatures in an encapsulating suit, it becomes a high priority to develop a method of determining the hydrations status of those involved. The best indicator of the hydrations status of entry team members is by comparing weights before starting the evolution and continuing to monitor their weight throughout the exercise. The weight change must be directly correlated to the loss of water. This loss can be replaced with forced hydration (hydrating until the entry weight is regained). Understanding the principles of the movement of water and the absorption within the body will aid medical personnel in determining the readiness of the team member to return to further suit work.

The ratio below can be used when utilizing a direct weight replacement method of rehydration.

8.35 lb	= 1 gal
3,780 ml	= 1 gal
1 lb	= 453 ml
3.78 l	= 1 gal
1 oz (fluid)	= 32 ml
15 oz (fluid)	= 1 lb

Temperatures within an encapsulating suit can reach well more than 100 °F. Core temperatures of a suited worker can approach 104–106 °F; heat loss/gain, along with fluid retentions, is a paramount concern when dealing with hazardous materials incidents. This remains true regardless of the level of suit needed for entry.

Treatment

Basic Life Support

- Assess vital signs paying attention to the temperature (oral or tympanic).
- Begin cooling efforts by placing cold packs in the axillary, the base of the neck, and groin.

  Note: The body attempts to cool itself by dilating surface blood vessels and allowing sweat to carry away heat. DO NOT apply cooling to large surfaces of the body (like chest and abdomen) as this will constrict the blood vessels and keep the excessive heat in the core of the body.

Advanced Life Support

Heat Cramps – result from a disproportionate loss of fluid and electrolytes from the body. In a hot environment, excessive sweating causes an imbalance in electrolyte concentration. This imbalance of electrolytes results in the skeletal muscles suffering from a deficit of these electrolytes (sodium, potassium, calcium, and magnesium). The cramping usually occurs in larger muscle groups but is most common in the abdomen, legs, arms, and fingers.

Initial Treatment

- Move patient to a cooler environment
- Administer oral fluids as tolerated (Gatorade)

Heat Exhaustion (with neurological signs or symptoms be prepared for instating the heatstroke protocol) – is one of the most commonly seen heat emergencies. It is caused by the loss of fluid and sodium (and other electrolytes) as a result of excessive sweating and an increase of core temperature up to 103 °F. If heat exhaustion is left untreated and allowed to progress, heatstroke may result.

Note: An increase of 1° of core temperature indicates an increase of 6% of the BMR. An increase of 6% of BMR will increase carbon dioxide production by 6%. A high capnography reading assists in diagnosing the heat-related injury and indicates a patient who is in danger of Heatstroke.

- Move patient to a cooler environment.
- Assess body temperature.
- Cool the body using cold compresses at the base of the neck, axilla, and groin.
- Observe for any signs of heatstroke.
- Assess capnography. Readings above 45 mm/Hg indicate a significantly overheated patient that will need rapid cooling.
- Start hydration and cooling
  - Normal Saline 20 ml/kg IV/IO
  - Slight cool saline if possible
- Administer oral fluids as tolerated (Gatorade)
- Monitor vital signs
  - Monitor LOC, EKG, and frequent VS as rapid changes are typical in heat-related injuries.

Heatstroke – is the complete failure of the thermoregulatory mechanism and can be observed by cerebral dysfunction. Symptoms include tachycardia followed by bradycardia, hypotension, rapid shallow respirations, seizures, decreased level of consciousness, eventually leading to coma. The estimated mortality from heatstroke is from 50% to 80%.

Treatment
- Rapidly move the patient to a cooler environment.
- Begin treatment immediately (this is a true emergency).
  - Remove extra clothing, including PPE, if the emergency involves an emergency responder.
  - Cool patient by applying cool packs in axilla, neck, and groin areas.
  - Start an IV of Normal Saline 20 ml/kg IV/IO. Cool saline if possible.
  - Hydrate until capillary refill time is less than two seconds.
  - Prepare for the possibility of seizure activity.
  - Continuously monitor LOC, EKG, and vital signs as they can change rapidly.
- Cardiac abnormalities are normal and usually do not require treatment.
  - If Ventricular dysrhythmias become constant and compromise cardiac stability, give:
    o Lidocaine 2 mg/kg IVP

Pediatric Considerations
  o Protocols used for adult patients are the same as those used for pediatric patients.

The Cold Environment

Exposure to Liquefied Gas and Cryogenics
The body tissue exposed to extreme cold can cause extensive damage. This section will not focus on environmental exposure to cold but exposure to chemical materials that are either transported in a cryogenic state or a liquefied gas state that, when released in an accident, the

danger from the cold material is great. These injuries may be of varying degrees, but the one injury that will be focused on is Frost Bite. These injuries occur almost instantaneously and can be devastating.

Cold injuries are separated into two categories: those that result from the freezing of tissue, such as frostbite, and those that involve nonfreezing injury to tissue, called chilblain and trench foot.

Chilblain is a mild form of cold injury and is usually associated with long exposure to a cold atmosphere against the bare skin. In the case of chilblain, the skin indicates damage by turning red, dry, and rough. Edema and skin sloughing are common later signs. The injury generally heals once the exposure to cold stops.

Trench foot or immersion foot has been a common wartime type injury. It receives its name from injuries commonly suffered during wartime activities. It is the result of feet (could be other parts) being submerged in very cold water or mud. This type of exposure results in a lack of circulation to the extremity, causing a white or cyanotic appearance of the skin and later severe swelling. Rewarming is necessary for a complete recovery that usually takes up to 10 days.

Exposure to Cryogenics. Emergency responders must be familiar with all types of chemical exposures, including cryogenics, those ultra-cold chemical gases that are kept in liquid form through refrigeration. They generate acute freeze injuries as a result of exposure. Cryogenic materials can penetrate most of the protective components used for encapsulating suits. The temperatures are so severe that even if the suit is not breached as a result of a frank exposure, tissue damage can still result because of conduction through the material into the skin. These exposures can cause surface tissue injuries or transverse the skin and injure tissues well below the surface.

What is cryogenic? Cryogenics is a field of science that deals with very low temperatures. In the hazardous materials sense, cryogenic is described as a gas that exists in a liquid form through a process of refrigeration and compression. Cryogenic temperatures range from absolute zero (in theory) as the lower limit and $-101\ °C$ ($-150\ °F$) as the upper limit. The gases usually stored or transported in a liquid cryogenic state are:

Oxygen	Hydrogen	Nitrogen	Neon
Argon	Carbon dioxide	Helium	

Frostbite Injuries

The injury that results from exposure to these cryogenics is usually an acute frostbite injury. Frostbite, which occurs as a result of tissue freezing, is a devastating wound that usually leaves scarred tissue and sometimes loss of extremities or appendages. Because the liquid portion of tissue is not water but electrolytes mixed in water, the temperature that is required to cause a freezing injury is several degrees less than $0\ °C$ ($32\ °F$). The tissue must reach a temperature of -3 to $-4\ °C$ for frostbite to be caused. The ambient temperature, therefore, must be at least $-6\ °C$, as the tissue temperature is influenced not only by the external temperature but also by the internal temperature. The most frequently injured areas due to frostbite are the feet, hands, nose, and ears.

Once the tissue is exposed to cold temperatures, ice formation results within the extracellular fluids, the formation of ice crystals increases the osmotic pressure in the extracellular area and draws the fluid from the intracellular space. The final results may involve several structures. Fire, the cells suffer severe dehydration, and if the exposure is of long enough duration, cell lysis results. Next, the formation of ice crystals can disrupt the vascular structures and surrounding tissues. The final result, once the area is rewarmed, can include vascular rupture and cellular death.

Rewarming of the injured tissue causes red blood cell sludging that leads to thrombus formation. The already compromised vessels supplying blood to the injured tissue may become blocked and, as a result of the injury, begin to leak. These conditions lead to a decreased circulation and, ultimately, necrosis of the tissue. Not unlike heart muscle injury, anything that compromises circulation to the area or decreases the concentration of oxygen in the blood will increase the area of injury.

Different tissues within the body respond to cold exposures differently. The tissues most sensitive to cold injuries include the nerves, blood vessels, and muscles. Moderately sensitive tissues include the skin and connective tissue, whereas very resistant tissues are the bones and tendons. Environmental conditions also factor in the severity of the injury. An increase in relative humidity and wind conditions improves the chances of injury. The potential for injury is also increased when the tissue is exposed to volatile liquids such as gasoline. Wet clothing or contact with metal also increases the chance of cold injury.

Other physiologic and environmental factors should be considered when judging how severe the cold injury can be. Nutritional status is an important factor. If the cells are starved for nutrition, then they have a little reserve to deal with a cold injury, and, as a result, the tissue damage will be markedly worse. Studies have also demonstrated that race may also be a factor. Some studies have identified blacks as having as much as six times more susceptibility to cold injuries than do whites.

Drugs and medications can also have a detrimental effect on cold injuries. Any drug that causes a constriction of peripheral circulation or altered thermal adaptation may increase the chances or contribute to the worsening of a cold injury. Vasoconstricting drugs such as caffeine or nicotine can cause hypoxia and rapid heat loss to an extremity. Vasodilators, like alcohol, contribute to a loss of core heat and cooling of the blood, eventually leading to hypothermia. Sedatives such as alcohol, barbiturates, and narcotics decrease metabolic activity and ultimately body heat production. As the body core temperature decreases, so does the ability to rationalize and problem solve, further contributing to longer exposure to environmental conditions. Any elevation inactivity increases core body temperature and lessens the chance of developing a cold injury.

Assessment

During the onset of frostbite, the sensation of pain diminishes because the pain impulses are no longer sent from the injured nerve fibers. At temperatures less than 110 °C, the pain impulses virtually cease and are replaced with a feeling of tingling or numbness. The tissues appear blanched, then become rock hard and develop a frosted appearance.

After rewarming occurs, the victims will complain of throbbing pain for three days to several weeks, depending on the severity of the injury. Sometimes blisters occur on the injured tissue one day to a week later. The frostbitten tissue will eventually turn black. During the next 3–4 weeks, the black tissue will gradually slough off, and a defined border separating viable tissue and dead tissue will become apparent. As the healing process continues, the dead tissue will continually separate, and eventually, a spontaneous amputation will occur.

Frostbite is rated in degrees, very similar to thermal burns. The evidence left by frostbite will determine the degree of injury suffered by the patients. For example, a frostbite injury affecting only the top layer of skin is deemed first degree. The second degree is a partial thickness injury. Third-degree involves a full-thickness injury, while a fourth-degree demonstrates complete necrosis and tissue loss to the bone or underlying structures.

Treatment

The treatment of cold injury involves rapid rewarming. This can be accomplished by submerging the affected part in the warm water of between 90 and 108 °F for 20–40 minutes. This can be safely done in the field as long as there is no chance of refreezing and the water temperature can be maintained. If the extremity becomes refrozen after thawing, the injury can become much worse. The rewarming process is very painful; therefore, analgesics should be given. During the rewarming process, flushing of the injured tissue will be seen, indicating that the blood vessel dilation has occurred. This is not necessarily a sign that all of the tissue will survive. This assumption cannot be judged until weeks after the injury, once the healing process is well underway.

Hyperbaric oxygen, as definitive therapy, has proven in some cases to decrease necrotic tissue formation and improve healing time. The use of hyperbaric oxygen on necrotic tissue healing is one of the recognized uses of hyperbaric oxygen.

Summary

A new student might read through this chapter and focus on the advanced life support options for those who are chemically injured when in reality conducting a good toxidrome assessment is probably the most important function. Understanding the body systems involved in an exposure, and developing a differential diagnosis is the most important process even before advanced treatments are ever considered. When limited information is available, the process of assessment is even more critical and providing supportive care may be the only option. Thus, the performance of the toxidrome exam is a systemic approach to body systems that we know are affected by a chemical exposure. When the information collected is detailed and complete, the medical care provider can confidently move forward with more advanced treatment.

This chapter provides detailed information on the routes of an exposure, the physiology affected by the exposure, and the treatment provided to change the detrimental physiology taking place. It is no longer acceptable in prehospital care to provide treatment without understanding the physiology of the illness. This is even more true for chemical exposure. With an understanding of the information presented, a medical care provider can move forward with confidence to provide the best care for the patient, whether that care is supportive or more detailed treatment for a specific poisoning.

3

Toxidromes

Introduction

This chapter could not possibly identify every chemical exposure. It is written to identify some of the most common chemicals found in industry and at victim's homes and workplaces. Toxidromes are groupings of chemicals that are similar in some way and may produce symptoms that are similar. But in the chemical world two similar chemicals can produce completely different symptoms.

Take for example cyanide and nitrates. Both are chemical asphyxiants. Both deprive the cells of oxygen but act in two different ways. Cyanide affects the cellular metabolism by blocking the use of oxygen in the cell. Nitrates work in the blood stream and modify hemoglobin to a non-oxygen-carrying compound called methemoglobin, which suffocates the cells by depriving them of oxygen. In addition, nitrates cause vasodilation and a drop in blood pressure. To the contrary, the hypoxia caused from cyanide exposure produces and initial high blood pressure.

This chapter guides the healthcare provider in the recognition of signs and symptoms so an appropriate differential diagnosis is developed and an appropriate care can then be provided.

In the beginning days of the original hazardous materials medical program, the authors created what was called "color-coded chemical treatment cards." These cards had the name of a chemical on one side and the current medical protocols for that chemical on the other. These cards were adopted by many agencies following the original program (see Figure 3.1).

Case Study – Silver Cyanide Exposure

A normal healthy 30-year-old male who worked in a jewelry factory was brought to the emergency department by fellow workers. They reported that 10 minutes earlier the victim accidently ingested cyanide. The source of the cyanide ingestion was an open beverage kept near a chemical process using silver potassium cyanide to polish silver.

Upon arrival the victim was in severe shock. He was immediately resuscitated and placed on a ventilator with intermittent positive pressure ventilation infusing 100% oxygen. His arterial blood gas indicated metabolic acidosis and was given sodium bicarbonate. His EKG displayed ST depression with right bundle branch block.

A Cyanide Antidote Kit was not available in the hospital pharmacy, so the workers were sent back to the factory to recover the Cyanide Antidote Kit kept in the clinic. In the

Hazardous Materials Medicine: Treating the Chemically Injured Patient, First Edition.
Richard Stilp and Armando Bevelacqua.
© 2023 John Wiley & Sons, Inc. Published 2023 by John Wiley & Sons, Inc.

Figure 3.1 The quick reference cards are used to provide a rapid reference to treatments of the more common exposures.

meantime, the victim began to convulse and the seizures were treated with thiopentone sodium (sodium pentothal). Once the seizures were under control, the victim's vital signs began to improve.

When the Cyanide Antidote Kit arrived, the victim was given amyl nitrite, sodium nitrite, and sodium thiosulfate. The victim's condition began to stabilize. SaO_2 improved to 96%, systolic blood pressure to 100–110 mm/Hg, and he regained consciousness.

After 24 hours the blood gas and vital signs were completely stabilized and the victim was extubated. After five days from his arrival in critical condition at the emergency department, he was discharged from the hospital.

Rapid identification of the toxin and efficient appropriate treatment was credited with his recovery.

Assessment Capabilities

Every emergency medical responder has been taught to take vital signs and make a basic assessment of how normal or abnormal findings affect the patient. This is a basic skill taught to even those responders providing basic first aid. Vital signs provide critical information about a patient's internal well-being. This section will take the basic vital signs a step further and teach the technician how to use those vital signs, advanced assessment tools, and basic knowledge of pathophysiology to help determine what kind of injuries are taking place after an exposure to a hazardous material.

Blood Pressure

Every healthcare provider understands that blood pressure varies depending on a patient's activities, situation, and injury/illness state. Blood pressure is regulated mostly by the nervous system but is also subjected to the endocrine system. Low blood pressure is called hypotension, and high blood pressure is called hypertension. Both have many causes that can be mild, minor, serious, or severe.

Increase in Blood Pressure

Blood pressure increases can be driven by a variety of causes. Smoking, alcohol, caffeine, core temperature, and even a full bladder can change a patient's blood pressure. Then there are the other health factors such as peripheral vascular disease, weight, diet, and of course, age and genetics. But these are not the focus of this text. Instead, this section will look at what physiological changes take place because of an exposure to a chemical that will change the blood pressure from what the patient normally experiences.

Chemicals that increase central nervous system (CNS) activity will typically raise blood pressure. The most common CNS simulators are in the drug category. Drugs like cocaine, amphetamine/methamphetamine, ecstasy, and caffeine top the list. Some hydrocarbons cause a temporary increase in CNS activity followed by depression. Gasoline is a good example. Another compound that also follows this trend is carbolic acid (phenol) that stimulates the CNS initially before causing perfuse CNS depression.

After exposure to a chemical that causes hypoxia, responders should initially expect to see the condition of high blood pressure. Any simple or chemical asphyxiant exposure has the ability to generate a significant increase in blood pressure as the body attempts to compensate for the lack of oxygen. There is one exception to this finding, and exposure to nitrites or nitrates. In this case, the loss of vascular tone that occurs due to nitrite/nitrate exposure causes a drop in blood pressure and not an increase as other chemical asphyxiants tend to do.

Decrease in Blood Pressure

A decrease in blood pressure can be caused by hypovolemia, loss of vascular tension (for various reasons), various medications, and a decrease in CNS activity. From a toxicity evaluation view, several different offenders may be the cause. First are those chemicals that reduce vascular resistance. The most prevalent ones are nitrites and nitrates. These chemicals are commonly found in fertilizers, paints, and dyes. Unfortunately, these exposures are often misdiagnosed as being a heat-related injury. When a victim who was working on landscaping or in their yard and is found unconscious, the natural first thought is a heat-related injury or a heart attack. Fertilizers, especially those high in nitrogen, exposed to the skin can cause perfuse vasodilation. The one quick assessment to see if the loss of blood pressure is related to heat or related to nitrates is to perform capnography. Heat exposure causes an increase in metabolism and thus an increase in the production of carbon dioxide providing a high capnography value. Nitrate/nitrites cause a decrease in cellular metabolism and a low capnography value.

Second are those chemicals that can cause a decrease in blood pressure related to plasma volume. Organophosphate, arsenic, ricin, carbamates, etc. that causes a decrease in circulating volume can be the culprit. Remember, if they are sweating, urinating, tearing, and producing excessive salivation, they are volume-depleted and will start to show signs of shock.

Third are those chemicals that cause CNS depression, such as hydrocarbon solvents and halogenated hydrocarbons. These decrease nervous system activity and can cause severe decreases in blood pressure. Be aware that if you find a low BP, go through the differential diagnosis to see what fits in the spectrum of exposure chemicals.

Pulse

Like blood pressure, many things can affect the pulse rate to be either higher or lower than what is considered to be normal. Blood pressure is one of them. Every healthcare provider has been taught that the pulse rate is often tied to blood pressure.

A decrease in blood pressure will certainly lead to an increase in pulse rate. This is common in blood loss situations such as trauma. But chemical exposures can also cause a drop in blood pressure. Those chemicals containing nitrates or nitrites dilate the blood vessels and, as a result, lower the blood pressure. The result is an increase in pulse rate.

Many chemicals cause either excitation/stimulation of the central nervous system or depression of the central nervous system. When the central nervous system is depressed, there is a drop in both blood pressure and pulse rate. When the central nervous system is stimulated, there is an increase in both. Central nervous system stimulation is seen with any of the stimulant drugs such as cocaine or methamphetamines. Depression is seen with marijuana, hydrocarbons, alcohol, and phenol.

It is interesting to know that any chemical that has the ability to rapidly raise blood pressure can cause reflex bradycardia. Both simple asphyxiants and chemical asphyxiants like carbon monoxide, cyanide, and hydrogen sulfide can cause this condition of reflex bradycardia. Reflex bradycardia occurs in response to the baroreceptor reflex that triggers to reduce a rapid rise in blood pressure. When there is an abnormal increase in mean arterial pressure, the baroreceptor reflex produces bradycardia as a means of decreasing cardiac output, which eventually will reduce blood pressure. This is explained in greater detail in the respiratory section "Rick's Hypoxic Syndrome."

A healthcare provider who understands the properties of the chemical offender may also be able to predict the changes in vital signs and treat them efficiently.

Toxidromes

Corrosives and Irritants Toxidromes

Corrosives and irritants are commonly found in homes across the county. Everything from drain openers, oven cleaners, pool chlorinators, and automotive batteries contain these hazardous chemicals (see Figure 3.2). This section will review some of the more common chemicals in this toxidrome.

Chlorine (Cl_2)

Agent Identification
Chlorine gas is a yellowish/green gas with a strong unpleasant and irritating odor. The odor is familiar as it is the same smell as chlorine bleach. Most people can smell chlorine at about 0.32 ppm, which is less than the permissible exposure limit (PEL) of 1 ppm. The PEL is the threshold limit of a chemical exposure that a person working an eight-hour shift, five days a week, can be exposed to without suffering ill effects. Over that amount is considered toxic. Chlorine is a strong oxidizing agent and can cause dramatic reactions with many common substances. Chlorine is heavier than air, so it has the ability to move over greater distances and concentrate in low-lying areas.

Figure 3.2 Corrosives and irritants can be found in both industry and in the household. They are commonly used for cleaning, purification of water, and to make plastics and synthetics.

History

Chlorine is a highly used chemical both in industry and at home. The United States produces more than 14 million tons of chlorine each year. Exposures to toxic levels are usually accidental and require an emergency response and rapid treatment. Chlorine incidents are most commonly seen during swimming pool maintenance or at water purification locations. Occasionally accidents and exposures are seen when some domestic cleaners are mixed with acid causing the release of chlorine gas. The biggest danger to multiple casualties occurs when a vehicle accident or railway incident causes the sudden release of large quantities of chlorine.

There has always been a concern about terrorists intentionally releasing chlorine in a heavily populated area. Chlorine is heavier than air and will tend to spread great distances in the wind and concentrate in low-lying areas. The U.S. Department of Homeland Security estimates that a terrorist attack using chlorine in a populated area could generate as many as 100,000 hospitalizations. This is not unheard of. On at least nine occasions since January 2007, terrorists have used attacks on chlorine transport and/or storage facilities in Iraq in an effort to cause mass casualties and mass fatalities.

Pathophysiology

Chlorine gas is water-soluble, and once in contact with the water-soluble mucous in the upper airways, it rapidly forms both hydrochloric acid and hypochlorous acid ($Cl_2 + H_2O = HCl + HOCl$) as it dissolves into airway surface mucous. Epithelial damage in these areas is a result of the dramatic change in the pH causing swelling and, in some cases, sloughing of tissue. In higher doses, the chlorine makes its way to the fine bronchioles and alveoli, causing damage there. The onset of chemically induced pulmonary edema may arrive as late as 24 hours after exposure.

Signs and Symptoms

Assessing a patient with an acute exposure to chlorine gas should be focused on mucous membranes but, most importantly, the upper airway. Generally speaking, dry skin will not react with chlorine, but moist skin will suffer the results of the gas-forming hydrochloric

Figure 3.3 Corrosives and irritants will attack the skin and eyes. If vaporized or as a gas they will immediately attack the respiratory system.

acid causing a burn. The medical care provider should focus on burns to the eyes and the respiratory system. Burns at the mouth and nose are signs that a more significant injury exists in the respiratory pathways. Rhonchi and wheezing are evidence that the upper respiratory system has been affected. The sound of rhonchi is related to an increase of mucous production or can signify the sloughing to surface tissue. Wheezing is caused by the narrowing of the bronchioles. The narrowing can be attributed to either swelling of the bronchial or squeezing of the bronchiole smooth musculature (bronchospasms) (see Figure 3.3).

Where Is Chlorine Found

Chlorine is used in industry and all types of water purification. It is transported in railcars and over the road tankers in a liquefied compressed gas configuration. Since chlorine is commonly used for the purification of water, many water treatment facilities will have liquefied chlorine gas in cylinders on-site and receive regularly scheduled shipments. These cylinders may be as large as the *Ton Cylinder* or as small as a 150 lb cylinder. Leaks from the cylinders can cause large plumes of toxic gas. In a high concentration, the gas appears to be a yellowish/green color, but as the gas becomes less concentrated but still in a toxic range, it will almost disappear. Chlorine is also found around large commercial pools. Most home/private pools use a dry chemical approach to maintain clear and clean water. The dry chemical has the capability of generating chlorine gas if it is allowed to get moist, but the chlorine generated is in lower concentrations capable of causing irritation but usually not significant injury.

In addition to its ability to clean water, chlorine is also used in the production of plastics and polyvinyl chloride (PVC). These plastic-based materials are made throughout this country.

One major concern about the transportation of liquefied chlorine is the fact that the expansion ratio from a liquid to a gas is very large. Chlorine liquid expands to chlorine gas at a ratio of 1 : 450–500. The permissible exposure limit for chlorine is 1 ppm. This means that one part chlorine mixed with 999,999 parts air is at the highest level of concentration in the air that will not cause permanent harm to a person if breathing for 8 hours a day and 40 hours a week. A leak from a 150 lb cylinder, commonly used at community pools, can create a huge plume and cover a wide area.

Decontamination

Decontamination is mostly achieved by removing a victim's clothing. Using a water rinse provides further decontamination. Flushing of the eyes with water, saline, sterile water, or lactated ringers is necessary if chlorine has entered the eyes.

Emergency Field Treatment
Basic Life Support

Ensure the patient is out of the contaminated area and has been made safe. Evaluate other systems for the possible effects of the chemical exposure (such as the eyes). Complete a rapid respiratory assessment and begin treatment:

- Pulse oximetry before and after oxygen administration
- 100% oxygen via NRB mask
- Assemble a nebulizer and administer 5 ml of sterile water

Advanced Life Support

- If burning persists, mix 2.5 ml of 8.4% sodium bicarbonate solution with 2.5 ml of normal saline (rendering a 4.2% mixture) and administer the 5 ml mixture through a nebulizer.

Note: Pediatric strength of sodium bicarbonate is 4.2% and can be used in full strength without titration.

- Administer methylprednisolone (Solu-Medrol) 125 mg, IVP
- If wheezing noted after the initial updraft of sodium bicarbonate:
  - Initiate an updraft of either ipratropium bromide (Atrovent) or/and albuterol (Proventil), dose nebulized
    - Albuterol 2.5 mg/3 ml via nebulizer and
    - Ipratropium bromide 0.5 mg/2.5 ml via nebulizer
  - These may be repeated 2 times

- Monitor heart rate. Contact medical control if HR over 150 bpm.
- Maintain adequate ventilation and oxygenation
  - Assess:
    - Oximetry
    - Capnography
    - EKG
  - Provide:
    - Oxygen by continuous positive airway pressure (CPAP) and set the positive end expiratory pressure (PEEP) setting above 10 cm of H_2O.
    - If rales or rhonchi are noted on auscultation and the oximetry or capnography reflect negative changes.
    - Maintain continuous positive airway pressure (CPAP) and set the PEEP above 10 cm of H_2O.
    - Consider NTG, morphine, or Lasix administration to reduce the pulmonary pressure and decrease the influx of fluid into the alveoli.

Ammonia (NH₃)

Agent Identification

Ammonia is a chemical compound that can take the form of a strong-smelling liquid or gas. It is highly irritating to tissue because it is a strong alkali and, even in low concentrations, can cause injuries to the skin, eyes, and respiratory system. People will notice the strong odor at levels ranging from 5 to 50 parts per million (PPM). Physiologic effects begin at around 25–50 ppm. The higher the concentration, the more serious the injury, up to and including death.

History

Ammonia is commonly used in fertilizer, refrigeration, and many cleaning products. The emergency responder is typically called out when a significant leak occurs involving anhydrous ammonia (pure ammonia gas). These leaks most often occur at production plants and cold storage facilities, pipelines, tank trucks, railcars, or ships and barges. Ammonia is commonly found in various industrial settings. In fact, the global production of ammonia in 2021 was 236,000,000 metric tons and is expected to be over 290,000,000 metric tons by 2030. It is shipped and used throughout the world.

Another exposure that seemed to occur often during the early 2000s was the results of stealing anhydrous ammonia to produce methamphetamines. The Hazardous Substances Emergency Events Surveillance (HEES) system from January 2000 to June 2004 recorded a total of 1791 events involving the theft of anhydrous ammonia, and 164 were confirmed to be involved with the production of methamphetamines.

Physiology

When anhydrous ammonia comes in contact with moist skin or a mucous membrane, it forms a strong alkali called ammonium hydroxide. This alkali will react with tissue causing liquefaction necrosis (saponification). The stronger the mixture, the more tissue is damaged. The alkali reacts with the tissue, allowing the alkali to gain access deeper into the underlying tissue destroying cellular membranes. Blisters often occur as the intracellular fluid is released into the extracellular space. In the upper respiratory system, this is critical as sloughing tissue and blisters have the ability to completely compromise the respiratory system and cause death.

In the eyes, exposure can also be critical. After causing liquefaction necrosis of the epithelial tissue, the chemical starts reacting with the stroma. Typically, the stroma will become cloudy and discolored, making the iris detail difficult to see. If the alkali reaches the endothelium, the injury creates permanent cloudiness and loss of visual acuity up to blindness.

Signs and Symptoms

Signs and symptoms include severe irritation to the eyes, nose, and throat. This results in difficulty breathing, wheezing, rales, and chest pain. If the victim is exposed to a high concentration of the chemical, it can gain access deeper into the respiratory system and pulmonary edema will ensue. Evidence of this will be displayed with the production of pink or bloody sputum. In addition, any areas of moisture on the skin will produce burns and blisters. If the exposure was direct contact to leaking liquefied anhydrous ammonia, frostbite might be present as a secondary issue.

Exposure to anhydrous ammonia, especially in high concentrations, is often fatal. After a severe exposure, injury to the eyes, lungs, and skin may continue to develop for 18–24 hours. Delayed pulmonary and ocular effects are possible.

Where Is Ammonia Found

Ammonia is easily mixed into water as a solution to be used as a cleaner. Window cleaners, oven cleaning foams, wax removers, toilet bowl cleaners, and other household cleaners often contain 10% ammonia. Chemicals containing ammonia should never be mixed with bleach (or many other acids). This will result in the release of a very dangerous chemical gas called chloramine. Chloramine is responsible for many injuries each year and the retirement of a number of firefighters who have been exposed during a response and received permanent lung injuries.

Commercial cleaners often contain 25–30% ammonia and are extremely dangerous because of their corrosivity and attraction to water. This concentration of ammonia solution is used in industry to etch metals like aluminum and copper. Anhydrous ammonia gas is used as a refrigerant, fertilizer, blueprinting, and for the manufacture of plastics, pesticides, and dyes.

Decontamination

Ammonia is extremely water-soluble. A person who is contaminated with ammonia should have all of their clothing taken off and washed with water. Getting water immediately on a contaminated patient is very important, even before their clothes are removed. Irrigation of the eyes is extremely important. The quicker eye irrigation starts, the better the outcome will be.

Emergency Field Treatment
Assessment/Treatment

Basic Life Support

Ensure the patient is out of the contaminated area and has been made safe. Evaluate other systems for the possible effects of the chemical exposure (such as the eyes). Complete a rapid respiratory assessment and begin treatment:

- Pulse oximetry before and after oxygen administration
- 100% oxygen via NRB mask
- Assemble a nebulizer and administer 5 ml of sterile water

Advanced Life Support

- Administer methylprednisolone (Solu-Medrol) 125 mg, IVP
- If wheezing noted after the initial updraft of 0.9% sodium chloride, then initiate an updraft of either ipratropium bromide (Atrovent) or/and albuterol (Proventil/), dose nebulized
  - Albuterol 2.5 mg/3 ml via nebulizer and
  - Ipratropium bromide 0.5 mg/2.5 ml via nebulizer
  - These may be repeated 2 times
  - Monitor heart rate. Contact medical control if HR over 150 bpm.
  - Maintain adequate ventilation and oxygenation

      o Oximetry
      o Capnography
      o EKG
- If rales or rhonchi are noted on auscultation and the oximetry or capnography reflect negative changes.
- Begin continuous positive airway pressure (CPAP) and set the positive end expiratory pressure (PEEP) above 10 cm of H_2O.

Phosgene, aka. Carbonyl Chloride

Agent Identification

Phosgene gas is either colorless or pale white and has an odor of newly mown hay. It is only slightly soluble in water and is primarily lipid soluble. Because of its poor solubility, during inhalation it bypasses the mucous of the upper respiratory system. It then travels deeper into the lower respiratory system and alveoli where it is slowly hydrolyzed to form hydrochloric acid damaging the fine bronchioles and alveoli. Phosgene also bonds with surfactant and removes the ability of surfactant to keep the alveoli open causing atelectasis. The latent effect of phosgene inhalation injury is chemically induced pulmonary edema (non-cardiogenic pulmonary edema).

History

Phosgene is an organic compound with the formula of $COCl_2$. Although phosgene is a colorless gas in low concentrations, it has a distinctive odor of newly cut hay or grass. It is used in industry to produce polyurethanes.

Phosgene was used in WWI as a chemical weapon where there are reports that it killed approximately 85,000 people. Small amounts of phosgene is formed during the breakdown and burning of organochlorine compounds.

Physiology, Signs, and Symptoms

Inhalation is the primary route of entry. One interesting fact is that the odor threshold for phosgene is five times higher than the threshold limit values. In other words, by the time a victim smells the gas, they are already in a very dangerous environment. The initial irritation from an exposure may be very mild and delayed causing a victim to ignore the minor symptoms resulting in a longer exposure. Phosgene is heavier than air and in confined areas can create an asphyxiating atmosphere.

Children are at a higher risk for serious exposure because of their faster metabolism, breathing, heart rate, and their short stature exposing them to higher concentrations closer to the ground.

Skin, eye irritation, and burns occur when phosgene contacts moist skin and may even create an irritated red swollen skin. In the eye it can damage the epithelial tissue causing it to slough from the surface of the eye and cause corneal opacification.

Where Phosgene Is Found

Phosgene is used to make plastics and pesticides. It is also formed when Freon is exposed to high temperatures or fires.

Decontamination

Generally, decontamination is accomplished by removing the victims clothing and washing their body with soap and water. Although phosgene is a gas, once skin is exposed it bonds to both the water (sweat) on the skin and the lipids (oils) found on the skin. The presence of continuous burning sensation on the skin indicates that further decontamination is still needed.

Emergency Field Treatment (Chloramine, Ammonia, and Phosgene)

Basic Life Support

Ensure the patient is out of the contaminated area and has been made safe. Evaluate other systems for the possible effects of the chemical exposure (such as the eyes). Complete a rapid respiratory assessment and begin treatment:

- Pulse oximetry before and after oxygen administration
- 100% oxygen via NRB mask
- Assemble a nebulizer and administer 5 ml of sterile water

Advanced Life Support

- Due to the solubility of phosgene expect that the result of the exposure will be the lower respiratory system and cause immediate or delayed chemically induced pulmonary edema and atelectasis.
  - Maintain adequate ventilation and oxygenation
    - Assess:
  - Oximetry
  - Capnography
  - EKG
    - Provide:
  - CPAP and set the PEEP setting at or above 10 cm of H_2O (or select the high setting) and 100% oxygen.
  - Provide advanced airway if needed.
  - If seizures occur post-exposure or the patient suffers from extreme anxiety give:
    - Diazepam (Valium) 10 mg IV/IO/IM. Maximum dosage 20 mg or
    - Midazolam (Versed) 2.5 mg IV/IO. Repeat at two-minute intervals to a maximum dosage of 10 mg or
    - Lorazepam (Ativan) 2 mg IM/IV/IO and can be repeated after 5–10 minutes
  - If Rales or Rhonchi are noted on auscultation and the oximetry or capnography reflect negative changes.
    - Maintain continuous positive airway pressure (CPAP) and set the positive end expiratory pressure (PEEP) setting above 10 cm of H_2O.
    - Consider NTG, morphine, or furosemide (Lasix) administration to reduce the pulmonary pressure and decrease the influx of fluid into the alveoli.
- If wheezing noted then:
  - Initiate a nebulized updraft of either albuterol (Proventil or Ventolin) or/and ipratropium bromide (Atrovent).
- Albuterol 2.5 mg (0.5 m or 0.5% diluted into 3 ml normal saline) give via nebulizer. May repeat three times at 20-minute intervals.

- Ipratropium bromide 0.5 mg (500 mcg) 2.5 ml via nebulizer, repeat at 20-minute intervals for a total of three doses. May be repeated up to three times at 20-minute intervals
- Consider the administration of Brethine/Terbutaline Subcutaneous Injection.
- Brethine (Terbutaline sulfate) 0.5 mg given subcutaneous injection. (0.5 ml of a 1 mg/ml solution).
  - Contact receiving hospital, poison control center, and provide **HazMat Alert.**

Pediatric Considerations (Chlorine, Chloramine, Ammonia, Phosgene)

A child's exposure can be much greater than an adult exposed to the same concentration. Children typically breathe much faster and their overall metabolism is higher making the exposure much more dramatic. All of the protocols stated above are applicable to a pediatric patient but expect more significant signs and symptoms.

Hydrofluoric Acid and Fluorine-Based Chemicals

Properties. Colorless, clear, fuming liquid or gas (hydrogen fluoride). Water-soluble. Vapor density of 3.0. Specific gravity of 1.2.

Explosive Limits. Will not burn. In contact with metal will form potentially explosive hydrogen gas.

Uses. Silicon and glass etching and frosting, semiconductor manufacturing, plastic and dyes manufacturing, floor stripping chemicals, and rust removal agents.

History

Hydrofluoric acid (HF) is not like any other inorganic acid. It is difficult to even compare it to hydrochloric or sulfuric acids because the injury caused by an exposure is not just a contact injury but also a system-wide toxic effect. Hydrofluoric acid is one of the strongest inorganic acids known. Because of its wide use and prevalence in industry, the probability of injuries is great. Hydrofluoric acid is also unique in the way it produces injury. Not only is it capable of producing severe corrosive effects on contact with tissue, but it also produces profound systemic effects that can be deadly.

Experience: Hydrofluoric Acid Spill   *Volusia County, Florida. A tanker carrying 4500 gal of hydrofluorosilicic acid (H_2SiF_6) broke in half, spilling its contents onto the interstate highway. The spill covered several 100 yd of the popular roadway. As a result, hundreds of autos drove through the spill, which was actively fuming during the hot summer day. Approximately 100 people reported to local hospitals, complaining of injury from the chemical. Several were held overnight, the worst of whom was a police officer who had responded to the spill and initially directed traffic in proximity to the spill and had direct contact with the liquid product. Interestingly, the police officer's skin wounds were treated with a solution made from Tums (antacid). Several days after the incident, all vegetation within a quarter-mile radius was brown or dead from the toxic fumes.*

Pathophysiology

The tissue injury generated from hydrofluoric acid is more like an alkali than it is a typical inorganic acid injury. Although hydrofluoric acid is an inorganic acid in the same class as hydrochloric and sulfuric acids, unlike other inorganic acids, hydrofluoric acid maintains its strong ionic bond between the fluoride and hydrogen, allowing it to penetrate deep into the sublayers of the skin and tissues. The injury typically produced by most inorganic acid burns is the result of the hydrogen ion rapidly breaking free (dissociating), causing a corrosive injury to the upper layers of skin. This type of injury produces immediate pain but tends to be superficial because of the coagulum formed by the injured tissue.

The injury generated from hydrofluoric acid may go unnoticed because of the lack of immediate pain. Unlike other inorganic acids, hydrofluoric acid maintains its strong ionic bond and slowly dissociates, allowing penetration into the subcutaneous tissue. Once deep in the tissue, the hydrogen ion separates from the fluoride causing a burn injury to the subcutaneous layers. As the tissue destruction progresses, blisters are formed on the surface of the skin. Typically blister production is seen more often in alkali injuries and not in acid injuries but in the case of hydrofluoric acid, blisters are common. The injury caused by hydrofluoric acid is twofold and, therefore, does not stop at tissue destruction.

The second part of the injury is much more severe and life-threatening. As with many other fluoride-containing chemicals, the fluoride ion seeks out and combines with all of the calcium and magnesium it comes in contact with. Bones and tissues provide the source of calcium and magnesium needed to produce calcium or magnesium fluoride. This process can result in severe decalcification and eventual destruction of the bone. Furthermore, the hydrofluoric acid that has not dissociated further penetrates tissue gaining access into the bloodstream. Once dissociated within the bloodstream, the hydrogen ion causes serum acidosis, and the fluoride binds with the calcium, causing hypocalcemia.

The bonding of the fluoride ion to the calcium found in the bloodstream causes an immediate threat to life. Skeletal muscles and cardiac muscle contraction are compromised as the result of serum hypocalcemia. The outcome of this dangerous condition is muscle tetany that can be followed by a sudden onset of asystole. This is displayed on an EKG as a failure of the ventricles to repolarize. The EKG will display widening of the QT interval until eventual loss of the T wave (repolarization of the ventricles) and asystole results. An exposure of 2.5% of the skin surface to a 90–100% concentration or a 10% skin exposure to a 70% concentration is enough to cause systemic hypocalcemia and death.

The fluoride ion also bonds with calcium on the cell membrane and increases permeability to potassium, which leads to spontaneous depolarization along nervous pathways, producing excruciating pain.

Weaker solutions cause an even more delayed onset. Lesions and pain may be delayed for up to 18–24 hours. One of the more common sites for this activity is the fingernail beds, which turn black and develop ulcerations with chronic low-dose exposure. The nail beds are one of the few areas lacking corneum stratum, the top layer of the epidermis, which provides a limited protective barrier against chemical exposure. This lack of protection allows hydrofluoric acid to penetrate easier and at lower concentrations.

Note: Recently there have been a rash of lithium battery fires. These batteries are common in automobiles, toys, and electronics. When these batteries have an internal short and fire ensues one of the most toxic gases released is hydrogen fluoride. When hydrogen fluoride enters the mucous membranes or respiratory system it is converted to hydrofluoric acid and causes serious injury that can be life threating.

Signs and Symptoms of Exposure

Hydrofluoric acid can affect any surface of the body and can enter through all routes. The symptoms displayed by the patient depend on the area of skin or tissue exposed to the acid. Routes of entry can include inhalation of vapors, skin exposure, and eye exposure.

Inhalation. Hydrogen fluoride is water-soluble, allowing the healthcare provider to predict the type of injury expected if the gas or vapor is inhaled. Typically, an inhalation injury that includes burning and swelling of the oral mucosa and upper airways may be immediately noticed. The delayed symptoms of pulmonary edema may start hours later and become very severe. Pulmonary edema has even been reported as late as two days after the initial inhalation injury. Inhalation provides an efficient route of chemical invasion into the circulatory system. Systemic reactions in this instance may have a rapid onset with a more severe reaction.

Skin Exposure. If the chemical comes in contact with the skin, initially, the patient may not have severe pain; therefore, they may believe that the injury is only minor. In concentrations of 20–50%, the onset of signs or symptoms can be delayed up to 24 hours. Strong concentrations of greater than 50% cause immediate, severe burning pain and evidence of acid burns much like that of hydrochloric acid. The skin may initially look blanched and feel firm to the touch. Up to two hours later, the skin will swell and turn gray in appearance. This initial injury will eventually turn into necrosed, blistered skin and develop ulcerations.

Eye Exposure. If hydrofluoric acid gains access into the eye, a sloughing of epithelial tissue and cloudiness of the cornea may be noted. As with any acid exposure, the injury may be severe if not immediately irrigated. Like skin injuries, hydrofluoric acid causes a deeper and more severe injury to the eye than would be expected for other inorganic acids. Due to the limited surface area involved in an eye exposure and the natural flushing that takes place once an exposure happens, systemic reaction, if present, would be more subdued.

Systemic Injury. When enough hydrofluoric acid comes in contact with the body, systemic effects take place. The fluoride begins to bind with the calcium in the bloodstream, causing perfuse hypocalcemia. This condition can be diagnosed on the EKG monitor and is evidenced by an elongated QT interval. As hypocalcemia progresses, the QT becomes longer and longer. It is critical at this point to ensure that a calcium bolus is given. If the QT gets long enough and repolarization fails to take place, the patient experiences cardiac arrest and asystole. Without a bolus of calcium chloride or calcium gluconate, the patient will die.

An inhalation, ingestion, or exposure of hydrofluoric acid of just 25 square inches of skin can cause severe hypocalcemia. If not recognized and accurately treated, this condition can cause death. An EKG indicated a prolonged QT interval is diagnostic for hypocalcemia. In these cases, IV infusion of calcium must be considered to reduce the life-threatening effects of hypocalcemia.

Where Hydrofluoric Acid Is Commonly Found

Hydrofluoric acid is used in a number of industries, usually involving manufacturing. It is used in the production of printed circuits, metal finishing, stainless steel, iron, and steel foundries, petroleum refining, and glassmaking. It is also found in over-the-counter items such as cleaning compounds, floor strippers, and rust removal agents. A form of hydrofluoric acid, hydrofluorosilicic acid, is used to fluoridate water in many communities.

Decontamination and Significant Danger to Rescuers

Rescuers must wear agent-specific protective suits and self-contained breathing apparatus if around the gas, vapor, or liquid. Victims of exposure must be provided with immediate decontamination. If ambulatory, the patient should be instructed to remove all clothing and move to a predetermined decontamination center.

Water is the initial decontamination solution of choice. If available, an Epsom salt solution (magnesium sulfate) or calcium hydroxide (lime water) are the most effective decontamination solutions. Maalox or Mylanta can be applied topically. Tums also contain calcium and magnesium and can be crushed, added to saline, forming a slurry that can be painted over the wound. If the hands have become exposed, then the nails should be trimmed back to the nail beds to facilitate adequate decontamination of the hands.

Treatment

Hydrofluoric acid burns require immediate and specialized first aid, much different from other chemical burns. If untreated or improperly treated, the patient can suffer permanent damage, disability, or even death. If, on the other hand, the burns are properly diagnosed, and proper treatment is rapidly started, the outcome is generally favorable. Treatment is geared toward tying up the fluoride ion, which essentially will limit and prevent further tissue damage and system responses.

Calcium gluconate is the drug of choice and is used in many different formats to provide calcium at the point of injury. Extra attention should be taken when the respiratory system or large areas of the skin (greater than 25 in.^{2}) are involved. A patient with hydrofluoric acid burns above the shoulders must be rapidly evaluated for injury to the respiratory system.

Assessment and treatment for HF exposure are highly dependent on the concentration and contact time. External exposures involving the eyes, skin, and respiratory system are usually followed by internal/systematic injury. All must be evaluated and, if necessary, aggressively treated.

Eye Injury Treatment (Hydrofluoric Acid)

Basic Life Support

- Immediately flush eyes with any mean possible.
- Once irrigation has been initiated, a nasal cannula can be used on the bridge of the nose to facilitate irrigation of both eyes. The eyelids may still need to be held open to ensure adequate irrigation.
- Ensure that all foreign materials or contact lenses are removed from the eye.

Advanced Life Support

- Eye irrigation should use normal saline and NOT a solution of calcium gluconate.
- Connect the saline bag and tubing to a Morgan Lens and bleed out any air.
- Insert two drops of Tetracaine ophthalmic analgesic solution into the eye. Before applying Tetracaine, determine if the patient is allergic to "caine" derivatives.
- Insert the Morgan lens(es) into the affected eye(s) and adjust the rate to ensure a continuous flow of solution out of the eye.

- Irrigate the eyes using the calcium gluconate that was prepared.
- Contact receiving hospital, Poison Control Center, and provide **HazMat Alert.**

Skin Burn Treatment (Hydrofluoric Acid)
Basic Life Support
- Immediately flush exposed area with large amounts of water

Advanced Life Support
- Prepare calcium gluconate gel by mixing 10 ml of 10% calcium gluconate into 2 oz of water-based lubricant (such as KY Jelly) and mix completely. This can be done by pouring both into an exam glove and mixed.
- Apply calcium gluconate gel over burned area and massage into exposure using a gloved hand.

Advanced Life Support
- If pain continues:
- Reduction or cessation of pain is an indication that the calcium is bonding with the fluoride.
- Contact receiving hospital and the Regional Poison Control Center.

Note: Tums mixed with water, saline, or water-based lubricant (K-Y Jelly), Maalox, or Mylanta (calcium blend) can be used in the absence of calcium gluconate or calcium chloride for topical treatment of HF burns. An Epsom salt bath solution (magnesium sulfate) may also be used as a skin surface treatment. Both calcium and magnesium bond with the free fluoride ions to prevent further injury.

Respiratory Injury Treatment (Hydrofluoric Acid)
Basic Life Support
- If respiratory exposure is suspected or if signs and symptoms are present, immediately begin humidified oxygen therapy using an NRB mask.

Advanced Life Support
- Mix 6 ml of sterile water into 3 ml of 10% calcium gluconate.
- Place the solution in the nebulizer and connect it to oxygen to provide effective fog.
- 9 ml is enough solution to provide two to three nebulizer treatments.
- If wheezing continues after treatment with calcium gluconate, administer:
  - Initiate a nebulized updraft of either albuterol (Proventil or Ventolin) and/or ipratropium bromide (Atrovent).
    - Albuterol 2.5 mg (0.5 ml of 0.5% diluted to 3 ml with sterile normal saline) via nebulizer. It may be repeated three times.
    - Ipratropium bromide 0.5 mg (500 mcg)/2.5 ml via nebulizer, then repeat at 20-minute intervals for a total of three doses (usually only one dose given in the field).
- Administer methylprednisolone (Solu-Medrol) 125 mg, IVP slowly.
  - Monitor heart rate. Contact medical control if the heart rate is over 150 bpm.
  - Consider the administration of Brethine (Terbutaline sulfate) 0.5 mg given by subcutaneous injection (0.5 ml of a 1 mg/ml solution).
- Contact receiving hospital and the Regional Poison Control Center.

Systemic Injury from Hydrofluoric Acid (Hypocalcemia)

- Continuously monitor VS and pay close attention to the EKG.
- Systemic injury is characterized by the presence of one or several of the below signs:
  ○ Cardiac dysrhythmias
  ○ Conduction disturbances
  ○ ST-Segment elongation on EKG (hypocalcemia)
  ○ Peaked T waves (indicating hyperkalemia)
  ○ Tetany
  ○ Seizures
- If the above symptoms are present, give:
  ○ 10 ml of Calcium Gluconate 10% IV
  ○ Repeat every three to five minutes to a maximum dose of 30 ml if the signs of systemic fluoride toxicity persist.

Note: Patients may be given larger doses of IV calcium for systemic toxicity; even large doses rarely cause hypercalcemia. Correction of acid-base status is paramount, but never give sodium bicarbonate and calcium through the same I/V line.

  ○ If calcium gluconate is not available, give calcium chloride 10% (9.3%) solution, 10 ml IV push slowly (very irritating to the vein).

Pediatric Considerations

- All topical and nebulized treatments remain the same for pediatric patients
- Pediatric dose of 10% calcium gluconate IV is:
  ○ Give 0.2 ml/kg IV or IO.
  ○ Repeat half the initial dose in three to five minutes if signs of systemic fluoride toxicity persist.
  ○ If calcium gluconate is not available, give calcium chloride 10% solution, 0.1–0.2 ml IV push slowly.

Phenol (Carbolic Acid)

Properties. A white, crystalline mass that turns pink or red if not pure or contaminated with water (moisture). Soluble in alcohol, water, ether, petroleum, and oils. Specific gravity of 1.07.
Combustible. Flashpoint at 172.4 °F.
Uses. Phenol is used as a general disinfectant, pharmaceutical, lab reagent, and wood preservative. Others are listed under where Phenol is Commonly Found.

History

Phenol/carbolic acid was first introduced as a disinfectant by Dr. Joseph Lister, who gained fame and known as the father of aseptic technique in 1865 with the use of phenol spray to disinfect the air in hospital wards and operating rooms. Dr. Lister theorized that many infections were spread from patient to patient by bacteria. The use of carbolic acid as a disinfectant greatly reduced the infectious rate between patients confirming his theory.

The technique worked so well that later in the 1800s, 5–10% phenol was being used for bathing patients who entered the hospital to kill any bacteria that would be brought into the hospital on their bodies. As a result of the popularity of this technique, phenol toxicity was first found in medical personnel. The toxicity resulted from medical personnel bathing patients, subjecting hospital workers to chronic low-dose exposure.

Today its popularity in smaller percentages is widespread. For example, Campho-Phenique® contains 4.5% phenol. Camphor and other similar substances interact with phenol, both to reduce its corrosive properties and to slow its absorption. Interestingly, phenol has also been used for facial treatments, coined "non-surgical face-lifts or chemical peels," where phenol is painted on the facial skin and a "controlled chemical burn" takes place, effectively removing the top layer of skin (corneum stratum) to reveal new, pink younger-looking skin.

An early antiseptic preparation containing phenol was Lysol. This Lysol has little resemblance to today's Lysol and was made by boiling heavy tar oils with vegetable oil in the presence of lye (sodium hydroxide). This preparation was toxic and frequently used in the 1930s for suicide attempts. At the time, it was called the "Fashionable Suicide."

Pathophysiology, Signs, and Symptoms

As mentioned in history, phenol has been recognized for its toxic properties since the 1800s. Little is actually known about the physiologic mechanisms of systemic phenol poisoning except through the symptoms displayed by a poisoned patient. Phenol is not only absorbed through the skin but, as a gas, dust, or fume, can also be inhaled through the lungs. Symptoms develop in 5–30 minutes. Even low concentrations can cause devastating results. There are reports of children dying after the application of compresses soaked with 5% phenol solutions being applied over open wounds.

Systemic phenol poisoning exhibits a variety of symptoms. Initially, an excitation stage may be witnessed. Later, more intense symptoms appear, including profound CNS depression, hypothermia, cardiac depression, loss of vascular tone, and respiratory depression. In test animals, seizures have been part of the excitation phase, but they have not been seen in human exposures.

In fact, the initial excitation phase may or may not be seen in human exposure. It is believed that the danger stage of phenol poisoning is not long-lasting and generally passes after about 24 hours.

In addition to the CNS depression seen in phenol poisoning, there is also an associated acid burn on direct contact to the skin. Like most acid burns, phenol denatures proteins found in the skin, causing the formation of coagulum and leaving a white to brown discoloration on the skin in the area of exposure.

Where Phenol Is Commonly Found

Phenol is most commonly found in over-the-counter disinfectants in very low concentrations. In high concentrations, it has various uses in the hospital. The pharmacies usually dilute the concentration prior to sending it out for use by the physicians. Other areas of use are listed below.

Compound	Use
Amyl phenol	Germicide
Creosol	Antiseptic
Creosote	Wood preservative
Quaiacol	Antiseptic

Compound	Use
Hexachlorophene	Antiseptic
Medicinal tar	Treatment of dermatologic conditions
Phenol	Outpatient podiatric surgery
Phenylphenol	Disinfectant
Tetrachlorophenol	Fungicide

Field Treatment and Decontamination

Treatment of phenol poisoning includes decontamination of the skin using soap and copious amounts of water. The decontamination procedure is only somewhat helpful because phenol has limited solubility in water. Further decontamination must be done using a mediator substance such as olive oil, mineral oil, or vegetable oil, then the soap and water decontamination should again follow. If only small amounts of water are used, absorption will increase, making the patient's conditions worse. Care must be taken by the rescuer to ensure that secondary contamination does not take place. The use of proper PPE is necessary to prevent secondary contamination. Remember, an exposure of 5% phenol has reportedly caused death in children. Supportive measures such as assisting ventilation and controlling ventricular ectopy and seizures are done as the need arises. Continuous cardiac and vital sign monitoring in conjunction with supportive respiratory care with 100% oxygen is mandatory.

Assessment/Treatment or Phenol

Basic Life Support

- Decontaminate initially with soap and water. If evidence of phenol is present after decontamination, use a mediating oil as described above and repeat a second soap and water wash.

  Note: There is no antidote to this exposure, but detailed decontamination will reduce the exposure and reduce the ongoing symptoms. Caution: using only small volumes of water increases absorption by expanding the surface area of exposure.

Advanced Life Support

- Support respiration, control seizures, and ventricular ectopy following the local medical protocols.
- Contact receiving hospital and the Regional Poison Control Center.

Pediatric Considerations (Phenol)

- The same protocols used for adults are appropriate for pediatric patient exposures.

Lacrimatory Agent Exposure

Lacrimatory agents are substances that intentionally increase the flow of tears. They are currently used as riot control agents and are primarily intended to incapacitate an individual without causing illness or permanent bodily harm. The compounds most commonly used by law enforcement and the military are chloroacetophenone (CN), 2-chlorobenzalmalononitrile (CS), and the newest compound, pelargonic acid vanillylamide (PAVA). All are commonly referred to as "tear gas" or "pepper spray." All of these chemicals contain some version of capsaicin that is derived from hot peppers. This is not a life-threatening chemical. Pepper spray is a fine solid material that is not water-soluble. Secondary contamination to the healthcare provider is a concern, and care needs to be taken to avoid effects to the care provider. Patients with pre-existing reactive airway disease (e.g. asthma) or who are allergic may present with more severe symptoms requiring immediate intervention.

Chemical Currently Being Used

Spray. Currently, most law enforcement agencies are using a chemical called PAVA. The spray contains about 0.3% solution of pelargonic acid vanillylamide (PAVA), which is a synthetic capsaicinoid (an analog of capsaicin) mixed in a solvent of ethanol. The handheld sprayer is propelled by nitrogen. PAVA replaced CS gas 2-chlorobenzalmalononitrile ($C_{10}H_5ClN_2$), commonly called tear gas used for years by both law enforcement and the military. PAVA is significantly more potent than CS gas and can be effectively sprayed about 13 ft.

Projectile. PAVA may also be dispersed in a projectile shot from a specialized pistol or rifle-style weapon. These "Pepperball Guns" have a maximum effective range of 60 ft. The projectile is called a pepper spray ball, pepper ball, or pepper spray pellet and is a frangible ball containing the PAVA powder.

These weapons are generally used as a stand-off weapon where a person is determined to be dangerous but deadly force is not warranted. These weapons allow law enforcement to use as many rounds as required to bring individuals, multiple persons, or crowds into compliance. Although these weapons are generally considered to be non-lethal when properly used, deaths have occurred when they have been fired at an inappropriate area such as the face, eyes, throat, or spine.

Effect

PAVA mostly affects the eyes causing blepharospasms (forced closing of the eyes) and severe pain. The pain from PAVA has been reported to be much greater than CS (tear gas) used previously. There are also reports of PAVA not working on people who are under the influence of drugs or alcohol. Exposure to moving air will generate a significant recovery from the effects of the chemical within 15–35 minutes.

PAVA and other capsaicinoids work by directly binding to pain receptors (transient receptor potential cation channel subfamily V or <u>TRPV1</u>) that normally produce the pain and sensation of heat. Victims report that the pain feels like scalding heat.

Treatment

The National Institute of Justice states that pepper spray (PAVA classification) is a naturally occurring substance found in the oily resin of cayenne and other varieties of peppers. Since this chemical is oil-based, water alone is not effective at removing it from the skin and eyes. There are two thoughts to treatment. First, numbing the eyes using an eye analgesic like Tetracaine or Ponticaine will remove the pain while the chemical is naturally removed from the eyes. The second treatment involves removing it from the eyes using whole milk or Milk of Magnesia. This is the least expensive and invasive of taking care of patients complaining about excessive pain or shortness of breath.

Since the agent does not cause significant tissue damage, the treatment is aimed at decontamination and relieving the pain caused by nerve stimulation.

Basic Life Support

- Provide immediate eye irrigation to reduce the effects of the chemical and decrease pain.
- Eye irrigation (decontamination) can include the use of whole milk to mix with the oleoresin capsicum and allow it to be washed from the eyes.
- Sudecon Wipes can be used to remove additional irritant spray from surface skin.

Advanced Life Support

- Before applying Tetracaine, determine if the patient is allergic to "caine" derivatives.
- Apply two drops of Tetracaine into each eye.
- When the blepharospasm is relieved, a detailed visual exam can be performed to evaluate eye trauma.
- Assess for clear lung sounds and blood pressure changes to ensure that sensitivity has not occurred.
- Contact receiving hospital and the Regional Poison Control Center.

Pediatric Considerations (Lacrimatory Agents)

- The same protocols used for adults are appropriate for pediatric patient exposures.

Asphyxiant Toxidromes

Asphyxiants are commonly found in every urban and rural environment. Some are found in household products, lawn and farming fertilizers, metal cleaning, polishing, and plating. Even faulty stoves, heaters, and running automobiles create it. See Figure 3.4 for some examples of areas that asphyxiants can be found.

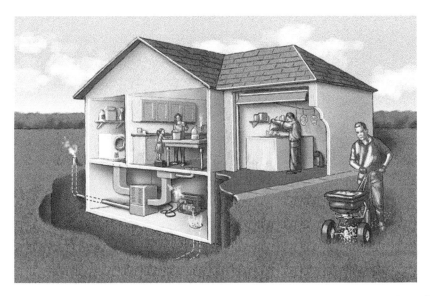

Figure 3.4 Simple asphyxiants can occur in any confined space where a gas replaces the oxygen in the area. Chemical asphyxiants can be a gas, like carbon monoxide, or a liquid/solid such as a cyanide salt (sodium cyanide). They are common in industry, storage facilities, and homes.

Effects of Hypoxia

When hypoxia occurs from either an exposure to a simple asphyxiant or a chemical asphyxiant (CO, CN, H_2S) (Figure 3.5), there are predictable signs and symptoms that occur.

The immediate symptoms are usually from the CNS that is the largest consumer of oxygen. Although the brain accounts for only 2% of the total body weight, it accounts for about 20% of the total oxygen consumption. Once hypoxia occurs, a victim will experience transient/delayed memory and short-term memory loss. In addition, they will suffer from fatigue, lethargy, and slow reaction time.

As hypoxia continues, there are notable changes to brain function and the following may be seen: fatigue, lethargy, and inattention. Hypoxia may also cause excitement, euphoria, disorientation, and uncoordinated movement, and headache.

Initially, hypoxia expands cerebral blood vessels and increases cerebral blood flow (Figure 3.6). As hypoxia-induced metabolic acidosis continues, there is an increase in cerebral vasoconstriction resulting in cerebral interstitial edema and an increase in both intracranial pressure and an increase in pressure in the cerebrospinal fluid. This process causes the signs and symptoms represented what is called in this book Rick's Hypoxic Syndrome during compensation (early stages of hypoxia) and decompensation (later stages of hypoxia),

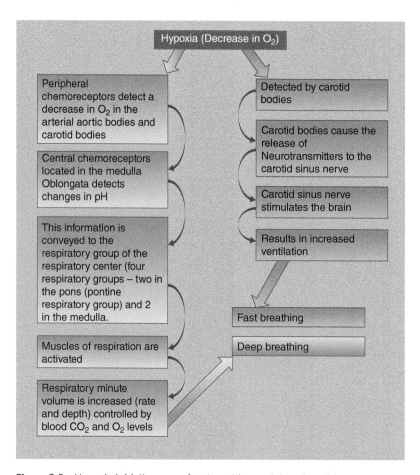

Figure 3.5 Hypoxia initially causes fast breathing and deep breathing.

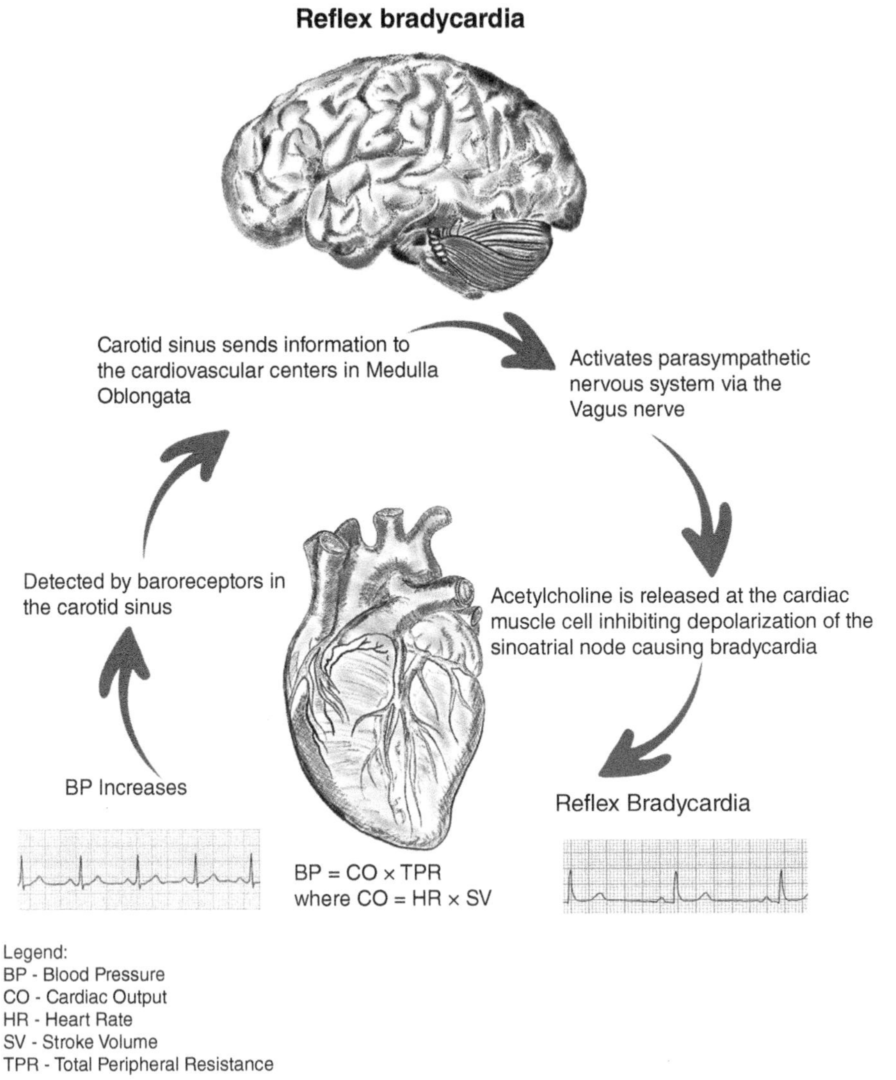

Reflex bradycardia

Carotid sinus sends information to the cardiovascular centers in Medulla Oblongata

Activates parasympathetic nervous system via the Vagus nerve

Detected by baroreceptors in the carotid sinus

Acetylcholine is released at the cardiac muscle cell inhibiting depolarization of the sinoatrial node causing bradycardia

BP Increases

Reflex Bradycardia

$BP = CO \times TPR$
where $CO = HR \times SV$

Legend:
BP - Blood Pressure
CO - Cardiac Output
HR - Heart Rate
SV - Stroke Volume
TPR - Total Peripheral Resistance

Figure 3.6 Reflex bradycardia is a complex process that occurs when there is a sudden rise in blood pressure.

causing the death of brain cells. Figures 3.5 and 3.6 describe the physiologic cause of both breathing and heart rate changes that occur during hypoxia.

Rick's Hypoxic Syndrome is a group of physiologic changes that take place early during a hypoxic event. The symptoms only appear for a brief period of time. During this time, the body is attempting to compensate for its lack of oxygen. Figure 3.7 represents both compensation and decompensation that occurs when the body becomes hypoxic.

But eventually hypoxia begins to affect the production of energy and neurologic function and decompensation occurs almost completely reversing the body's compensation effort (see Figure 3.8). This syndrome occurs with simple asphyxiation, carbon monoxide, cyanide, and hydrogen sulfide poisoning but does not occur with nitrite/nitrate poisoning because of the vasodilating effects from nitrates. The below flow chart displays this physiologic based syndrome.

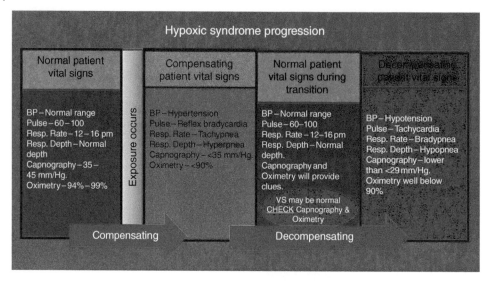

Figure 3.7 Rick's hypoxic syndrome occurs when there is an exposure to a chemical or simple asphyxiant and there is a sudden loss of oxygen being used by the cells. This condition is only temporary and may last only minutes. When the vital signs reverse, this is an indication of severe cellular hypoxia and cellular death.

Hypoxic syndrome signs and symptoms

Compensatory/early sign

Respiratory compensation
  Rapid breathing
  Deep breathing
  Shortness of breath

Cardiovascular compensation
  High blood pressure
  Reflex bradycardia
  EKG changes indicating hypoxia

Decompensation/late sign

Respiratory decompensation
  Slow breathing
  Shallow breathing
  Apnea

Cardiovascular decompensation
  Low blood pressure
  Fast pulse rate
  Intraventricular irritability (multifocal PVCs and ST segment changes)

Figure 3.8 Hypoxic syndrome signs and symptoms.

Simple Asphyxiants

Simple asphyxiants cause a decrease in the amount of oxygen entering the external respiratory system. The normal oxygen concentration in the air is approximately 21%. Occupations Safety and Health Association (OSHA) states that any atmosphere containing less than 19.5% oxygen is an oxygen-depleted atmosphere. It is at this level that decision-making abilities are significantly decreased and may hinder a person's ability to escape from the environment. They displace oxygen, resulting in lower concentrations reaching the alveoli and ultimately supplying the cells. The end consequence is cellular hypoxia, acidosis, and eventually death. Examples of simple asphyxiants are carbon dioxide, nitrogen, and halogenated extinguishing systems.

Experience; Death at McDonald's: "Five Lousy Feet"

An 18-year old McDonald employee and a delivery truck driver were killed at a McDonald's in Sanford, Florida, last week.

Saturday afternoon, the carbon dioxide delivery driver arrived to fill the restaurant's carbon dioxide tank, according to police. He took the supply hose through an emergency door and into a children's play area but could not reach a fill valve in an adjacent storeroom because the door was locked, as reported by a Sanford police Sargent.

Officials think that an 18-year-old McDonald's worker scaled a 10-ft wall into the storeroom, which had no ceiling, to open the door from the inside, but it had a deadbolt lock that required a key. The delivery man then apparently fed the hose over the wall to the 18-year-old. Soon the 18-year-old was overcome by the gas, and the delivery man apparently climbed over the wall and also was overcome. Another restaurant employee found them.

Assessment/Treatment for Simple Asphyxiants
Basic Life Support
- Remove patient from the environment.
- Immediately administer 100% oxygen; if unconscious secure airway and deliver 100% oxygen.

Advanced Life Support
- If conscious, apply CPAP with a PEEP setting of ≥10 cm of H_2O
- If unconscious, apply PPV using a PEEP valve set at greater than 10 cm of H_2O
- Secure an advanced airway if needed.
- Start IV of 1000 cc normal saline at a KVO rate.
- Follow general medical protocol – treat symptoms.
- The goal is to maintain oximetry >92% and capnography between 35 and 45 mm/Hg.
- Contact receiving hospital and provide **HazMat Alert.**

Pediatric Considerations (Simple Asphyxiants)
- The same protocols used for adults are appropriate for pediatric patient exposures.

Chemical Asphyxiants

Chemical asphyxiants are those chemicals that interfere with either the transportation of oxygen in the body or directly interfere with the usage of oxygen by the cells. Unlike simple asphyxiants that displace oxygen with another gas, some of the chemical asphyxiants can be ingested or absorbed through the skin and not just an inhalation hazard. All asphyxiants cause some similar symptoms. Figure 3.9 represents some of these similar symptoms.

Carbon Monoxide Poisoning

Properties. Colorless, odorless gas. Soluble in water, benzene, and alcohol. Specific gravity of 0.96716.
Explosive Limits. 12–75%.
Uses. Carbon monoxide is used for organic synthesis and metallurgy.
Other Names Include. Carbon monoxide, carbonous oxide, carbon(II) oxide, carbonyl, flue gas, monoxide.
Toxic Limits. OSHA PEL 50 ppm/NIOSH TWA 35 ppm, IDLH 1200 ppm

Figure 3.9 Asphyxiation causes cyanosis, shortness of breath, high blood pressure, and reflex bradycardia. Some asphyxiates can cause inaccurate pulse oximetry readings. Depending on the asphyxiate the skin may be red or cyanotic in color.

History

Carbon monoxide poisoning was first described by John Scott Haldane, MD, in 1919. Doctor Haldane was an industrial, medical physician whose responsibilities including caring for workers building tunnels under waterways in the Midwestern United States. Because these tunnels were kept under pressure to keep the water from flowing in and flooding the tunnel; the ventilation was somewhat compromised. During this time, early gasoline-powered tools were used to dig and move the earth from the tunnel. These pieces of equipment were normally vented to the outside, but many leaked and created an unsafe environment for the workers. It is interesting to note that there were few checks and balances and no federal agency overseeing the safety of the workers. OSHA was not established until 1970 when President Nixon signed it into law. Through Doctor Haldane's experience caring for these workers, he composed several research papers describing both the symptoms and treatment for carbon monoxide poisoning. He is also well known for his study on the bends that are typically suffered by scuba divers. Because his workers were under pressure in the tunnels and would sometimes leave too quickly without allowing time for the body to off-gas the pressurized nitrogen, he had the opportunity to both study and treat many victims of the bends. The current treatment for both the bends and for carbon monoxide that are used today began its roots in Doctor Haldane's research.

Pathophysiology

Carbon monoxide (CO) is an odorless, colorless, and nonirritating gas. It is primarily produced as a byproduct of incomplete combustion, but it is also produced in industry for use in manufacturing, health assessment equipment, and laboratories. The most common exposures are related to internal combustion engines, smoke from fires, and industrial facilities. The gas can only enter the body via the respiratory system, where it crosses the alveolocapillary membrane and enters the bloodstream. Once in the bloodstream, its effects are systemic.

Although CO is odorless and colorless, some victims of acute exposure report an acidic taste. It is important for those working the fire or emergency medical service to have a high suspicion of the presence of the gas related to how it is used. Emergency responders should be able to identify common signs and symptoms displayed by a victim and determine emergency responses necessary to keep themselves and their crew safe. There are many cases of emergency responders responding to basic first aid calls only to become injured or overwhelmed from the CO because it was not recognized as an exposure event.

Some fire-based EMS agencies have gone to the extent of providing a CO-only detector on their EMS equipment bags, so the responders are alerted if they enter a dangerous environment containing CO.

Year after year, news stories about families being killed by malfunctioning heating systems within their homes are found in both newspapers and TV. Others are killed when smoke from a house fire builds up and hinders the occupant's escape. Although smoke contains many toxic chemicals (reviewed in the smoke inhalation section of this book), it is usually carbon monoxide and cyanide that kill victims of fire. Approximately 12,000 Americans die each year from fires; 80% of those deaths are attributed to exposure to toxic gases, including CO.

Note: Methylene chloride (dichloromethane) is a halogenated hydrocarbon found in products used in both industry and around the home. Most commonly, methylene chloride is found in items such as paint strippers, insecticides, aerosol propellants, and Christmas tree bubble lights. Once it gains access into the body (most commonly inhalation but can be absorbed through the skin, eyes, and GI system), it is rapidly absorbed and then slowly metabolized into carbon dioxide (70%) and carbon monoxide (30%). Because methylene chloride is fat-soluble (polar solvent), it concentrates in the fat cells, specifically the liver, where it gradually releases, causing a long-term exposure to carbon monoxide. Exposure to paint strippers containing methylene chloride has led to the death of workers from carbon monoxide poisoning.

The symptoms of acute exposure to CO may appear suddenly and without warning. The concentration of this gas within the body and the effects experienced are directly related to the physical activity, metabolic rate, and susceptibility of the victim. The percentage of gas in the environment is also a determination of how quickly symptoms appear. Today, many EMS organizations are carrying Masimo, Rainbow Technologies that, through the use of a finger light refractory device, can determine the amount of carboxyhemoglobin (the combination of carbon monoxide and hemoglobin) in the body. In addition, because of the color and opacity of carboxyhemoglobin (bright red and transparent), the pulse oximetry will display a false high reading, most often at 100%. Many times, the sign of bright red skin is not noticeable until the carboxyhemoglobin is at a dangerously high percentage.

The effects of CO poisoning cause the victim to rapidly become disoriented and lose muscular coordination. This is the main contributing factor to the high loss of life found in residential fires where victims are found in unusual places, like closets and bathrooms, even when they are well aware of their surroundings and escape routes.

Once CO is inhaled, it crosses the alveolocapillary membrane and enters the bloodstream. In the bloodstream, it rapidly combines with hemoglobin, forming a strongly bonded compound called carboxyhemoglobin (COHg). Hemoglobin's affinity to CO is about 200 times greater than it is for oxygen. The bond between hemoglobin and CO is also much stronger than the bond between hemoglobin and oxygen, making the problem even more difficult to treat.

Carbon monoxide is also attracted to and binds with an enzyme called cytochrome oxidase. The enzyme found in the cells is necessary for cellular respiration and the production of energy. The bonding of CO to cytochrome oxidase does not occur early in the exposure event. In fact, it only occurs when the cells are deprived of oxygen. Typically, cytochrome oxidase has a great affinity to oxygen. This affinity to oxygen is nine times greater than it is for CO. So, the process of CO bonding to cytochrome oxidase only occurs when there is an absence of oxygen related to CO-induced hypoxia. Some toxicologists believe that this mechanism is responsible for the symptoms noted during CO poisoning and not the widely believed physiology caused for the lack of oxygen transport to the cells. Because of the distinctive manner in which carbon monoxide affects the body, it is classified as a chemical asphyxiant (prevents the usage of oxygen by the cells).

Carbon monoxide causes injury to the tissues by combining with hemoglobin (HG) to form a non-oxygen-carrying compound, carboxyhemoglobin. The first signs and symptoms displayed are caused by the organs with the highest oxygen consumption, specifically the CNS and the myocardium. The complaints are related to hypoxia suffered by these organ systems. The most commonly reported symptoms are headache, dizziness, and agitation, related to the effects of hypoxia on the CNS. Other symptoms such as chest pain can be related to myocardial hypoxia. Sudden exposures to very high concentrations may cause almost immediate syncope and cardiovascular collapse.

The sign of cherry red skin reported on carbon monoxide victims is related to the color of the victim's blood. Carboxyhemoglobin is bright red and transparent to light. Once a victim is poisoned by CO, both the arterial and venous blood is bright red giving the patient's skin the characteristic red appearance. This sign is not always reliable. The absence or presence of this sign should not be relied upon to determine a field diagnosis. Depending on the natural color of the victim's skin, the redness associated with CO poisoning may be present with carboxyhemoglobin levels of 30% or not seen until concentrations reach 60%.

The bright red and translucent appearance fool the pulse oximeter into reading carboxyhemoglobin as oxyhemoglobin. Therefore, the oximeter reading on a patient suffering from carbon monoxide poisoning should not be relied upon to estimate the oxygenation status of the patient. In fact, a patient severely poisoned with CO will read unusually high with an expected SaO_2 reading at 100%.

The only way of determining the oxygenation of cells in the field is to analyze the Capnography reading. When aerobic (with oxygen) metabolism is underway, there is a normal production of carbon dioxide. Normal carbon dioxide production is read on exhalation and should be between 35 and 45 mm/Hg. If CO poisoning is well underway, the production of carbon dioxide will be severely reduced, giving a low capnography reading (below 35 mm/Hg).

Carbon monoxide is also a normal end product of metabolism, so expected levels of COHg are found in the blood of healthy individuals. These levels are usually around 1% for healthy non-smoking persons. The level for moderate smokers is higher and is usually around 4–6%. Heavy smokers can approach 10%. Symptoms for a non-smoker usually do not appear until levels of about 10% are found in the blood with the initial symptom being a temporal headache.

Concerns

Several important factors influence the actual levels of carbon monoxide within the blood of an exposed victim. Pregnant patients are an extreme concern to caregivers. Pregnant patients exposed to carbon monoxide have shown levels of carboxyhemoglobin twice as high in the fetus than that of a mother's level. The half-life of total carboxyhemoglobin in fetal blood is around 15 hours, and even with oxygen therapy, carboxyhemoglobin takes five times longer to reduce.

Children also exhibit sensitivity to the exposure of CO. They appear to be especially vulnerable because of their higher metabolic rate that causes a faster heartbeat and more rapid breathing. In addition, children have a reduced blood volume as compared to an adult.

Patients with heart disease or chronic lung diseases such as chronic obstructive pulmonary disease (COPD), pulmonary fibrosis, or asthma are also at a much higher risk. These patients already suffer from decreased oxygen-carrying capability and/or decreased circulation, and even minor exposure can be devastating to this group.

Concentrations, exposure times, metabolic rates, and the physical activity of those exposed all play a part in determining the severity of the effects. The most recent American statistics state that from 1999 to 2014 there were 24,890 CO poisoning deaths. This included 6653 accidental and 18,231 intentional with an average of 1555 deaths per year. This number does not reflect the number of lives lost in smoke inhalation and fire situations.

Signs and Symptoms

The following concentration of carboxyhemoglobin elicits certain symptoms. These particular symptoms are generalizations and not found in all patients at the represented percentages. At best, they are approximations of predictable events.

Other symptoms caused by CO poisoning can mimic neurologic conditions such as multiple sclerosis, parkinsonism, bipolar disorders, schizophrenia, and hysteria. Cardiovascular effects are related to rapid hypoxia experienced by the myocardial tissue. Those victims with previous coronary artery disease may suffer angina pain at carboxyhemoglobin levels less than 10%. In healthy younger patients, CO poisoning lowers the ventricular fibrillation threshold, and if hypoxia continues, cardiac arrhythmias are the most frequent cause of death (see Figure 3.10).

The carbon monoxide level in a structure fire rapidly reaches well over 10,000 ppm within only minutes. This is important to note because the permissible exposure level (PEL) is only 35 ppm and the immediately dangerous to life and health (IDLH) is 2000 ppm. A firefighter not wearing or improperly using a self-contained breathing apparatus for 10 minutes can reach a carboxyhemoglobin level of 28%. A free burning fire generates less carbon monoxide than a smoldering fire. Many fire departments enforce air pack regulation during the active firefighting phase but do not enforce the use of air packs during the overhaul phase when

Carboxyhemoglobin levels

CO levels	Signs and symptoms
0–10%	No symptoms
10–20%	Tightness across forehead and headache
20–30%	Headache and throbbing temples
30–40%	Severe headache with N/V and dim vision
40–60%	Coma and convulsions
>60%	Cardiovascular collapse and respiratory failure

Figure 3.10 Although every person reacts slightly differently, these are the general signs and symptoms suffered at the percentage of carboxyhemaglobin levels.

CO levels are the highest. In reality, the highest percentage of toxic gas is generated during the overhaul phase. Air packs are warranted even more during this dangerous time.

Where Carbon Monoxide Is Typically Found

Carbon monoxide can be generated from anything that burns, including internal combustion engines. There are also other processes in industry that generate carbon monoxide. Below is a list of areas where carbon monoxide is very common:

Internal Combustion Engines. A typical passenger vehicle produces about 7% carbon monoxide. Some literature suggests that in the past years, up to 50% of the annual suicides were committed using carbon monoxide from internal combustion engines. Today, with the advent of catalytic converters, the level of carbon monoxide exhausting a vehicle has been reduced. Lawnmowers, blowers, string trimmers, generators, and other gasoline-powered machinery that do not employ the use of a catalytic converter still produce large amounts of carbon monoxide.

Faulty Gas and Kerosene Stoves. These stoves, when not ventilated properly or when the air/fuel mixture is not correct, generate extensive carbon monoxide and are responsible for many deaths each year.

Charcoal and Sterno™ Fire. Use of these fires in unventilated areas has been responsible for a number of deaths in recent years. The use of charcoal fires to intentionally generate carbon monoxide with the intent of committing suicide has seen a significant increase over the past five years.

Industry and Laboratories. Some industries and laboratories produce carbon monoxide, and others use it in chemical reactions. It can be purchased in high-pressure cylinders for industrial use. There has been an influx of home inventors creating a hydrogen/carbon monoxide mix to use for fuel. This is not only a dangerously flammable fuel but is also toxic.

Fires. Fire not only produces carbon monoxide but a number of other dangerous chemicals. Carbon monoxide is flammable and, under the right conditions, causes flashover or backdrafts.

Methylene Chloride is also called dichloromethane. It is a widely used solvent in home and industry. Found in paint strippers, aerosols, and other home products. Methylene chloride can be absorbed through the skin, eyes, ingested, or inhaled. Once in the body, the liver provides Stage 1 detoxification, changing 30% of the product into carbon monoxide, poisoning the patient.

Decontamination and Danger to Responders

Providing decontamination to a victim of carbon monoxide is usually not necessary. The gas can only be inhaled, not absorbed through the skin. Usually, the clothing of a victim will not hold enough of the gas to injure a medical provider, but it is always a good practice to grossly decontaminate (remove the clothing) the victim involved in any type of exposure. Fire victims who have been removed from burning structures have many chemicals, including cyanide, formaldehyde, and hydrogen chloride, within their clothing. Therefore, it is very important to, at the minimum, remove the clothing of these victims.

Field Treatment

Remove the patient from the toxic atmosphere and administer 100% oxygen. Provide continuous positive airway pressure (CPAP) or attach a positive end expiratory pressure (PEEP) device to a bag valve mask device and set the PEEP at $8-10\,cm/H_2O$ with 100% oxygen being supplemented. It is important to note that:

1) Carboxyhemoglobin is <u>halved in room air in six hours</u> (This is called the half-life).
2) Carboxyhemoglobin is <u>halved when breathing 100% oxygen in 1.5 hours</u>.
3) Carboxyhemoglobin is <u>halved when breathing 100% oxygen while in a hyperbaric chamber in less than one hour</u>.

Basic Life Support

- Immediately administer 100% oxygen if intubated or as high a concentration of oxygen as possible with a non-rebreather mask if conscious. If unconscious, secure the airway to deliver 100% oxygen.
- Capnography to determine appropriate cellular metabolism (between 35 and 45 mm/Hg)

Advanced Life Support

- If the patient is unconscious, secure an advanced airway.
- Provide CPAP if conscious and PPV if unconscious utilizing a PEEP set at ≥10 cm of H_2O.
- Start IV to KVO.
- Treat unconscious patients by evaluating glucose levels and administration of naloxone (Narcan) 2 mg IVP, IO, IM, or intranasal every three to five minutes up to 8 mg.
- Contact receiving hospital and the Regional Poison Control Center.

Note: A carboxyhemoglobin of 9% in a pregnant female can be enough to cause fetal hypoxia. Although the patient may not display critical signs, the fetus may be in critical condition. Immediate transport to a medical facility is necessary.

Note: Pediatric patients will always have more severe effects compared to adults in the same environment due to their higher metabolism.

Pediatric Considerations (Carbon Monoxide)

- Apply CPAP after exposure to carbon monoxide to reduce the half-life of carboxyhemoglobin.
- When providing CPAP to pediatric patients, always start the PEEP setting at low pressures of 5 cm H_2O. Then increase it in increments of 1 cm H_2O, as tolerated by the patient.
- In patients less than 12 years old, the maximum PEEP should be 15 cm H_2O.
- For patients 12 years old and above, the maximum PEEP could be increased to as high as 20 cm H_2O if tolerated.

Definitive Treatment and Follow-Up If a medical hyperbaric chamber is located within the area, all attempts should be made to transport the patient to that facility. Hyperbaric oxygen (Dr. Haldane's treatment) has been proven to provide the best reduction of carboxyhemoglobin. At a pressure of 2.5 atm and 100% oxygen, carboxyhemoglobin concentration is reduced to 1/2 in less than one hour. Providing any level of oxygen to the alveoli under pressure utilizing CPAP or hyperbaric oxygen increases the oxygen-carrying capability of the blood by forcing more oxygen into the solution of the blood when the hemoglobin is bound with carbon monoxide.

Cyanide Poisoning – Hydrogen Cyanide, Cyanide Salts, and Cyanide Containing Gases

Cyanides are any chemical compound containing a cyano group; CN. This group consists of a carbon that is triple-bonded to a nitrogen atom. Inorganic cyanides are the salts of hydrocyanic acid and are highly toxic. Sodium cyanide (NaCN) and potassium cyanide (KCN) are examples of these.

Organic cyanides usually contain the term nitrile in their chemical name. Here the CN group is covalently bonded to a carbon-containing group. An example would be a bonded methyl (CH_3) group with CN producing methyl cyanide (acetonitrile).

Hydrocyanic acid is also known as hydrogen cyanide (HCN). This is the acid that killed the historically significant chemist Karl Wilhelm Scheele. HCN is a liquid but is highly volatile and is used in industry to produce acrylonitrile, one of the toxic products found in ABS (acrylonitrile/butadiene/styrene) plastic.

Properties Include:

- Soluble in water
- Commercial-grade materials can be 96–99.5% pure
- The specific gravity of liquid is 0.688
- The vapor density of gas is 0.938
- Explosive limits: 6–41%

Uses: Cyanide is used in the manufacture of acrylonitrile (vinyl cyanide), acrylates, adiponitrile, cyanide salts, dyes, chelates, rodenticides, and insecticides. There are many more uses for this versatile chemical that are too numerous to list.

Hydrogen cyanide is a colorless gas, while sodium cyanide or potassium cyanide (cyanide salts) are in a crystal form. Many books and authorities state that the smell of cyanide is like a "bitter almond," but in reality, this smell is produced when the body is attempting to detoxify cyanide and, as a result, forms benzaldehyde. Benzaldehyde has a distinctive smell of almonds and give actual almonds their characteristic smell. Benzaldehyde can be found in most grocery stores, being sold as an "Artificial Almond Extract" that is used in cooking recipes. Therefore, the body of someone poisoned with cyanide may smell like bitter almonds because of the presence of benzaldehyde, but the chemical itself does not smell of almonds.

History

The use of cyanide as a poison date back to the early Egyptians and Romans. It is well known for its use in executions and homicides. The use of both apricot and peach pits being crushed and added to food or drink has been documented throughout history. Peach and apricot pits, along with some 150 other plants, contain cyanide.

There are a number of more recent events that bring focus on the effects of cyanide poisoning. On November 18, 1978, 909 members of the Peoples Temple under the leadership of a religious fanatic named Jim Jones (also called Jonestown) died in a mass suicide. While in San Francisco, the Temple's activities came under scrutiny after a scathing newspaper article. This eventually led to the Temple's relocation to Guyana. Even after moving the Temple and its members out of the United States, the government kept up the investigation. After a visit (investigation) by the federal government lead by Congressman Leo Ryan, Jim Jones ordered the murder of the congressional party, including Congressman Ryan. He then commanded that all Temple members drink from a vat containing grape-flavored drink laced with cyanide. Nine hundred seven members of the congregation died as a result. This was the largest mass suicide ever conducted and except September 11, 2001, which was the largest loss of civilian life during one event.

In another event that took place in 1982, nine people lost their lives to cyanide poisoning when a person took Tylenol capsules and placed potassium cyanide in them. The bottles were then placed back in the original package and returned to shelves in stores to be sold. As a result, nine people died from cyanide poisoning. Some of the nine deaths were from this event, while several others were from copycat events. These poisoning led to a dramatic change in pharmaceutical packaging standards that included the tamper-resistant seals on both the bottles and boxes of over-the-counter pharmaceuticals.

Pathophysiology

Cyanide is one of the most rapid-acting poisons. It gains access into the body most often through inhalation but can also be ingested and absorbed through the skin and mucous membranes. It can cause death within minutes in higher concentrations. The speed of the poisoning and death is dependent on the quantity, route of entry, exposure time, and the activity level of the victim. Remember that increased metabolism increases the effects of a poisoning, so children are especially at risk. The speed at which cyanide gas works has been witnessed during executions (when gas chambers were used in the United States). Typically, a convicted prisoner dies within seconds of an exposure to cyanide gas.

Following an exposure, cyanide rapidly enters the cells, bonding with an enzyme, cytochrome oxidase. The bonding of cyanide to cytochrome oxidase prevents the transport of electrons from cytochrome to oxygen. This bonding results in a disruption of the electron transport chain, and the cell can no longer produce adenosine triphosphate (ATP) for energy and the production of carbon dioxide as a waste product. The enzyme adenosine triphosphatase is necessary for cellular respiration (the use of oxygen to convert glucose and oxygen to energy – aerobic metabolism). By binding to cytochrome oxidase, cyanide stops aerobic metabolism by not allowing oxygen to be carried throughout the metabolic process. Without cytochrome oxidase, the cell cannot utilize oxygen from the bloodstream. The process eventually causes the cells to have limited functionality under anaerobic metabolism. This inadequate anaerobic state ultimately causes a decrease in cellular energy production, metabolic acidosis, cellular suffocation, and death. It is interesting that the half-life of cyanide in the body is only about an hour, but during a true exposure, death takes place well before detoxification or excretion can take place.

During a quick assessment, performing a capnography reading will show a significant reduction in exhaled carbon dioxide, indicating levels well below 35 mm/Hg. This can help with diagnosing CN poisoning. Remember, simple asphyxiants, CO, and H_2S will also cause a decrease in carbon dioxide production and will also display a low CO_2 reading using capnography.

Cyanide can cause death in an oral dose of fewer than 5 mg/kg. This dosage can be explained better by saying that a 150 lb patient will die ingesting a dose of about seven drops of concentrated cyanide. Inhalation of cyanide gas in a concentration of 0.3 mg/l is fatal within seconds.

Many sources suggest that the presence of a bitter almond smell (also reported as a garlicky or oniony smell) may be noted either emitting from the patient's body or on their breath. The ability to detect a bitter almond smell is a sex-linked recessive trait, and only 60–80% of the general population can smell the aroma. The remaining cannot detect the odor due to a sensory deficit. The deficit is greater in men than in women by a ratio of 3 : 1.

Because a small amount of cyanide is present in various foods and other environmental exposures (like cigarette smoke), the body naturally produces an enzyme named rhodanese that converts cyanide to thiocyanate that is renally excreted. The enzyme works through a chemical combination of sulfur and cyanide. Unfortunately, this enzyme is only present in

Blood cyanide levels and associated signs and symptoms

Serum cyanide levels	Signs and symptoms
0–0.2 µg/ml	Normal (non-smoker)
0.1–0.4 µg/ml	Normal (smoker)
0.5–1.0 µg/ml	Anxiety, confusion, unsteadiness, tachypnea
1.0–2.5 µg/ml	Headache, palpitations, dyspnea, depressed level of consciousness, atrial fibrillation, ectopic ventricular beats
2.5–3.0 µg/ml	Loss of muscular coordination, convulsions, reflex bradycardia, respiratory depression, coma
Greater than 3.0 µg/ml	Apnea, cardiovascular collapse, asystole

Figure 3.11 Although every person reacts slightly differently, these are the general signs and symptoms suffered at each serum cyanide level.

small amounts and responds too slowly to assist in detoxifying cyanide during a true exposure.

Signs and Symptoms

The patient may present with a wide variety of signs and symptoms because cyanide poisoning affects virtually all of the cells in the body. The most sensitive target organ is the CNS, where the urgent need for oxygen is first sensed. Early effects can include headache, restlessness, dizziness, vertigo, agitation, and confusion. Later signs are seizures, coma, and death (see Figure 3.11).

Since it is impossible to determine the serum cyanide level in the field under an emergency situation, it might be more beneficial to recognize the early signs of cyanide poisoning so preparations can be made for the late signs, especially as they relate to the respiratory and cardiovascular system.

To Put 3.0 up/ml in Perspective:

- Average US human weighs 178 lb
- 60% of the average human is water
- $178 \times 0.60 = 107$ lb of water
- $107/8.33 = 12.8$ gal
- 3785 ml/gal
- $3785 \times 12.8 = 48,619$ ml in the average human
- One grain of salt weighs an average of 300 up.
- $48,619 \times 3$ µg/ml (3 µg/ml = cyanide fatal dose) = 145,857 total µg
- 145,857 µg/300 µg/grain of salt = 486 grains of salt
- A typical McDonalds™ salt packet contains about 270,000 µg of salt
- So, about half of a salt packet from McDonalds filled with sodium or potassium cyanide would be a lethal dose

Compliments of Lieutenant Robert Cruthis, Seminole County Fire Department

Hypoxic Syndrome: Hypoxic Response to Chemical Asphyxiants (With the Exception of Nitrates and Nitrites)

Early (during compensation)	Late (decompensation)
Respiratory effects	
Tachypnea	Decrease in respiratory rate
Hyperpnea	Respiratory depression
Dyspnea	Apnea
Cardiovascular effects	
Flushing	Hypotension
Hypertension	Acidosis
Reflex bradycardia	Tachycardia
Arterioventricular (AV) nodal or intraventricular rhythms	ST changes and cardiovascular collapse EKG-ST segment changes

Definitive Diagnosis

Red Blood Cell or Plasma Cyanide Levels   This test takes considerable time, and this test is not generally helpful if an active poisoning is taking place. This test does provide subsequent confirmation and can be used for both documentation or for therapeutic treatment during the follow-up assessments.

Arterial and Venous Blood Gases   In the case of cyanide poisoning, the oxygen tension will be at a normal level or high while the venous oxygen tension will also be high, and the arteriovenous oxygen difference will be less than 10%.

Field Assessment   During an active cyanide poisoning, the pulse oximeter will read abnormally high, even as high as 100%, while the capnography reading will read abnormally low, usually less than 35 mm/Hg. This is also true for CO and H_2S poisonings. So this cannot be diagnostic for cyanide but can be used to determine the differential diagnosis.

Where Cyanide Is Commonly Found

Cyanide is found in industry either combined with other chemicals or in a relatively pure form. It is also used in commercial pest control and used in metal plating, polishing, and used to join two dissimilar metals. Probably the most common poisoning seen today by emergency responders involves victims of smoke inhalation. Many products in use today will emit toxic, cyanide-containing gases when burned. The plastic commonly found in automobiles and aircraft is ABS plastic. The ABS stands for acrylonitrile, butadiene, and Styrene. Acrylonitrile is actually vinyl cyanide, and when it burns an extensive amount of cyanide is released.

Approximately 150 plant species also contain a cyanide-producing agent called cyanogenic glycoside (amygdalin). Amygdalin was produced and distributed in health food stores under the trade name Laetrile. Laetrile was marketed as a cancer treatment but was subsequently withdrawn after many deaths were related to its use and therapeutic drug testing proved that there were no benefits to its use. There remain many people who believe that there is a conspiracy by drug manufacturers to keep Laetrile away from those who would benefit from its use. In fact, actor Steve McQueen used Laetrile for the treatment of his colorectal cancer only to die from it a short time afterward.

A number of cyanide poisonings were reported from the use of Laetrile, and it currently remains off of the market in the United States. There is no doubt that people in the United States are still able to obtain Laetrile through mail order and the chance of seeing a cyanide patient as a result of its use is high.

Most seeds or pits of cyanide-producing plants contain both amygdalin and an enzyme named emulsion. Once amygdalin is mixed with emulsion, the reaction yields hydrocyanic acid. Therefore, just crushing these seeds and adding water releases the cyanide. As few as one peach or apricot pit or 15–60 crushed seeds can cause significant cyanide toxicity.

$$C_{20}H_{27}NO_{11} + 2H_2O = 2C_6H_{12}O_6 + C_6H_5CHO + HCN$$
$$amygdalin + water = glucose + benzaldehyde + cyanide$$

The following is a list of areas that commonly either produce or use cyanide.

Industry	**Pest control**
Metal cleaning	Insecticides
Metal polishing	Fumigants
Electroplating	Rodenticides
Metal heat treating	**Fires**
Synthesis of plastics and rubbers	Polyurethane
Soil sterilization	Polyacrylonitriles
Fertilizers	Nylons, wools, and silks
Clandestine PCP (phencyclidine) labs	

Decontamination of Patients

Cyanide and its components are largely water-soluble. A victim exposed to cyanide in the liquid, solid, or gaseous form should be decontaminated by completely removing the victim's clothing, followed by a complete washing with soap and water or a water-based decontamination solution. Decontamination consisting of, at least, the removal of contaminated clothing should be done on all smoke inhalation victims. This step is necessary to reduce secondary contamination to emergency care providers, transporters, and the hospital staff.

Emergency Medical Field Treatment

Natural detoxification of cyanide in low concentrations takes place quickly. In fact, the half-life of cyanide in the body is about one hour. The initial treatment of small doses of cyanide is focused on keeping the patient alive with as much supportive treatment as necessary as the natural detoxification process works. In more severe cases of exposure to high concentrations, aggressive treatment is necessary.

Since respiratory arrest develops quickly, establish a good patent airway. Then as quickly as possible, begin advanced treatment utilizing either the traditional Cyanide Antidote Kit (Pasadena or Lily Kit) or the CyanoKit (by Meridian).

Basic Life Support

- Support respirations and provide 100% oxygen via a non-rebreather (NRB) mask.
- If the patient is not breathing, immediately begin CPR and other supportive measures.

Advanced Life Support

- If the patient is conscious, apply CPAP at 100% oxygen with the PEEP setting ≥ 10 cm/H_2O.
- If unconscious or not breathing, provide an advanced airway and positive pressure ventilation. Consider the use of a PEEP valve on the BVM with a setting of 10–15 cm/H_2O.
- Prepare to give one of the two available cyanide antidotes.
  - The CyanoKit is preferred for confirmed <u>cyanide</u> poisoning or smoke inhalation:
    - CyanoKit (hydroxocobalamin 5 g)
    - The CyanoKit will **NOT** work on a patient poisoned with hydrogen sulfide.
  - For <u>Hydrogen Sulfide or Cyanide</u>:
    - Utilize sodium nitrite IV to bond with the sulfide ion, which will reduce the cellular metabolic toxicity of the poisoning.

Note: Never give the nitrite-based components of the Lilly or Pasadena Cyanide Antidote Kit to smoke inhalation patients who may already be compromised with elevated levels of carboxyhemoglobin and methemoglobin.

CyanoKit – Hydroxocobalamin (Preferred Treatment for Cyanide Poisoning)

In 2006, the FDA finally approved treatment for Cyanide Poisoning that has been used in Europe (specifically France) for decades. The use of vitamin B_{12a}, or hydroxocobalamin, has shown to be an efficient and less toxic way to treat cyanide poisoning. In the United States, the hydroxocobalamin is found in 5 g vials packaged in the "CyanoKit." Hydroxocobalamin has a number of advantages over the traditional Cyanide Antidote Kit.

Note: Hydroxocobalamin is vitamin B12a. This is not to be confused with vitamin B12 that is found in many energy drinks. Vitamin B12 is cyanocobalamin and will not treat cyanide poisoning.

Hydroxocobalamin works by bonding with cyanide to form cyanocobalamin, which is really excreted. It is usually chosen over the traditionally nitrate-based Cyanide Antidote Kits because it is relatively few side effects, does not compromise the oxygen-carrying capability of the blood, and does not cause a drop in blood pressure. The reported side effects do include erythema, headache, gastrointestinal distress, itching, and non-specific IV site reactions. Some people will also experience a rise in blood pressure. Other side effects include a facial rash that may appear as late as a week after infusion and discoloration of the urine that can last up to a couple of days.

Because of its rapid effect and less serious side effects, hydroxocobalamin is the choice for smoke inhalation patients who are suspected of having cyanide exposure.

Generally speaking, there should be no other infused drugs through the same IV line that is being used for hydroxocobalamin. But, just to make this clear, there has also been no published data that identifies any incompatibility of hydroxocobalamin with any other drug. If no other IV line is accessible, the hydroxocobalamin line can be used only after the line has been flushed and efforts are taken to avoid mixing any drug with hydroxocobalamin. The adult dose is:

Give 5 g Hydroxocobalamin IV Infusion Over 15 minutes.

- Start a dedicated IV line.
- Reconstitute the 5 g vial with 200 ml of 0.9% sodium chloride. (Lactated Ringers and D5W can also be used if sodium chloride is not available). Renders 25 mg/ml.
- To mix the reconstituted drug, invert or rock the vial. Do not shake.
- Administer 5 g over 15 minutes (~15 ml/min). An additional 5 g dose may be administered if necessary.

(Lily or Pasadena) Nitrite-Based Cyanide Antidote Kit (Used for hydrogen sulfide or (Cyanide if the CyanoKit Is Not Available)

Nitrites (sodium nitrite) convert hemoglobin into methemoglobin. Methemoglobin then competes with cytochrome oxidase for the cyanide ion, actually attracting the cyanide away from the cytochrome oxidase. Methemoglobin is formed when the ferrous iron (Fe^{+2}), located in the hemoglobin is oxidized (adds an electron) to form ferric iron (Fe^{+3}). This change in the valence of the iron atom within the hemoglobin attracts the cyanide ion, freeing the cytochrome oxidase to again participate in aerobic cellular metabolism. The methemoglobin bonding with the cyanide forms cyanmethemoglobin, which is then renally excreted. The last step is to infuse sodium thiosulfate, which acts as a cleanup agent by changing the remaining cyanide into a relatively harmless substance, thiocyanate.

Cyanide Antidote Kit (2 Step Process)

Note: Amyl Nitrite was the first step in the nitrite-based antidote, but it has been removed from the market.

- Step 1 – establish an IV of normal saline and immediately give:
  - IV Sodium nitrite ($NaNO_2$) 10 ml of a 3% solution over two minutes while closely monitoring the patient's blood pressure.
  - Sodium nitrite converts approximately 20% of the circulating hemoglobin to methemoglobin. Methemoglobin bonds with both cyanide and sulfide and are excreted in the urine.
- Step 2
  - IV Sodium thiosulfate ($Na_2S_2O_3$): 50 ml of a 25% solution over 10 minutes.
  - If after 30 minutes the symptoms of Cyanide poisoning still exist give a second dose of both sodium nitrite and sodium thiosulfate at 1/2 the initial dose.

Note: The use of sodium nitrite cause approximately 20% of the hemoglobin to be converted to methemoglobin. Therefore, 20% of the hemoglobin will not carry oxygen. Securing an advanced airway and providing 100% oxygen is critically important.

- Contact the receiving hospital and Regional Poison Control Center.

Pediatric Consideration (Cyanide/Hydrogen Sulfide)

CyanoKit – Hydroxocobalamin (Preferred treatment for Cyanide Poisoning)

- The safety and effectiveness of the CyanoKit have not been established in the pediatric population (per package insert). The CyanoKit has been successfully used in the non-US market at a dose of 70 mg/kg to treat pediatric patients.

Cyanide Antidote Kit – Lilly or Pasadena Kit

- Use with great caution in children – Sodium nitrite ($NaNO_2$) 10 ml of a 3% solution administer 0.33 ml/kg of a 3% solution over 10 minutes.
- Sodium thiosulfate ($Na_2S_2O_3$) 50 ml of a 25% solution over 10 minutes. Monitor BP.
- Children – Administer 1.65 ml/kg up to 50 ml over 10 minutes.

Hydrogen Sulfide Poisoning

Properties:

- Colorless
- Has a strong odor

- Water and alcohol soluble
- Specific gravity 1.189
- Explosive limits: 4.3–46%

Hydrogen sulfide is used in industry for the purification of hydrochloric and sulfuric acids and as an analytical reagent. Found naturally in below-grade confined spaces and any location where there is rotting organic matter.

History

For many years, hydrogen sulfide poisoning was the leading cause of death related to toxic inhalation in the workplace. The OSHA's 29 CFR 1910.146 – Permit Required Confined Space standard was developed primarily because of the dangers of hydrogen sulfide in confined and below-grade spaces. Today it has fallen to second but still poses a severe risk in the workplace.

Hydrogen sulfide is formed during the decompensation of organic materials. It has a distinctive odor of "rotten eggs" that may only be present briefly under higher concentrations. The rotten egg odor can be detected at very low concentrations. Some research states that concentrations as low as two parts per billion (ppb) can be detected through its odor. One of the characteristics of hydrogen sulfide gas is its ability to cause paralysis of the olfactory sensors (olfactory fatigue). In other words, in a high enough concentration, a person will soon not smell anything at all and not realize that they are in danger. The victim may only receive a brief whiff of the chemical before a total numbing of the smelling sense develops. That is why so many workers have fallen victim to the effects of this potent poison.

History has proven that two or more victims result when an incident involving hydrogen sulfide takes place. The typical scenario involves the first worker falling unconscious in a confined space without prior warning. When the victim's coworkers see him collapse, they rush in to rescue, resulting in the death of both.

Those who have spent many years working in the sewers or around low doses of the gas develop chronic conjunctivitis that they have termed "gas eye" or "sewer eye." Some say that the workers use the severity of conjunctivitis caused by irritating properties of hydrogen sulfide to determine if the affected worker should take a day off or at least work in an area without a concentration of the gas.

Pathophysiology

Hydrogen sulfide acts on the body as a respiratory irritant. In low doses, it affects the upper respiratory system with bronchospasms, localized irritation, and edema and causes conjunctivitis and many other nonspecific complaints, such as dizziness, nausea, and headache.

In higher concentrations, the respiratory irritations go deeper into the respiratory system causing chemically induced pulmonary edema. Susceptibility to these symptoms varies among individuals and may be increased due to multiple previous exposures, causing a sensitized-type reaction.

Both high and low levels cause cellular asphyxia in much the same way that cyanide does. Once hydrogen sulfide enters the bloodstream, it finds its way to the cells, where it combines with the enzyme cytochrome oxidase. Cytochrome oxidase is responsible for providing the catalyst for the transfer of electrons to oxygen. Hydrogen sulfide stops the transfer of electrons and the ability of oxygen to move through the process of forming ATP for the production of cellular energy. Aerobic respiration becomes very inefficient, and anaerobic respiration causing acidosis and eventually cellular hypoxia and death. Hydrogen sulfide is a stronger inhibitor of cytochrome oxidase than cyanide. For this reason, hydrogen sulfide is considered to be more toxic than cyanide.

Signs and Symptoms

The signs and symptoms presented by a patient who has had significant exposure to hydrogen sulfide are almost exactly the same as a victim poisoned with cyanide. Both cyanide and hydrogen sulfide cause cellular hypoxia in all organ systems of the body; therefore, the symptoms can be very diverse. The organ systems that use the most oxygen are the first to demonstrate the symptoms. Organs such as the brain (CNS) and heart are affected early in the poisoning. Any deficits noted in these areas should stimulate the medical provider to deliver care immediately.

The early signs of hydrogen sulfide poisoning are a direct result of the hypoxic state suffered by the brain. The brain, not receiving oxygen, sends messages to the respiratory center to breathe faster (tachypnea) and deeper (hyperpnea). The cardiovascular system responds by increasing blood pressure. The sudden increase in blood pressure results in reflex bradycardia. As hypoxia gets more severe, irritability is noted in the heart in the form of AV nodal and intraventricular dysrhythmias.

The later signs are the result of prolonged cellular hypoxia. As the brain stops sending out signals to breathe faster and deeper, a slower respiratory rate will be noted. The same goes for vascular tension that relaxes in the absence of signals to constrict, and the result is hypotension. Soon acidosis and cardiovascular collapse take place, and the patient, without treatment, will die.

Hydrogen sulfide is one of the most rapid-acting poisons and can cause death within minutes. The emergency care provider must aggressively provide supportive and ALS antidote as soon as possible if the patient is to benefit.

Where Hydrogen Sulfide Is Commonly Found

The most common site where hydrogen sulfide is found involves areas where organic materials are breaking down. These include septic tanks, sewers (both sanitary and non-sanitary), wells, tunnels, and mines. It is also prevalent in gas and petroleum pumping operations. Occasionally, hydrogen sulfide is found in laboratory settings. Hydrogen sulfide should be monitored for anytime a rescue of any type is done within a confined space or below-grade setting.

Decontamination and Significant Danger to Rescuers

Decontamination of the skin is not necessarily due to poor cutaneous absorption. Removal of clothing should be sufficient to remove a chance of secondary contamination, although rescuers have been known to lose consciousness while giving mouth-to-mouth ventilation to a poisoned victim.

Field Treatment

The use of nitrites is recognized as an antidotal treatment for hydrogen sulfide poisoning. Nitrites attract the sulfide from the cytochrome oxidase and reactivate aerobic metabolism. This reaction forms sulfhemoglobin (very similar to methemoglobin and also does not carry oxygen), which is then quickly filtered by the kidneys and eliminated from the body. Because the physiology is very similar to cyanide toxicity, the treatment is very much the same. Treatment consists of the following drug and dosages.

Basic Life Support

- Support respirations and provide 100% oxygen via a non-rebreather (NRB) mask.
- If the patient is not breathing, immediately begin CPR and other supportive measures.

Advanced Life Support
- If the patient is conscious, apply CPAP at 100% oxygen with the PEEP setting $\geq 10\,cm/H_2O$.
- If unconscious or not breathing, provide an advanced airway and positive pressure ventilation. Consider the use of a PEEP valve on the BVM with a setting of $10-15\,cm/H_2O$.

(Lilly or Pasadena) Nitrite-Based Cyanide Antidote Kit (Used for Hydrogen Sulfide)
- Using the Cyanide Antidote Kit for Hydrogen Sulfide requires only one step.
  - Establish an IV of normal saline and immediately give:
    - IV Sodium nitrite ($NaNO_2$) 10 ml of a 3% solution over two minutes while closely monitoring the patient's blood pressure.
    - Sodium nitrite converts approximately 20% of the circulating hemoglobin to methemoglobin. Methemoglobin bonds with both cyanide and sulfide and are excreted in the urine.

Note: The use of sodium nitrite cause approximately 20% of the hemoglobin to be converted to methemoglobin. Therefore, 20% of the hemoglobin will not carry oxygen. Securing an advanced airway and providing 100% oxygen is critically important.

- Contact receiving hospital and the Regional Poison Control Center.

Pediatric Consideration (Hydrogen Sulfide)
Cyanide Antidote Kit – Lilly or Pasadena Kit
   Use with great caution in children – Sodium nitrite ($NaNO_2$) 10 ml of a 3% solution administer 0.33 ml/kg of a 3% solution over 10 minutes.

Definitive Treatment and Follow-up Care
Several sources suggest the use of hyperbaric oxygen to enhance the elimination of hydrogen sulfide. The treatment of hydrogen sulfide toxicity is not one of the identified uses of HBO, and therefore, until significant proof is established recognizing its benefit, it will not become a part of standard care.

   Hospital follow-up care should include monitoring the patient for aspiration pneumonia and late-developing chemically induced pulmonary edema. Pulmonary edema may occur up to 48–72 hours after the initial exposure. Also, if nitrites were used as an antidote, a methemoglobin level should be drawn.

Nitrites, Nitrates, Nitrobenzene Poisoning

History
Nitrites (NO_2), nitrates (NO_3), organic, and inorganic nitrogen compounds are found in our environment in many different products and forms from colognes to paints and fertilizers. Although commonly found in both home and work environments, poisonings more commonly occur due to intentional misuse or recreational purposes. A pharmaceutical, amyl nitrite, a yellow liquid that was available in glass pearls wrapped in cloth much like an ammonia capsule, began the trend of using nitrates for recreational purposes. Although not available anymore as a pharmaceutical, a similar product called isobutyl nitrite is sold over the counter and over the internet in a product called RUSH.

   There are a number of recreational uses for Rush. It is used to enhancing the "high" experienced by drug users. It is used in conjunction with other drugs such as marijuana so the user can experience a different kind of high. More commonly, Rush is also used to enhance sexual pleasure.

Rush can be purchased in a number of locations, such as head shops where drug paraphernalia is sold or in sex shops, but the most prevalent area of purchase is from the internet. The production and sale of nitrites are primarily uncontrolled.

Most users of Rush fall into one of two categories. Juvenile drug abusers experiment with nitrites to compound the effects of other drugs. The other group, primarily men, use the drug to enhance sexual activity. The abusers report that the use of nitrites just before orgasm increases the intensity and duration of the orgasm.

There has been a recent trend to use nitrates to commit suicide. Sodium nitrate can be purchased from some of the most popular shopping websites for about 15 U.S. dollars. It is listed on some suicide sites as a painless means of committing suicide. The over consumption of sodium nitrate causes a perfuse loss of blood pressure and creates a hemoglobin that cannot transport oxygen. Sodium nitrate is used as a fertilizer and food preservative.

The signs and symptoms of nitrate poisoning may be subtle, but with a suspicion of the poisoning, knowledge of pathophysiology, and good assessment skills, the poisoning can be rapidly identified and treated.

Pathophysiology

Exposure to nitrogen compounds can occur through absorption of the skin, mucous membranes, respiratory system, and gastrointestinal tract. The most efficient routes of access are through the gastrointestinal and respiratory systems. In industry, nitrogen compounds are found most commonly in the solid form, making inhalation of dust the primary route of poisoning in the workplace.

The initial exposure to a nitrogen compound is the loss of blood pressure. Nitrogen compounds cause a relaxation of vascular smooth muscle causing dilation of the vessels. The loss of vascular tone results in a dramatic loss of blood pressure.

Once absorbed into the bloodstream, these nitrogen compounds combine with hemoglobin and change the iron molecule, ferrous iron (Fe^{+2}), into ferric iron (Fe^{+3}). The conversion of the iron molecule changes the hemoglobin into a non-oxygen-carrying compound called methemoglobin. This change causes the poisoned patient to become hypoxic.

The condition of hypoxia caused by methemoglobin is termed methemoglobinemia. The color of the blood (actually the red blood cells), while in this condition, changes from bright red to chocolate brown and is easily assessed during a blood draw. Even blood vigorously shaken in an oxygen-rich atmosphere will remain chocolate brown in color.

Another diagnostic clue comes from the pulse oximeter. The oximetry unit works on the principle that oxygen-rich hemoglobin is bright red and transparent. Because of this principle, a patient suffering from increased methemoglobin, which is dark brown and opaque, will show an inaccurately low reading. Normally, a 1% methemoglobin level is found in healthy individuals. Significant signs and symptoms do not appear until levels at or above 10% are formed.

Nitrates and nitrites are also used medicinally because of their ability to relax smooth muscles, thus causing vasodilating effects. Nitroglycerin, a nitrate, and amyl nitrite are two examples of pharmaceutical use of these chemicals. Furthermore, the conversion of ferrous iron into ferric iron is used during the treatment of cyanide poisoning. The original Cyanide Antidote Kit contained amyl nitrite and sodium nitrite that during a single dose of each converts up to 25% of the hemoglobin into methemoglobin. The change of the electrical valence of the iron atom during this process attracts the cyanide ion out of the cell, allowing the cell to return to aerobic metabolism.

Masimo® has developed a technology (they call it the "Rainbow") that allows first responders and even hospital emergency department personnel the ability to measure both carboxyhemoglobin and methemoglobin levels through a finger probe similar to the oximetry. If the technology is available, these levels can be assessed and treatment options selected based on the findings. This field technology is expensive, so many emergency medical agencies have not been able to purchase it, but, over time, the price will become more reasonable, and all prehospital advanced life support providers will have the capability.

Signs and Symptoms

The toxic effects of exposure to nitrogen compounds are exhibited in two different ways. First, the vasodilating effects are evidenced by the following signs and symptoms, which are usually short-lived and often corrected with only simple treatments.

1) Throbbing headache and fullness of the head are due to dilation of the meningeal vessels.
2) Flushing of the neck and face are signs of cutaneous capillary and vasodilation.
3) Dizziness and syncope are due to cerebral ischemia related to profuse vasodilation.
4) Tachycardia, sweating, and pallor are responses of the sympathetic nervous system to hypotension.

Second, evidence of methemoglobin presents with signs and symptoms related to methemoglobinemia.

Percentage related to normal hemoglobin (%)	Signs and symptoms
10–15	Mild cyanosis in extremities but usually no other symptoms
20–30	Shortness of breath, changes in mental status, and changes in vital signs
Approximately 50	Lethargy
Greater than 60	Cardiac collapse and death

Cyanosis may be caused by low levels of methemoglobin. The early appearance of cyanosis will be seen in the nail beds, mucous membranes, and commonly on the lips. These low levels may not generate other cyanotic-type symptoms seen in the CNS. In contrary, cyanosis caused by high levels of methemoglobin and hypoxia will cause symptoms such as confusion, anxiousness, and be displayed as a severely dyspneic patient.

Because of the darkness of the blood color generated even with low levels of methemoglobin, the pulse oximeter will be almost immediately affected, providing a false low reading. This tends to mislead the prehospital care provider even in the absence of other CNS symptoms. If the care provider suspects that methemoglobin is affecting the oximetry reading, an immediate capnography reading should be obtained. Methemoglobin does not affect the accuracy of the capnography reading, and it will give critical information to the healthcare provider concerning the real condition of the patient related to the level of methemoglobin found in the blood.

Where Are Nitrogen Compounds Found

Organic nitrogen compounds are used in the manufacturing of dyes, paints, polishes, photographic chemicals, crayons, food preservatives, and fertilizers. For those who intentionally misuse the chemical for the physiological effects, nitrogen compounds are purchased at head shops and sex stores and, of course, available on the internet.

Except for intentional misuse, fertilizers account for the highest percentage of nitrogen compound poisonings. Well water contaminated with fertilizers is often reported as a cause of nitrogen compound poisoning. Those most susceptible to poisonings of this nature are neonates. Fetal hemoglobin is more susceptible to oxidations by nitrates and nitrites. Furthermore, neonates also lack the natural ability to reduce methemoglobin. In children and adults, the body produces an enzyme called methemoglobin reductase, which effectively changes methemoglobin back into hemoglobin by converting ferric iron back into ferrous iron.

Field Treatment

Treatment of nitrogen compound poisoning initially involves the removal of the patient from the chemical then removing the chemical from the patient. This is accomplished by decontamination efforts using soap and water.

Basic Life Support

- Immediately administer 100% oxygen if conscious; if unconscious secure an airway to deliver 100% oxygen.
- Do not rely on the pulse oximeter as the darker colored methemoglobin will create a false low reading.

Advanced Life Support

- Masimo/Rainbow technology can assess the level of methemoglobin and provide guidance for further treatment
- If conscious, place the patient on CPAP with the PEEP setting at or above 10 cm of H_2O.
- Assess cellular respiration by evaluating the end-tidal CO_2 (capnography).
- If unconscious or -not breathing, secure an advanced airway and use PPV with a PEEP valve set above 10 cm of H_2O.
- Start 2 – IVs of 1000 cc normal saline utilizing large-bore catheters.
- If hypotensive, position the patient and reassess the blood pressure.
- If needed, provide a fluid challenge.
- If still unable to maintain BP, start a dopamine drip or norepinephrine to maintain a systolic pressure of greater than 90 mm/Hg.
  - Dopamine:
    - Mix 400 mg in a 250 ml bag of 0.9% D5W = 1600 mcg/ml.
    - Provide an IV infusion rate of 5 μg/kg/min.
    - If this does not improve the BP, the infusion rate can be increased to 10 μg/kg/min.
  - Levophed (norepinephrine):
    - 8–12 μg/min titrated to maintain systolic BP higher than 90 mm/Hg.
- If symptomatic and methemoglobinemia can be confirmed, then:
  - Administer methylene blue, 1–2 mg/kg IVP over five minutes. (methylene blue may momentarily affect the pulse oximeter because of the opaqueness of the drug. The dose may be repeated but not to exceed a maximum dose of 7 mg. Must be used with caution or not used in patients with G6PD deficiency).

Methylene blue activates an enzyme methemoglobin reductase, which then reduces (Fe^{*3}) ferric iron back into (Fe^{+2}) ferrous iron and again enables the hemoglobin molecule to carry oxygen. If methylene blue is not given, the methemoglobin is identified by the kidneys as a foreign substance and filtered out in the urine. This leads to a longer recovery because of the significant loss of hemoglobin after the poisoning.

Note: Methylene blue will turn secretions and urine green in color.

Note: Methylene blue must be used with caution or not considered at all in patients with G6PD deficiency. G6PD deficiency is a genetic abnormality that results in an inadequate amount of glucose-6-phosphate dehydrogenase (G6PD) in the blood. Hemolytic anemia develops when red blood cells are destroyed faster than the body can replace them, resulting in reduced oxygen flow to the organs and tissues.

• Contact receiving hospital and Regional Poison Control Center.

Pediatric Consideration (Methemoglobin) The same protocols used for adults are appropriate for pediatric patient exposures.

• Provide 1–2 mg of 1% methylene blue IV push over five minutes.
• It is not recommended to provide more than 2 doses in a prehospital setting.

Cholinergic Toxidrome

Organophosphate Insecticide Poisoning

These insecticides can enter through all routes. They bind with the enzyme acetylcholinesterase that is necessary to remove acetylcholine (neurotransmitter) from the synapse. The chronic presence of acetylcholine in the neuropathway causes overstimulation of primarily the parasympathetic nervous system (although the CNS and somatic systems are affected to a lesser degree).

There are many different classifications of pesticides. Some are for insects, some are for wild animals such as coyotes, and some for rodents such as mice and rats. But the pesticide that seems to be one of the most commonly involved in accidental (and sometimes intentional) poisonings is organophosphate insecticides. In fact, 80% of all hospitalizations related to pesticide poisonings are from organophosphate insecticides.

Organophosphates have gained such great popularity because of their effectiveness as a nuisance bug killer. Organophosphate's popularity is related to its relatively unstable chemical structure, allowing them to break down in the environment and not contaminate soil for extended periods of time. Once in the body, these insecticides do not persist in the body tissue like other insecticides used in the past, such as dichlorodiphenyltrichloroethane (DDT), which was an organochloride.

In fact, organophosphates have mostly replaced the use of DDT as an insecticide worldwide (see Figure 3.12). The use of DDT was outlawed in the United States in 1970 and further outlawed in both Mexico and Canada with the signing of the North America Free Trade Agreement (NAFTA) in 1994. Interestingly, although DDT was outlawed in the United States, it was still produced here for other countries even after its use was outlawed in the United States. Probably the most toxic organophosphate insecticide is one of the first synthesized in the 1800s called tetraethylpyrophosphate (TEPP). In fact, TEPP is still in use and is available as a commercial insecticide.

Organophosphate first gained notoriety during WWII when the German military developed and tested various versions of organophosphate as antipersonnel nerve agents. The Germans developed military nerve agents such as Tabun, Sarin, and Soman, all were developed to cause rapid illness after very small doses. The United States and allied countries joined in the development of these agents after WWII, and VX, one of the most potent nerve agents, was developed by both the United States and Great Britain.

Figure 3.12 Organophosphates and carbamates are used in farming and home pest control. They can be purchased at local hardware stores and building supply centers. In developing countries, organophosphates are commonly used to commit suicide.

As recent as September 2020, military nerve agents have been used to kill people. There are many reports of government-sponsored attacks on groups and individuals with various nerve agents. The article below demonstrates the most recent attack.

Experience: Novichok Nerve Agent Used Against Russian Dissident Has Dark History

Los Angeles Times
LAURA KING STAFF WRITER
SEPTEMBER 2, 2020
3:42 PM

The German government on Wednesday announced "unequivocal" proof that hospitalized Russian anti-corruption crusader Alexei Navalny, who fell gravely ill August 20 while aboard a domestic flight, was poisoned by a military-grade agent from the banned Novichok group.

Nearly two weeks after being stricken, Navalny, 44, remains in a medically induced coma, his condition described as stable. Amid an international outcry, he was brought to Berlin two days after his hospitalization in Siberia. Russian doctors who initially treated him denied any sign of poisoning, suggesting – ludicrously, critics said – a metabolic disorder such as low blood sugar.

The German government's conclusion, based on toxicology tests conducted by a specialized military laboratory, heightened already intense suspicions that the Kremlin was behind the attack on one of President Vladimir Putin's most prominent critics because of Novichok's state origins, distinct toxicological footprint, and the difficulty of making it.

Armed with medical proof of the sophisticated nerve agent's use, Germany demanded an explanation. Putin's government, which has previously scoffed at the notion it played any role in Navalny's collapse, responded with trademark nonchalance.

Novichok was last in the headlines in 2018, when it was identified as the poison used against turncoat Russian spy Sergei Skripal in an assassination attempt attributed to Russian intelligence agents. He and his daughter Yulia, who was visiting him at his home in the quiet English town of Salisbury at the time of the attack, nearly died, but both eventually recovered.

In today's political climate, it would not be unexpected to find radical hate groups using these chemicals for terrorism. In fact, this was exactly the case in Tokyo during March 1995.

Experience: Tokyo Subway, Site of an Attack Using Sarin Nerve Agent

Tokyo, Japan, March 20, 1995. A terrorist attack left at least eight dead, 17 critical, 37 serious, 984 treated with antidotes, 4073 treated as outpatients with an estimated total of 5510 Tokyo residents involved. These victims were exposed in a subway station to a chemical warfare agent named "Sarin." Sarin (2-[fluoro(methyl)phosphoryl]oxypropane, $C_4H_{10}FO_2P$ or [9(CH₃)2CHO(CH₃)FPO]) is a wartime nerve agent of organophosphate makeup that has a strong cholinesterase-inhibiting effect that is toxic by inhalation and absorption. The symptoms suffered by victims of this attack are similar to but more severe than parathion. Witnesses stated that a strong odor was present and described the odor as acidic. The Sarin was intended to be released under the seat of government but was released before reaching its destination.

Until the last few years, the United States had large stockpiles of nerve agents. But these agents have been steadily destroyed through incineration, and now they are mostly gone from American soil. There is much speculation that other countries still have large quantities of nerve agents stored and ready for possible use.

There are many commercial grade organophosphate pesticides that are extremely toxic and can cause serious symptoms if a person was to be exposed. Organophosphate pesticides such as parathion are used in the agricultural industry. It is readily absorbed through the skin, eyes, respiratory, and gastrointestinal systems. The organophosphate pesticides typically available to the public include Malathion and Diazinon; both are considered to be much safer for home use because they have been synthesized to have poor absorption qualities and low oral toxicity.

One other interesting side note concerning these insecticides, organophosphate insecticide poisonings are the number-one way to commit suicide in the world. Although it is not preferred in the United States (gunshots make up 51% of the suicides), it is easily available in almost every country, and virtually everyone in contact with it understands that it is an extremely toxic substance capable of causing death.

Pathophysiology

As previously described, organophosphates can enter through all routes. The effects on the body are systemic, affecting the nerve conduction pathways. In general, organophosphates display a large range of differences in their toxicities, which are due to the difference in the chemical's ability to penetrate skin or absorb through the oral route.

Organophosphate chemicals cause harm by bonding with an enzyme called acetylcholinesterase. This enzyme is responsible for removing the neurotransmitter acetylcholine from the synaptic junction. Acetylcholine is one of the numerous neurotransmitters used by the body. It is found in the CNS, somatic system, sympathetic nervous system and is most prevalent in the parasympathetic nervous system. After an exposure occurs, all of these systems display symptoms that are described below.

Acetylcholine is the primary and most important chemical transmitter located at the synaptic junction of the parasympathetic nervous system. As an electrochemical impulse is transmitted through a nerve cell, it reaches a junction (synapse) between two nerve cells. For the impulse to be transmitted to the next cell, acetylcholine is released. The acetylcholine then traverses across the junction, stimulating the conduction at the distal end of the synapse, thereby continuing the electrical impulse. Once it has fulfilled its function, an enzyme, acetylcholinesterase, is released to break down the acetylcholine by hydrolysis into two inert chemicals. All of these chemical reactions take place millions of times a day in the body, with each only lasting a fraction of a second.

Note: One of the most common and consistent symptoms are pin point pupils (miosis) but much less common is fully dilated pupils that occasionally occur after an exposure to an organophosphate.

Organophosphates cause harm by binding with the enzyme acetylcholinesterase, inhibiting it from functioning. The results are overstimulation and excitation of the nerve impulse because the synapse becomes flooded with acetylcholine. After the phase of the excitation, which lasts variable amounts of time, the nerve cell becomes paralyzed and stops functioning.

The nerve's most prominent symptoms come from those systems using acetylcholine as a neurotransmitter. They are:

- The central nervous system is causing anxiety, restlessness, convulsions, coma, respiratory and circulatory depression.
- The somatic (musculature) nervous system is causing fasciculation, cramps, weakness, paralysis, and cardiac arrest.
- The sympathetic nervous system is causing tachycardia and hypertension.
- The most prominent symptoms are displayed by the parasympathetic nervous system. These symptoms include sweating, constricted pupils, lacrimation, salivation, wheezing, diarrhea, bradycardia, urinary incontinence, hypovolemic hypotension, bronchial constriction, excessive bronchial excretions. See Figure 3.13 for the extensive symptoms common with organophosphate exposure.

Signs and Symptoms

The acronyms SLUDGEM and DUMBELS are all used to describe the symptoms found in a victim of organophosphate poisoning:

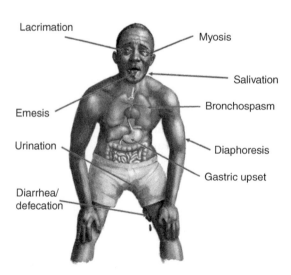

Lacrimation

Myosis

Salivation

Emesis

Bronchospasm

Urination

Diaphoresis

Gastric upset

Diarrhea/ defecation

Figure 3.13 Organophosphate poisoning is commonly displayed by several obvious symptoms remembered by using the acronyms DUMBELS or SLUDGEM.

Salivation	Diarrhea
Lacrimation	Urination
Urination	Miosis (pinpoint pupils)
Defecation	Bronchospasm (wheezing)
Gastrointestinal	Emesis
Emesis	Lacrimation
Miosis	Salivation

Figure 3.14 The killer Bee's of organophosphate poisoning are bradycardia, bronchorrhea, and bronchoconstriction. These are the symptoms that present the greatest immediate danger to a victim of organophosphate poisoning.

Bronchoconstriction

Bronchorrhea

Bradycardia

The most common cause of death in organophosphate exposure are, what is called, the *Killer B's*. These include Bradycardia, Bronchorrhea, and Bronchoconstriction (see Figure 3.14).

Location of Organophosphate Insecticides

These chemicals are found in large quantities in farming and industrial settings. They have also been the insecticide of choice for the home setting for many years, although carbamate insecticides are also popular. Those developed for home use are found in lower concentrations and are formulated to have poor skin and respiratory absorption. Regardless, both commercial and home insecticides have caused significant exposures. The pesticides get to their destination by truck and rail, which increases the possibility of an accident in all communities.

Decontamination and Significant Danger to Rescuers

Typically, the patient presents with the strong scent of pesticides on his breath and some body fluids. Protection with rubber gloves and respiratory gear is necessary until the patient is stripped and appropriately decontaminated. Soap and water are usually enough to decontaminate, but if the odor remains, a second washing may be required.

There are many beliefs concerning the body fluids of a victim of organophosphate poisoning. First, vomit, urine, and feces may all be contaminated with the organophosphate chemical. On the other hand, sweat and tears are probably not. There is no clear path for the organophosphate chemical to infuse into these body fluids. Therefore, the best route for secondary contamination is through direct contact with vomit, urine, and feces. Of course, direct contact with any of the body fluids should be avoided.

The other misinformation that is often presented is the vaporization of the chemical into the air from the setting of an ambulance or hospital treatment room. When dealing with commercial or over-the-counter organophosphate pesticides, the danger from inhalation of the insecticides is very small. Insecticides are formulated to be sprayed on crops or household plants to kill insects. It would make no sense to place the chemical on a plant only for it to evaporate in a matter of minutes. It is most efficient if it will remain on the plant for some time after application. So, the organophosphate is formulated to have a very low vapor pressure and very high boiling points so that the chemical will not evaporate and will stay in place for an extended amount of time.

Organophosphates are very concentrated chemicals. These concentrates are then added to a carrying substance to increase the overall volume. The carrying agent will evaporate off of the crop, leaving behind the insecticide. The carrying agents are usually xylene or toluene. Both of these chemicals evaporate at about the rate of water, easily mix with organophosphate, and effectively increase the volume of the chemical. In fact, the odor associated with insecticides is really the odor of xylene or toluene.

For Example (Parathion):

- Parathion: VP is 0.000,04 and BP is 707 °F
- Toluene: VP is 280 and BP is 231 °F
- Xylene: VP 8.29 and BP 282 °F

With this example, it is easy to see how the carrying agents are vaporizing in an ambulance or treatment room, but the organophosphate is staying on the floor. Both toluene and xylene are CNS depressants and will cause dizziness, a staggered gate, and slurred speech if there is an extended exposure. These are often misdiagnosed as organophosphate poisoning, and those complaining of these symptoms are given atropine and 2-PAM antidotes when, in reality, they were never needed. The quickest treatment for xylene or toluene is fresh air, NOT atropine. The danger of secondary exposure comes from direct contact with the organophosphate chemical and not from the vaporization of the chemical once it has been spilled or applied.

Note: In the 1970s it became a popular drug practice to huff model airplane glue. These users were called "glue sniffers.". Before model glue was reformulated in the 1980s the "high" that was achieved from huffing glue was from xylene. Model glue contained high concentrations of xylene. This is the same carrying agent found in organophosphate insecticides and is responsible for the secondary effects experienced by pre-hospital and hospital emergency care providers. Not organophosphate poisoning. The determining diagnostic finding is the change in pupils. If a care provider is complaining of symptoms after carrying for an organophosphate poisoning patient the pupils should be the determining factor to give atropine or not. If the pupils are normal is size and reactivity, the symptoms are not related to organophosphate exposure but instead xylene exposure and they only need fresh air to recover.

Treatment

Atropine is the first medication utilized for the treatment of organophosphate poisoning. It is administered in high doses while under close cardiac and vital sign monitoring. Often patients who have significant signs and symptoms from organophosphate poisoning will be hypoxic from increased production of upper airway mucous and severe congestion. Great caution should be used in the administration of atropine, and it should not be given to a hypoxic patient. Oxygen needs to be administered prior to giving high doses of atropine. Giving a high dose of atropine to a hypoxic heart may cause ventricular fibrillation.

Atropine blocks the release of acetylcholine into the synaptic junction while the body is naturally metabolizing the organophosphate. Atropinization must be maintained until all of the absorbed organophosphates have been metabolized and the body has again produced sufficient quantities of acetylcholinesterase. Acetylcholinesterase is produced slowly at a rate of 1% a day. This means that a loss of 30% of the acetylcholinesterase due to bonding with organophosphate will take 30 days for the body to replace all of the lost enzymes. That is why it is so critical to begin the second step of treatment as soon as possible.

The second step of treatment is the administration of IV (or IM) Pralidoxime (also called Protopam or 2-PAM). Early doses of this drug will reduce the need for extended administration of atropine to reduce the release of acetylcholine into the synapse. Pralidoxime is the actual antidote to organophosphate poisoning, where atropine is really a first-aid treatment. Pralidoxime has three desirable effects:

1) First and foremost, it frees and reactivates acetylcholinesterase
2) Second, it detoxifies the organophosphate chemical
3) Third, it has anticholinergic effects (it helps atropine block the release of acetylcholine)

Therefore, both atropine and pralidoxime should be given to any organophosphate patient displaying effects from the poisoning. It is also important to give the pralidoxime as soon as possible after the exposure. The bond between the organophosphate and the acetylcholinesterase enzyme is a weak bond at first. Over time the bond gets stronger (ages). If the bond is allowed to age, the pralidoxime will not break the bond. As a result, the bonded organophosphate and acetylcholinesterase will be moved into the circulatory system and filtered out by the kidneys. This results in the loss of the acetylcholinesterase enzyme. Remember, it is only replaced by the body at a rate of 1% per day. The following case demonstrates the importance of giving pralidoxime early in the poisoning.

Experience: Malathion Overdose Treated Without Protopam

One case of parathion overdose required the use of 2950 mg of atropine over the first 24 hours of treatment. Over the next 24 days, the patient received 19,590 mg of atropine just to keep his symptoms in control. This was because there was no pralidoxime given during the treatment period (Golsousidis and Kokkas 1985).

Treatment

Decontaminate the patient to stop further absorption of the organophosphate through the skin. Exercise care to protect responders from secondary contamination.

Basic Life Support

- Immediately provide 100% oxygen using a non-rebreather mask.
- Be aware of excessive mucous production and suction as needed.
- Assess vital signs paying particular attention to both blood pressure and respiration.
- Ensure the use of PPE to avoid secondary contamination from vomit, urine, or feces.
- Significant secondary exposure from airborne organophosphates is rare unless the exposure is from a weaponized form of the chemical.
- If dermally exposed, start immediate decontamination to terminate the exposure.

Advanced Life Support

- Start IV with normal saline and give:
  - If symptomatic, give atropine 2–6 mg IVP until Atropinization (drying pulmonary secretions, relieve bronchoconstriction, normalize heart rate/BP) occurs.
  - There is not a maximum dose; use what it takes to dry pulmonary secretions, relieve bronchoconstriction, and reverse hemodynamically significant bradycardia. Use

extreme caution in a hypoxic patient (giving atropine to the hypoxic heart may stimulate critical arrhythmias).

○ Atropine should ONLY be given if muscarinic (DUMBELS or SLUDGE) effects are seen. It is NOT effective for nicotinic effects (muscle fasciculation/weakness/paralysis).

○ Pralidoxime (2-PAM, protopam chloride) IVP 1 g over no less than five minutes to remove insecticide from acetylcholinesterase and treat nicotinic symptoms.

   ▪ Pralidoxime may also be given IM as well (300 mg/ml – give 600 mg in 2 ml). May give a second dose after 15 minutes.

○ If seizures occur post-exposure or the patient experiences excessive anxiety, give:

   ▪ Diazepam (Valium) 10 mg IV/IO/IM. Maximum dosage 20 mg.

   ▪ Midazolam (Versed) 2.5 mg IV/IO. Repeat at two-minute intervals to a maximum dosage of 10 mg.

- If atropine and pralidoxime are given via a Mark 1 kit or DuoDote kit:

   ○ 1 Mark 1 = 1 DuoDote.

   ○ For mild symptoms (eye pain, miosis), no dosing is needed.

   ○ For moderate symptoms (sweating, systemic symptoms without intubation, twitching vomiting, weakness), give two kits and repeat in five minutes.

   ○ For severe systemic symptoms with intubation (unconscious, seizures, apnea, flaccid paralysis, significant bronchorrhea) give three kits.

   ○ If seizures follow the exposure, give:

     ▪ Midazolam (Versed) IV/IO 2.5 mg. Repeat at two-minute intervals to a maximum dosage of 10 mg or

     ▪ Lorazepam (Ativan) 4 mg IV at a rate of 2 mg/min. May repeat in 5–10 minutes.

- If the patient is wheezing or if capnogram indicates a shark fin pattern:

   ○ Initiate a nebulized updraft of either albuterol (proventil or ventolin) and/or ipratropium bromide (atrovent).

     ▪ Albuterol 2.5 mg (0.5 ml of 0.5% diluted to 3 ml with sterile normal saline) give via nebulizer. It may be repeated three times.

     ▪ Ipratropium bromide 0.5 mg (500 mcg)/2.5 ml via nebulizer repeat at 20-minute intervals for a total of three doses (usually only one dose given in the field).

   ○ Administer methylprednisolone (Solu-Medrol) 125 mg, IVP slowly.

     ▪ Monitor heart rate. Contact medical control if HR is over 150 bpm.

   ○ Consider the administration of Brethine/Terbutaline Subcutaneous Injection.

     ▪ Brethine (Terbutaline sulfate) 0.5 mg given subcutaneous injection (0.5 ml of a 1 mg/ml solution).

Pediatric Considerations (Organophosphate)

- Atropine 0.05 mg/kg IV/IO push or IM, min 0.1 mg, max 5 mg, initial dosing should be given as soon as possible (1 g mixed in 20 ml N.S. = 50 mg/ml solution).

- Pralidoxime (2-PAM), 25 mg/kg IV or IM (max 1 g IV, 2 g IM) over 5–10 minutes. Dose may be repeated in 30–60 minutes (one to two doses) for weakness or high atropine requirements.

- If seizures occur post-exposure or the patient suffers from extreme anxiety, consider the administration of:

   ▪ Diazepam (Valium) 0.2 mg/kg IV/IO/IM.
   or

   ▪ Midazolam (Versed) 0.1 mg/kg IV/IO. Repeat at two minute intervals to a maximum dosage of 0.6 mg/kg.
   or

   ▪ Lorazepam (Ativan) 0.05 mg/kg IV, can be repeated after 10–15 minutes.

Mark 1/DuoDote

- Mark 1 and DuoDote should NOT be used in children as the atropine dose is too high. The pralidoxime 600 mg (Mark 1 kit) can be used. Use the AtroPen auto injector for pediatrics.

AtroPen Auto-Injector

- Use the 1 mg Atropen to deliver an initial dose and follow with a second dose if symptoms are not controlled.

Carbamate Poisoning

Carbamates are another group of insecticides that cause exposure of injuries. Although not as dangerous as organophosphates, they still have the ability to stimulate severe symptoms. These pesticides are formulated from carbamic acid and have effects on the body very similar to those of organophosphates. Like organophosphates, carbamates inhibit acetylcholinesterase, causing a build-up of acetylcholine in the synaptic junction. They can enter the body through inhalation, ingestion, and dermal exposure. Unlike organophosphates that cause irreversible inhibition of acetylcholinesterase (unless treated with Protopam), carbamates' bond with acetylcholinesterase is temporary and short-lived. Carbamates also poorly penetrate the CNS; therefore, CNS depression and seizures are rarely found in these poisonings.

Some of the most common carbamates are sold under these trade names: Temic, Matacil, Vydate, Isolan, Furadan, Lannate, Zectran, Mesurol, Dimetilan, Baygon, Sevin. There are many other carbamate insecticides that are not listed. The important point is that if a responder can research the chemical and the formula of the insecticide cannot spell out CHOP, then, more than likely, the insecticide is a carbamate.

Treatment

Initial treatment is supportive, such as maintaining the airway and decontamination. Atropine is again the drug of choice. Pralidoxime (Protopam, 2-PAM) is not indicated in carbamate poisoning as the bond between acetylcholinesterase and carbamates is transient and time-limited. Usually, the poisoning only lasts from 6 to 12 hours.

Only atropine is used for the treatment of carbamate poisoning. Pralidoxime (also called Protopam or 2-PAM) is not used as the bond between carbamates and acetylcholinesterase is not permanent and is transient in nature. While treating carbamate poisoning, the dosage of atropine is greatly reduced and given for a much shorter period of time. Therefore, the use of a DuoDote is not necessary. In addition, because the carbamate is unable to enter the CNS, seizures are not seen in carbamate poisoning. In inhalation exposures, the healthcare provider must still assess the respiratory system and determine if bronchial constriction is occurring, and treat that condition as needed. The following are the recommended treatment protocols.

Basic Life Support

- Immediately provide 100% oxygen using a non-rebreather mask.
- Be aware of excessive mucous production and suction as needed.
- Assess vital signs paying particular attention to both blood pressure and respiration.
- Ensure the use of PPE to avoid secondary contamination from vomit, urine, or feces.
- Significant secondary exposure from airborne carbamate is rare.
- If dermally exposed, start immediate decontamination to terminate the exposure

Advanced Life Support

- Start IV with normal saline and give:
  - ○ If symptomatic, give atropine 0.4–2 mg IVP until atropinization (drying pulmonary secretions, relieve bronchoconstriction, normalize heart rate/BP) occurs.
  - ○ Atropine should ONLY be given if muscarinic (DUMBELS or SLUDGE) effects are seen.
- If the patient is wheezing or if capnogram indicates a shark fin pattern:
  - ○ Initiate a nebulized updraft of either albuterol (Proventil or Ventolin) and/or ipratropium bromide (Atrovent).
    - ■ Albuterol 2.5 mg (0.5 ml of 0.5% diluted to 3 ml with sterile normal saline) give via nebulizer. It may be repeated three times.
    - ■ Ipratropium bromide 0.5 mg (500 mcg)/2.5 ml via nebulizer repeat at 20-minute intervals for a total of three doses (usually only one dose given in the field).
  - ○ Administer methylprednisolone (Solu-Medrol) 125 mg, IVP slowly.
    - ■ Monitor heart rate. Contact medical control if HR is over 150 bpm.
  - ○ Consider the administration of Brethine/Terbutaline Subcutaneous Injection.
    - ■ Brethine (terbutaline sulfate) 0.5 mg given subcutaneous injection. (0.5 ml of a 1 mg/ml solution).

Pediatric Considerations

- Children less than 12 years: 0.02–0.05 mg/kg Atropine IV every 10–20 minutes until the drying of mucous membranes is seen, then repeat dose every 1–4 hours for at least 24 hours.
  - ○ Max single dose for children: 0.5 mg.
- Children 12 years of age or older: 1–2 mg IV initially. If no response, the dose is doubled every 5–10 minutes until the mouth is dry.

Hydrocarbons and Derivatives Toxidrome

Hydrocarbon Toxicity

Hydrocarbon exposures are common among the public and those working in industry. These chemicals are found in fuels, including gasoline and diesel fuels. They are also prevalent in products such as turpentine, furniture polish, kerosene, and household cleaners (see Figure 3.15).

Hydrocarbons include all compounds that are composed of (contain) carbon and hydrogen, but the ones that will be focused on here will include those derived primarily from petroleum. These include petroleum distillates and include straight-chain hydrocarbons and aromatic (containing the benzene ring) hydrocarbons.

Hydrocarbon gases have the ability to displace oxygen, causing an asphyxiating atmosphere. Non-halogenated hydrocarbons are also flammable or combustible, some creating an explosive atmosphere. Exposure to these hydrocarbons causes CNS depression and an anesthetic state. In addition, hydrocarbons can cause myocardial excitation and sensitization and lower seizure threshold. Epinephrine and other catecholamines should be avoided post-exposure as the heart is sensitized to these hydrocarbon based chemicals.

The types of exposures most often seen include accidental ingestion, intentional recreational abuse, accidental inhalation, dermal exposure, and intentional ingestion for the purposes of committing suicide. Complications related to hydrocarbon exposure are aspiration pneumonitis after ingestion of the chemical, followed by the CNS and cardiovascular complications.

Figure 3.15 Fuels and solvents fit into the hydrocarbon and derivatives toxidromes. Ethylene glycol (automotive antifreeze) also fits in this toxidrome category. These are common is every garage across America.

Pathophysiology

Hydrocarbons that are volatile (vaporize quickly) are much easier to inhale. Highly volatile chemicals are more likely to be inhaled or aspirated. Simple petroleum derivatives such as kerosene, gasoline, and furniture polish are examples of hydrocarbons that are easy to aspirate.

A certain group of hydrocarbons, halogenated hydrocarbons, are more likely than others to be absorbed and cause systemic poisoning. This chemical family includes carbon tetrachloride, dichloromethane, and trichloroethylene. The effects depend on the chemical's toxic potential. The toxic potential is directly related to both the dose and the chemical's physical properties: solubility, volatility, and viscosity.

The following halogenated hydrocarbons present additional toxic concerns. For example, the halogenated methane derivatives (which have also been given alternate names) are the cause of more common exposures. They include:

- Dichloromethane (Methylene chloride) – is found in many cleaning solutions and paint removal mixtures. Exposure to this chemical causes carbon monoxide poisoning because the liver, during phase two detoxification, creates carbon monoxide as the water-soluble material placed back in the bloodstream to be eliminated via the kidneys.
- Trichloromethane (Chloroform) – is a strong neurotoxin that has been used in the past as an anesthesia drug.
- Tetrachloromethane (Carbon tetrachloride) – was used in dry cleaning, as a fire extinguishing chemical, and as a refrigerant until its toxic properties were noted. Acute exposure to this chemical causes neurologic toxicity, and prolonged exposure causes significant damage to both the liver and kidneys. The overall use of this chemical was significantly restricted in the twentieth century because of its toxic effects and the environmental impact on the ozone.

Chemical compounds that are lipid-soluble are able to cross the blood–brain barrier and affect the CNS. Halogenated hydrocarbons such as methylene chloride, chloroform, carbon tetrachloride, and aromatic hydrocarbons such as benzene, toluene, and xylene are easily absorbed through the respiratory system or gastrointestinal system and rapidly lead to CNS toxicity.

Cardiac Effects

One of the major concerns after an exposure to halogenated hydrocarbons is the effects on the myocardium. Of course, dysrhythmias are a major concern. It is known that an exposure to halogenated hydrocarbon sensitizes the myocardium to certain neurotransmitters like epinephrine and dopamine that can lead to ventricular arrhythmias with little to no warning. These chemicals are also thought to inhibit calcium influx and the sodium/potassium channels after depolarization. This leads to enhanced automaticity and intraventricular and nodal-based rhythms.

Direct myocardial damage from continued exposure also puts a patient at risk for sudden cardiac arrest. Prolonged repeated exposure leads to structural damage that may include edema, hemorrhage, rupturing of myofibrils, interstitial fibrosis, and myocarditis. Sudden death has been reported as a result of coronary vasospasms following an acute inhalation.

CNS Effects

After inhalation, these hydrocarbons rapidly enter the bloodstream. Most are CNS depressants and mimic a patient who is suffering from alcohol intoxication. Euphoria is common, much like what is seen in both alcohol and narcotic use. If exposure continues, lethargy, headache, and coma follow. The most pronounced effects are seen in the young and adolescents with developing brains. Hydrocarbon exposure has been linked to memory deficits and learning disabilities found in adolescents who inhale hydrocarbons for recreational purposes.

Emergency Medical Care

Patients with an altered level of consciousness will first need to have their airway stabilized and be provided supplemental oxygen. Conducting an assessment of the patient's current oxygenation status is important using both a pulse oximeter and performing capnography. Aggressive intubation and ventilation using CPAP will keep the alveoli open and increase the circulating oxygen volume.

If there is wheezing present or there is an indication on the capnography that bronchiole construction is taking place, treatment to both relax the respiratory smooth muscles and reduce inflamation should be considered. Providing an updraft of a bronchodilator and an IV of steroids is recommended.

Decontamination is indicated in the case of skin exposure. Provide decontamination utilizing soap and water as soon as possible to reduce the exposure. Early decontamination will also assist in reducing the vapor inhalation that continues when a patient is contaminated.

Signs and Symptoms

Signs and symptoms include multiple system effects:

- *Eyes.* Blurred vision, dilated pupils, and eye pain
- *Skin.* Irritation
- *Cardiovascular.* Palpitations, cardiac sensitivity, EKG changes, myocardial excitation, low blood pressure
- *Neurologic.* Confusion/anxiety, decreased LOC, slurred speech, seizure, staggered gate
- *Respiratory.* SOB, Poor SaO_2/$EtCO_2$, cough, rales, slow respirations, rhonchi (see Figure 3.16 for detailed signs and symptoms).

Treatment

Basic Life Support

- Provide a complete assessment, including cardiac monitoring and respiratory status.
- If an inhalation injury has taken place, provide 100% oxygen utilizing a non-rebreather mask.
- Decontamination of the patient should be performed for dermal exposures.
- Position patient to prevent aspiration in case of vomiting.

Advanced Life Support

- Maintain adequate ventilation and oxygenation
  - Assess:
    - Oximetry
    - Capnography
    - EKG
- If acute bronchoconstriction (wheezing):
  - Initiate a nebulized updraft of either albuterol (Proventil or Ventolin) or/and ipratropium bromide (Atrovent).
  - Albuterol 2.5 mg (0.5 m or 0.5% diluted into 3 ml normal saline) give via nebulizer. May repeat three times at 20-minute intervals.
  - Ipratropium bromide 0.5 mg (500 mcg) 2.5 ml via nebulizer, repeat at 20-minute intervals for a total of three doses. It may be repeated up to three times at 20-minute intervals.

Note: Albuterol should be used with great caution as it can exacerbate myocardial irritability.

- Administer methylprednisolone (Solu-Medrol) 125 mg, IVP slowly.
  - Monitor heart rate. Contact medical control if HR is over 150 bpm.
- Consider the administration of Brethine/Terbutaline Subcutaneous Injection, if cardiac rate or irritability is of concern
  - Brethine (terbutaline sulfate) 0.5 mg given subcutaneous injection (0.5 ml of a 1 mg/ml solution).
- If oxygen saturation is below 92%, provide oxygen via CPAP.
- Provide advanced monitoring evaluating end tidal CO_2 ensuring a reading between 35 and 45 mm/Hg and monitoring the waveform for the occurrence of a shark fin pattern.

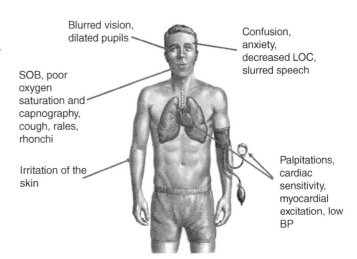

Figure 3.16 Hydrocarbons are CNS depressants displaying symptoms similar to alcohol intoxification.

Blurred vision, dilated pupils

Confusion, anxiety, decreased LOC, slurred speech

SOB, poor oxygen saturation and capnography, cough, rales, rhonchi

Irritation of the skin

Palpitations, cardiac sensitivity, myocardial excitation, low BP

- If Rales or Rhonchi are noted on auscultation, and the oximetry or capnography reflect negative changes.
  - Maintain CPAP and set the PEEP setting above 10 cm of H_2O.
  - Consider NTG, morphine, or lasix administration to reduce the pulmonary pressure and decrease the influx of fluid into the alveoli.
- Provide advanced airway if needed.
- If tachycardia occurs secondary to hydrocarbon exposure:
  - Administer Adenocard (adenosine)
    - 6 mg of adenosine is given rapid IV push followed by a 20 ml syringe bolus of 0.9% sodium chloride (normal saline).
    - If no conversion within one to two minutes, give 12 mg IVP, repeat a second time if necessary (30 mg total).
  - Consider diltiazem (Cardizem) 0.25 mg/kg IV/IO bolus over two minutes.
    - A second bolus of 0.35 mg/kg may be used if necessary.
  - If superventricular tachycardia occurs, give:
    - Esmolol (Brevibloc) 0.2–0.5 mg/kg IVP q5 minutes PRN or
    - Procainamide 5–15 mg/kg IVP with maintenance drip of 20–50 µg/kg/min. (Loading dose: 15–18 mg/kg administered as a slow infusion over 25–30 minutes or 100 mg/dose at a rate not to exceed 50 mg/min repeated every 5 minutes as needed to a total dose of 1 g).
- If seizures occur post-exposure or the patient suffers from extreme anxiety, administer:
  - Diazepam (Valium) 10 mg IV/IO/IM. Maximum dosage 20 mg or
  - Midazolam (Versed) 2.5 mg IV/IO. Repeat at two-minute intervals to a maximum dosage of 10 mg or
  - Lorazepam (Ativan) 2 mg IM/IV/IO and can be repeated after 5–10 minutes.
- If evidence of suicide attempt, overdose, and/or respiratory depression in the setting of a decreased level of consciousness administer (with suspect opioid history):
  - Naloxone (Narcan)
    - Adult dose: 2 mg IVP, IO, IM, or intranasal every three to five minutes up to 8 mg.
- Contact receiving hospital, Poison Control, and provide **HazMat Alert.**

Pediatric Considerations

- Monitor both cardiac and respiratory status.
- If bronchiole construction (wheezing) occurs, administer:

Note: Administration of a nebulized Inhalation Solution is not recommended for pediatrics less than two years old.

- ***Albuterol (Proventil or Ventolin).*** 2–12 years old less than 15 kg: give one 3 ml unit-dose vial of 1.25 mg by nebulizer (5–15 minutes).
  - 13 and older get an adult dose
- ***Ipratropium Bromide 0.02% (Atrovent).*** Administered at an adult dose by a nebulizer, but the number of overall doses is reduced to two. Only one should be given in the field.
- Monitor the EKG closely for cardiac irritability (ectopy).
- ***Solu-Medrol (Methylprednisolone).*** 1.5 mg/kg not to exceed 60 mg dose, IV. Give only one dose in the field.

Note: The correct size mask must be available to provide CPAP to a pediatric patient. If the patient is not breathing, you must have a pediatric BVM with an integrated PEEP valve or an inline supplemental PEEP valve available.

Toxic Alcohols

The toxic alcohols include ethylene glycol, propylene glychol, methanol, and isopropyl alcohol. Toxicity from these alcohols is usually related to the ingestion for the purposes of intoxication or suicide. In the case of ingestion of ethylene glycol, the body forms glycolic acid and oxalic acid, causing acidosis and forming calcium oxalate crystals in the urine leading to kidney injury or stroke, and ultimately resulting in systems failure.

Glycols cause the formation of crystals in tissues and severe metabolic effects may result in irreversible brain damage. The formation of these crystals depletes the calcium found in the blood and affects the repolarization of the heart muscle. Hypocalcemia can be worsened by the administration of sodium bicarbonate.

Methanol causes the formation of formic acid, which leads to severe acidosis and permanent damage to the retina. All produce CNS depression, with isopropyl producing the most. Isopropyl alcohol will cause the production of acetone in the body and can result in severe CNS depression and hypotension with significant ingestions. Actions of methanol and ethylene glycol will be delayed 4–12 hours with the exception of the onset of CNS depression, which will have a rapid onset.

Treatment
Basic Life Support
- Support respirations and provide 100% oxygen via a non-rebreather (NRB) mask.
- If the patient is not breathing, immediately begin CPR and other supportive measures.

Advanced Life Support
- Assess and treat any cardiac dysrhythmia.
- Administer 50 mEq of 8.4% sodium bicarbonate (50 ml) IV.
- Administer thiamine 100 mg IV for ethylene glycol exposures
- Administer pyridoxine 1 mg/kg IV for ethylene glycol exposures
- Administer folic acid 50 mg for methanol exposures
- If an elongated ST interval is noted on the EKG (indicating hypocalcemia), administer 10 ml/10% calcium chloride or 10 ml/10% calcium gluconate IV.
- The antidote is fomepizole (Antizole). If fomepizole is unavailable, ethanol 10% 8 ml/kg IV may be used. If either is unavailable, give ethanol 20% 4 ml/kg PO (Whisky, Vodka, or Gin may be substituted in an emergency situation). If transport time is short, this therapy should be reserved for a healthcare facility. In cases of severe acidosis, renal failure, etc., from late presenting patients, this modality will not be of much help.
- Activate a HazMat Alert to the receiving facility.

Etiological Toxidrome

Overview

Biohazards are communicable diseases that are a threat to the life and health of emergency response personnel unless precautions and protective measures are exercised. In the 1990s, the term *biohazard* became popular to identify those organisms that invade and infect a host's body. The risk of communicable diseases cannot be seen. However, understanding communicable diseases will help emergency response personnel know how they can better protect themselves and how infectious diseases are transmitted during patient care and biohazard cleanup (see Figure 3.17).

Figure 3.17 Etiologic agents causing infections are a concern for emergency responders and hospital staff. We have seen outbreaks from anthrax, Ebola, and various corona viruses that all become a concern to those offering emergency care.

For many reasons, infectious diseases have become part of the hazardous materials of the future. Each year new biological agents or new strains of old biological microorganisms are affecting the population. Legionella pneumonia, AIDS, and drug-resistant tuberculosis are just a few examples.

In 2014 four cases of Ebola virus (a hemorrhagic fever viral disease) occurred in the United States. In total, there were 11 cases in the United States, but seven were medically evacuated from other countries. Nine of them contracted the disease outside of the United States then traveled into the United States. Two of those died. Two contracted the virus in the United States after treating an Ebola patient. Both of those recovered.

In 2019 Novel Coronavirus better known as COVID-19 (also called SARS-CoV-2) began infecting people in China. It rapidly spread worldwide and became the fastest and most virulent viral pandemic ever seen in the World. At the writing of this book, the world is still dealing with efforts to end the pandemic without the end in real sight.

Our environment is filled with tiny microscopic living things called microorganisms. The living organisms that affect the providers of emergency care are divided into two categories: viruses and bacteria. Bacteria are single-celled microorganisms that are plantlike but lack chlorophyll. Viruses are strands of RNA or DNA with a protein covering that requires a host cell to multiply and establish an infection.

These microorganisms develop a relationship in humans that can be beneficial or detrimental to a person's health and well-being. The host–parasite relationship, as it is sometimes called, works in a beneficial way in the digestive system, where the bacteria *Escherichia coli*, among others, work to break down food for digestion within the intestines. Sometimes the same microorganisms that are beneficial cause disease (see Figure 3.18).

Infection is the result of either bacteria or viruses gaining access into the body and multiplying. This invasion overwhelms or bypasses the defensive barriers that are always present in a healthy host. Infections are caused by primary or opportunistic pathogens.

Primary pathogens can, unaided by other opportunities, invade a healthy body, bypassing defensive mechanisms, and establishing an infection. They are aggressive and can enter through many different routes. Measles and gonorrhea are examples of primary pathogens.

Figure 3.18 Responders need to understand the difference between bacterial and viral pathogens and learn to practice precautions to keep them safe from the dangers both may present.

Opportunistic pathogens are usually unable to penetrate the defense mechanisms found in a healthy individual. Unlike the primary pathogens, opportunistic pathogens cause infection or disease by taking advantage of broken defense mechanisms. This breakdown can be as simple as an open wound or as complex as a compromised immune system. The breakdown of defense mechanisms occurs during injury, preestablished illness, drug and alcohol abuse, old age, and so forth. A staph infection is an example of an opportunistic infection. Staph is commonly found in our everyday environment and does not normally affect us adversely. However, staph often infects those with a compromised defense mechanism. Vulnerable populations such as those in prisons, nursing homes, homeless shelters, and daycare centers provide the source of diseases for everyone else. These areas are the same ones that have daily visits from emergency medical services.

Adhesion is the first stage of microbial infection. It is the establishment of a pathogen on a surface of tissue that is in contact with the external environment. After adhesion, invasion into the epithelial cell layer then into the tissue. Infection occurs when the pathogen invades surrounding tissue. Figure 3.19 demonstrates the pathogen invasion process.

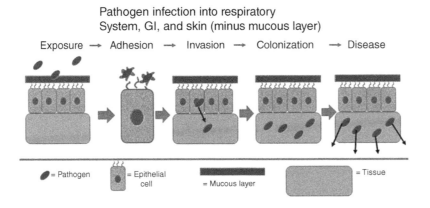

Figure 3.19 The process of pathogen invasion to infection is a multistep process.

There are several infectious diseases that are of special concern in the emergency environment. This includes those who are on-scene emergency medical responders and those who treat patients in the emergency room where there are limits to infection control practices. Generally, in-patients are diagnosed, and appropriate precautions are in place when infectious diseases are identified, but these do not typically exist in the emergency room and definitely not in the prehospital setting. The disease of concern are:

- Vancomycin-resistant *enterococci* (VRE)
- Methicillin-resistant *Staphylococcus aureus* (MRSA)
- *Clostridium difficile* (*C. Diff*)
- Necrotizing fasciitis

Vancomycin-Resistant Enterococci (VRE)

Enterococci are normal flora (bacteria) found in the mouth, GI, and female genitalia. VRE generally spreads from one person to another through contact. They are not usually spread through the air. Healthy individuals are not as likely to get this infection, but it does occur. Typically, those at greatest risk are people with weakened immune systems.

Occasionally, these bacteria will establish an infection in the urinary tract or other mucous membranes but more often in an open wound.

Some *enterococci* have developed a resistance to the antibiotic normally used to kill the bacteria (vancomycin).

Symptoms
If VRE has infected a wound, the area of infection will be red and tender. In addition, it will be warm to the touch. If it causes a urinary tract infection, the patient will complain of back pain, burning upon urination, urinary urgency, and frequent urination. Some patients will develop diarrhea, feel weak and sick, and suffer from fever and chills

Diagnosis
If VRE is suspected, a physician will send a wound, blood, urine, and/or stool sample to a lab. The lab will, in turn, grow the bacteria in media then test to determine what antibiotics will kill the bacteria. This testing process takes several days to weeks.

Treatment
VRE is difficult to cure because it does not respond to many antibiotics. Antibiotics will be prescribed that are either taken orally or by IV. The antibiotics may be a mixture of several different antibiotics in order to kill the infection. If the colonization of the bacteria occurs, the infection may return at any time, especially when the person becomes immunocompromised. Prevention is always the key. Frequent washing of hands, keep cuts and scrapes clean and covered, and avoid direct contact with someone diagnosed with VRE.

Methicillin-Resistant *Staphylococcus aureus* (MRSA)

MRSA is a bacterium responsible for several difficult to treat infections in humans and is resistant to beta-lactam antibiotics, including:

Penicillins (methicillin, dicloxacillin, nafcillin, oxacillin, etc.) and cephalosporins.

Initial presentation is usually characterized by small red bumps that resemble pimples or spider bites and is accompanied by a fever. 75% of these infections are localized to skin and soft tissue

Pathophysiology

MRSA is a *Staph. aureus*, a gram-positive coccus that has developed resistant qualities to typical antibiotics used to treat the infection. Once an infection is established, it produces a range of toxins that cause systemic disease processes such as toxic shock syndrome.

Research has found that up to 80% of people are eventually colonized with *Staph. aureus*. Most only have colonization intermittently, and only 20–30% have persistent colonization. Colonization rates are higher in healthcare workers, those with diabetes, and patients on dialysis. About 2 in 100 people carry MRSA, mostly in their nostrils.

Those at high risk of developing an infection from the colonization are:

- Diabetics
- IV drug abusers
- Those with long-term intravenous catheters
- Trauma victims

Signs and symptoms

- *Skin.* Swelling, warmth, redness, and pain in infected skin.
- *Blood and Deep Tissue.* Fever of 100.4 or higher, chills, malaise, dizziness, confusion, muscle pain, swelling and tenderness, chest pain, cough, and lingering wounds.

Treatment

Incision and draining of the wound may be necessary if an abscess occurs.

Pharmacological treatment includes:

- Clindamycin
- Tetracyclines (doxycycline and minocycline)
- Rifampin (in conjunction with other antibiotics)
- Linezolid

Clostridium Difficile (C. Difficile)

Overview

Clostridium difficile is also known as *Clostridioides difficile*, is often called *C. difficile* or *C. diff*. Once this bacterium establishes an infection, it can cause symptoms ranging from diarrhea to life-threatening colon inflammation. This bacterial commonly affects older adults but can infect any age after an overtreatment with antibiotics. In recent years, *Clostridium difficile* infections have occurred more frequently, are severe, and become more challenging to treat.

Intestines typically contain approximately 100 trillion bacterium and about 2000 different kinds of bacteria; many helps protect the body from infection. When high-dose antibiotics are taken, they tend to destroy the good bacteria along with the bad. Without the helper bacteria, *C. difficile* can quickly grow and establish a severe infection. The antibiotics that can cause a *C. difficile* infection include:

- Fluoroquinolones
- Cephalosporins
- Penicillin's
- Clindamycin

Symptoms

Many people carry *C. difficile* in their gut but never develop the infection. Signs and symptoms generally begin between 5 and 10 days after starting a course of antibiotics that dangerously lower the normal flora found in the intestines. A mild to moderate infection causes watery diarrhea three or more times a day for two or more days. This may be accompanied by mild abdominal cramping and tenderness.

More severe infection is characterized by watery diarrhea 10–15 times a day, abdominal cramping and pain, rapid heart rate, fever, blood in the stool, nausea, dehydration, weight loss, swollen abdomen, kidney failure, and increased white blood cell count. In addition, the colon can become inflamed, enlarged (called toxic megacolon), and septic. These symptoms often require a patient to be admitted into ICU.

Clostridium difficile is found in many locations, including soil, air, water, and human/animal feces. Spores from *C. difficile* bacteria are passed from feces to food from surfaces and dirty hands. The *C. difficile* spores can persist in the environment for weeks or months, making it very easy to transmit from an object to a person. This is called a fecal/oral route.

Once an infection is established, the *C. difficile* produces toxins that attack the lining of the intestines. The toxin produces inflammatory cells, destroys tissue, and leaves decaying cellular debris in the colon, causing watery diarrhea.

Since the year 2000, there has been an aggressive new strain of *C. difficile*. This strain produces far more toxins than other strains and is resistant to many medications. In addition, it has infected people who have not been medicated with antibiotics. This presents many concerns to the emergency response community who can get infected without being immunocompromised and without a history of being over-medicated with antibiotics.

The majority of *C. difficile* infections are caused for persons who have recently been in a hospital, nursing home, or long-term care facility. *C. difficile* spreads on unwashed hands and equipment like beds, sinks, stethoscopes, and thermometers.

Other Risk Factors

- Women are more likely than men to get *C. difficile* infection
- Ten times greater for people aged 65 or greater

Complications from *C. Difficile* Include

- Dehydration occurs from severe diarrhea.
- Kidney damage and ultimately failure from dehydration.
- Toxic megacolon disease from the toxins created by the bacteria.
- Bowel perforation occurs from cellular damage created from the infection and bacteria invasion.
- Ultimately, death occurs if unable to get the disease under control.

Prevention

- Avoid unnecessary antibiotics
- Hand-washing
- Use contact precautions
- A thorough cleaning to remove spores

Necrotizing Fasciitis

Overview

Necrotizing fasciitis is an invasive infection that usually begins as a skin infection, then moves to the underlying tissue and eventually to musculature (fascia). This invasive infectious disease is caused by several different types of bacteria and generally spreads rapidly. The infection may cause flu-like symptoms, redness at the site of the condition, and pain. The healthcare provider must make a rapid diagnosis and then provide prompt treatment to prevent the rapid spread, multiple organ failure, and death.

Symptoms

Symptoms can begin within hours of an injury. The infection causes pain and tenderness over the affected area. The flu symptoms may include fever, sore throat, stomach ache, nausea, diarrhea, chills, and body aches. Redness around the initial wound (infection) is a common sign. The area may become swollen, shiny, discolored, and hot to the touch. Ulcers and blisters are seen in many cases.

If the infection spreads or goes untreated, the patient will become dehydrated, develop a high fever, fast heart rate, and low blood pressure. Pain may become less intense as tissue and nerves are destroyed by the infection. If untreated, necrotizing fasciitis will lead to shock and death.

Cause

Necrotizing fasciitis can attack anyone. The types of bacteria that can cause necrotizing fasciitis includes:

- Methicillin-resistant *Staph. aureus* (MRSA)
- Klebsiella
- Clostridium
- *Escherichia coli*

Half of the cases of necrotizing fasciitis begin as a streptococcal bacteria infection. Half of the diagnosed infections occur in young and healthy individuals. Many times, it develops after a traumatic wound that causes an opening in the skin. It can also develop after minor trauma where there is no break in the skin. In addition, it can occur after surgery and even after a minor injury such as an insect bite. A weakened immune system increases the chances of developing this infection as well.

Treatment

A rapid diagnosis and prompt use of IV antibiotics are critical to treating the infection. Surgery is usually necessary to remove dead tissue caused by the infection. As the tissue becomes damaged from the infection, blood supply also is restricted, making IV antibiotics of limited help. Early surgery to eliminate damaged and infected tissue will assist in the healing process and reduce the need for amputation. Hyperbaric oxygen therapy may be used to heal tissue that has limited blood supply because of the damage to the arteries and capillary beds.

There is a 25% fatality rate with necrotizing fasciitis due to complications related to septicemia and organ failure. The speed of diagnosis and treatment influences outcome.

Means of Entry

Communicable diseases are transmitted by direct and indirect routes. Both forms of transmission involve either viruses or bacterial, which must be in a sufficient quantity. Blood-to-blood is the most direct method of contracting either viruses or bacteria. Other means of transmission are extremely efficient, such as a blood-to-mucous membrane. For example, the exposure of a patient's blood or other body fluids into the mouth or eyes is a danger that emergency responders must be cognizant of and protect themselves from.

Viruses are transmitted in a host, such as blood or body fluids (referred to as bloodborne or airborne) and cannot multiply outside of the living cell. Once the virus enters the body, it begins to reproduce, whereas bacteria are transmitted outside of the host on objects such as equipment and can multiply outside of the body, not needing a host to multiply or reproduce. Some pathogens can enter and affect the body only through open wounds, others can gain access if inhaled, and others only if swallowed or exposed to other mucous membranes. Emergency medical care providers should understand the different routes of entry so that protective measures can be taken to prevent the transmission of disease.

Virulence

Virulence refers to the amount of a particular microorganism required to cause an effect in the host (in other words, the toxicity of the pathogen). The smaller the dose (fewer microorganisms) needed to infect a patient, the more virulent the organism is. Virulence is based on three effects: infectiousness, invasiveness, and pathogenicity. Infectiousness is the ability, once a pathogen gains access into the body, to initiate and maintain an infection. Invasiveness is the ability of the pathogen to progress further into the body once an infection is established. Pathogenicity is the ability of the pathogen to injure the body once the infection is established. Because of the extended use of antibiotic medicines and the ability of microorganisms to change their resistance to them, the virulence of many of the more common infections has become much greater.

Emergency care providers are at a higher risk of becoming infected with pathogens simply because of the uncontrolled environment in which they work. Emergency medical providers often work on scenes that have poor lighting, are grossly contaminated with blood and other body fluids, and may be extremely unsanitary. Many times, these providers meet patients who are uncooperative, unruly, or violent. Bloodborne and airborne pathogens pose the highest risk in these environments.

When concerns about communicable diseases are discussed, usually human immunodeficiency virus (HIV) or hepatitis B virus (HBV) are the topics. Unfortunately, airborne diseases are also on the rise today. COVID 19 was initially identified as a large droplet transmission but seeing how quickly the disease spread; many believe that this is more like an airborne transmittable disease.

Exposure

If exposed to a communicable disease, remove all contaminated clothing and personal protective equipment and clean the exposed area. The area should be scrubbed with soapy water. If this type of cleaning is immediately impractical, then use a waterless antimicrobial cleaning agent or alcohol product containing over 70% alcohol. Afterward, and as soon as practical, wash the areas with soapy water. If the area exposed does not allow for scrubbing, such as a mucous membrane, then remove as much of the material as possible and irrigate with water repeatedly to remove the infectious material. This includes vigorous irrigation of the eyes if they are exposed to blood or body fluids.

Never use chlorine bleach products or any other strong cleaning agents on skin that may have been contaminated, believing that it is the best way to kill bacteria or viruses. These cleaning agents work well on porous and nonporous surfaces to kill bacteria and viruses, but the effects are different on the skin. Household bleach (sodium hypochlorite, $NaClO$) damages the first layer of the epidermis and worsens future contamination. Damage to the epidermis, as a result of bleach and other strong cleaning agents, is evidenced by the liquefaction of fats in the skin layer (saponification), similar to the effects of an alkali burn. The damaged skin has a slick or soapy feel and allows bacterial contaminates to enter much more easily. There is no remarkable documentation of an exposure to intact skin causing a communicable disease, but it is reasonable to suspect that a bacteria or virus could gain access if the skin is injured in this way.

Significant exposure is normally defined as exposure by a contaminated needle stick or sharp puncture by other instruments in which blood-to-blood contamination takes place. Significant exposure may also be an exposure where blood or body fluids come in contact with an open wound, mucous membrane, or eyes. Each emergency healthcare agency must have a written protocol to follow if exposure takes place. OSHA's 29 CFR 1910.1030 Occupational Exposure to Bloodborne Pathogens identifies rules that must be followed on a federal level. Even some state government agencies have produced documents regulating exposure protocol. If a significant exposure has occurred during patient contact, the exposure must be documented and follow-up care identified.

Radiological Toxidrome

Overview

Radiation and radioactive sources are used in medical testing and in industry to determine road density and to gain radiologic imaging of walls and construction projects. High concentrations are found in radiologic fuels for power plants and military vessels. All present a danger to exposure. See Figure 3.20 for location of source of radiation.

Radiologic emergencies can be addressed from two different aspects, those involving wartime activities and those that happen during peacetime accidents. This section primarily addresses peacetime accidents, although many of the principals involved can be adopted to

Figure 3.20 Radioactive sources are used in industry for a variety of reasons ranging from determining the density of roadways to treatment of tumors. There is great concern in the country for the potential of using radiation in a dispersal device.

wartime radiologic emergencies. This text's approach is toward recognition and identification of the problem rather than field treatments. It is recognized that there is little in the way of field treatment that can be accomplished after an exposure takes place. Therefore, this section will focus on recognition of the hazard, the level of hazard, and a decontamination of those that are affected by the radiation incident.

When radiation emergencies exist, the term "radiation" refers to ionizing radiation. Radiation is a broad term that describes energy transmission. There are other types of radiation such as ultrasound, radio waves, microwaves, cell phones, and infrared. These are examples of nonionizing radiation. These should not be confused with ionizing radiation that causes tissue damage.

Simply put, ionized radiation is energy that disrupts the integrity of the atom producing a particle or ray of energy that is capable of causing harm to living cells and tissues. Ionized radioactivity can be produced by reactors but is also naturally occurring (also called orphan isotopes) and can be found in and around the planet.

We are constantly being bombarded with minute amounts of radiation emitting from space. None of our senses can detect the presence of this radiation. Furthermore, the lack of understanding concerning how it affects biological tissue and the physics behind radiation has caused fear of this hazard event.

Types of Radiation

There are four common types of ionizing radiation, known as alpha, beta, gamma, and neutrons (X-rays and medium- to high-ultraviolet light fall into this group) that are involved in radioactive decay (spontaneous transmutation). These particles also release energy and are the sources of hazard within this type of emergency.

Alpha Particles

Alpha particles are positively charged particles that are the weakest of the ionizing radiation. These particles are made up of two neutrons and two protons and are relatively large, moving more slowly than beta particles. Alpha particles can be propelled about 4 in. from their source. These particles are easily stopped by a thin piece of paper, and therefore these particles cannot penetrate even light clothing. If alpha particles come in direct contact with the skin, they may only penetrate the top layer of tissue, made up of dead cells, so they pose little threat to the outside of the body. However, alpha particles pose a greater health hazard if they are inhaled, ingested, or enter the body through an open wound. Thus, they represent an internal exposure hazard.

Beta Particles

Beta particles are positively or negatively charged particles. This is depending on the neutron to proton ratio which produces a negative beta particle, and when the neutron to proton ratio is too small, a positive beta is released called a positron. They are higher in speed, 7000 times smaller (about the mass of an electron), and have much more penetrating power than alpha. Beta particles in general have many penetrating capabilities and can readily penetrate skin tissue if the responder or public is close enough to the source.

Most beta particles can travel about 30 ft, with some traveling as far as 100 ft in the air, and can penetrate 0.1–0.5 in. into the skin. They can be blocked by a thin layer of metal or other dense material. Some particles can even be blocked by heavy clothing. Because the beta particles can penetrate and cause damage to the tissues, once striking or penetrating the skin, they are designated as internal and external hazards. Ingestion, inhalation of beta particles is a severe medical emergency and a health physicist should be notified.

Gamma Rays

Gamma rays are not particles but emissions of energy generated from a nucleus that is unstable due to excessive energy levels produced by the ratios of protons, electrons, and neutrons. This excessive energy is emitted as a photon, which is a pure package of energy. This electromagnetic energy has a considerable deep penetrating ability and should be considered extremely dangerous. Gamma rays pass completely through the body, causing damage through and through. Gamma rays are referred to as penetrating radiation. They are deemed to be an internal exposure hazard.

Neutrons

These are particles that are released during transmutation as a part of the decay process. The number of neutrons released is dependent on the isotope and consist of a free neutron. They move at speeds well below the speed of light but faster than gamma emissions and are effectively stopped by hydrogen-rich material such as plastics, concrete or water. They ionize atoms and have the capability to turn non-radioactive material into radioactive.

X-Rays

X-rays are another type of electromagnetic radiation capable of penetrating the body tissues. They are produced on the site of use as the need arises. The technique used to produce X-rays involves bombarding a metallic target with electrons, which are then directed to the point needed. There is a danger of overexposure to X-rays, but this danger is usually due to carelessness and not a radiation accident. If the X-ray equipment is involved in an accident, such as a fire or explosion, there is no danger to radiation exposure.

Measuring Radioactivity

Understanding the measurement of radioactivity comes with an understanding of what is activity and what is the measurement of effect. The activity of the radioactive material is the amount of radiation or activity that is being released by the isotope. This activity is measured in curies (Ci) or becquerel (Bq). This is a unit of measurement of radioactivity as compared to 1 g of radium. More specifically, it is the amount of a substance in which 37 billion radioactive decays per second or 3.7×10^{10} becquerels, which undergo radioactive disintegration. These are units of activity measurement and have to be converted into a level of effect or dose. Becquerels are used in place or curies within the International System of Units.

When an individual is exposed to radiation, it is the amount of radiation that travels through the air and is measured as a roentgen (R) or coulomb/kilogram (C/kg). Once the individual is exposed to the energy that is being released, an absorbed dose is reached (or exceeded). At this point the body is absorbing the energy starting to affect biological tissue. This is measured in radiation absorbed dose (RAD) or gray (Gy). The effective dose combines the absorbed dose and the subsequent medical effects. This combination comprises the dose equivalent or radiation roentgen equivalent man (rem) or Sievert (Sv). The Sievert is becoming the standard of measurement to describe the biological dose equivalents.

Medical severity	Radiation measurements				Symptoms						
	Gy	Sv	Dose equivalent (rem)	Absorbed dose (RAD)	Nausea vomiting	Weakness	Headache	Fever	Hemorrhage	Diarrhea	Leukopenia
	0.01	0.01	1	1							
	0.05	0.05	5	5							
	0.1	0.1	10	10							
Supportive	0.25	0.25	25	25							
Supportive	0.50	0.50	50	50							
Supportive	0.75	0.75	75	75	X						
Supportive	0,85	0.85	85	85	X						
Serious	1	1	100	100	X	X					
Emergent	2	2	200	200	X	X	X	X	X		
Emergent	3	3	300	300	X	X	X	X	X		
Critical	4	4	400	400	X	X	X	X	X		
Lethal critical	5	5	500	500	X	X	X	X	X		
Lethal critical	12	12	1200	1200	X	X	X	X	X	X	X

Principles of Protection

Once a general understanding of radiation has been developed, the principles of radiation protection are easier to understand. The four basic principles to protect yourself, and any victims involved, are that of time, distance, shielding, and quantity.

Time. The time factor involving protection from radiation is simple: <u>*Limit your time of exposure.*</u> The longer the exposure to a source of radiation, the greater the damage to the body tissue. If an extended amount of time needs to be spent in an area affected by radiation, such as rescue or extrication, then a plan should be made to rotate personnel through the area, in effect limiting individual exposure.

Protective action guidelines

Guidelines	Activity	Condition
5 rem (50 mSv)	All occupational exposures	All reasonably achievable actions have been taken to minimize dose.
10 rem (100 mSv)[a]	Protecting valuable property necessary for public welfare (e.g. a power plant)	Exceeding 5 rem (50 mSv) unavoidable and all appropriate actions taken to reduce dose. Monitoring available to project or measure dose
25 rem (250 mSv)[b]	Lifesaving or protection of large populations	Exceeding 5 rem (50 mSv) unavoidable and all appropriate actions taken to reduce dose. Monitoring available to project or measure dose

[a] For potential doses >5 rem (50 mSv), medical monitoring programs should be considered.
[b] In the case of a very large incident, such as an IND, incident commanders may need to consider raising the property and lifesaving response worker guidelines to prevent further loss of life and massive spread of destruction.

Distance is the second protection principle. <u>*Stay as far away as possible.*</u> By doubling the distance away from the source, the exposure is decreased by a factor of four. For example, an exposure of 16 mR/h at a distance of 3 ft would be reduced to 4 mR/h at a distance of 6 ft. But the opposite also holds true, and exposure of 8 mR/h at a distance of 4 ft would be 32 mR/h at 2 ft. Therefore, it is important to realize that exposure to radiation is always greatly reduced by moving away from the source. (R = roentgen, a measurement of exposure to gamma or X-rays, mR/h, represents the amount of exposure to produce a certain number of ions in $1 cm^3$ of air in one hour. 1 m/R = 1 mR/h or one-thousandth of a roentgen.)

Shielding is the third principle of protection. <u>*Keep something between you and the source.*</u> Shielding utilizes the fact that the denser the materials, the more radiation is blocked by them. Lead is the most common shield used around X-ray equipment because of its high density. At the site of a radiation emergency, this principle can be put to use using an automobile or a hill of dirt. As mentioned earlier, heavy clothing will stop all alpha particles and some beta particles. Plastics can assist with shielding of alpha, beta, neutrons but gamma will still penetrate plastic materials. Lead shielding, although not practical for rescue, is effective against gamma rays and X-rays.

It may be difficult to use this principle, especially when performing a rescue. But, utilizing this and the other two protection principles, time and distance, will minimize the exposure.

Quantity. The last protection principle is that of quantity. Exposure is directly related to the amount of radiation being emitted from the source. Any means necessary to limit or reduce the amount of radiation emission will reduce the amount of exposure. It may be difficult to immediately reduce the amount of the spill (exposed radioactive material), but this principle also holds true for the victim dirty with radioactive particles. *Remove contaminated clothing.* The victim's clothing should be removed and gagged, then taken from the immediate area, therefore limiting the number of materials in the direct vicinity. It should never be placed in the ambulance and transported with the patient to the hospital.

Location of Radiation and Common Sites for Accidents

Accidents involving radioactive material can be divided into six different categories:

1) Transportation of radioactive materials.
2) Medical uses of radioactive substances.
3) Research facilities are utilizing radioactive substances.
4) Industrial uses and nondestructive testing.
5) Nuclear reactor sites.
6) Isotope production facilities (radioactive fuel).

Types of Injuries

Although victims of radiation accidents rarely show immediate signs or symptoms of radiation injury, the effects of radiation on the body's cells include interference with cellular division, damage to the chromosomes, damage to the genes (DNA mutations), neoplastic transformation (cancer), and cellular death. Although the mechanisms causing these changes are not fully understood, it is thought to be the end result of chemical alterations that are caused by radiation as it transverses the cell. The rescuer must be aware of the possibility of a radiation injury by evaluating the scene and looking for any evidence of a radioactive source. Radiation injuries are found in one or a combination of either external irradiation, contamination, or incorporations. Figure 3.21 shows the systems typically affected by a radiologic exposure.

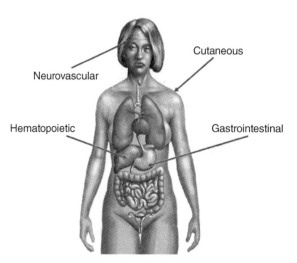

Figure 3.21 Although signs and symptoms of radiation exposure rarely show immediate signs, the internal damage to multiple systems could be devastating at high-dose exposure.

External Irradiation   External irradiation is the injury that occurs when penetrating radiation (gamma or X-ray) passes through the body from an external source. The intensity of the injury depends on the amount of exposure to the radiation. Once the victim is externally exposed, he does not become radioactive and can be handled by the rescuers without fear of receiving a secondary radiation injury from the patient.

Contamination   When a patient becomes partially or wholly covered with radioactive material, it is referred to as a contamination injury. The contamination can be in the form of a solid, liquid, or gas. The contamination can enter the body through the skin, lungs, open wounds, or the digestive tract. Therefore, this injury can be an internal injury, external injury, or both. Rescuers handling the contaminated patient can, themselves, become contaminated and need to exercise the principles of time, distance, shielding, and quantity. If external contamination exists, decontamination of the patient must be done prior to transportation of the patient to the hospital.

Incorporation is the uptake or combining of the radioactive material by the tissues, cells, and target organs (the organs affected by the injury). Based on its chemical properties, the radioactive material seeks out the areas that easily incorporate the type of radiation materials involved, such as radium to the bone or iodine to the thyroid. Medicinal radioactivity uses this principle to treat and diagnose these target areas. Incorporation cannot occur unless contamination has occurred, and therefore, incorporation is always a combination of internal and external injury.

Irradiation (from gamma) of a part or even the whole body does not usually require immediate emergency treatment. This type of injury usually causes signs and symptoms days, weeks, or months after the exposure takes place. Consequently, a contamination injury (from alpha and beta) is handled as a medical emergency. Unless treated promptly, the contamination will lead to an internal injury and eventual uptake into cells and tissues, causing permanent injury sometimes identified years after the incident.

Depending on the type of radiation, how much of the body was exposed, and the dosage determines how much radiation sickness the victim will suffer. The effects of radiation on the body are varied depending on the dosage. The skin may swell, blister, redden, flake, or itch. The breathing may be affected because of the swelling or damage at the bronchioles and alveoli. Other widespread damage may be evidenced by permanent or temporary sterility, loss of menstruation, reduction of sperm count, damage to blood vessels, cancer, and genetic damage. Figure 3.22 provides a detailed description of the damage that results from an exposure to ionized radiation.

Recovery from large doses of radiation may require months or years. Recurrent or chronic problems such as chromosomal damage and reproductive difficulty may last a lifetime.

Acute Radiation Syndrome   Acute Radiation Syndrome – (ARS) is a grouping of symptoms and injuries that appear within 24 hours of an exposure to ionizing radiation. The dose of radiation received by the victim is significant enough to cause cellular degradation and damage to the DNA. This occurs after whole or partial body irradiation of greater than 100 RAD. There are four sub syndromes to the overall acute radiation exposure that can occur within the first 24 hours. They are:

- ***Neurovascular (N)***. Effect on the tissues that make up the CNS. Can occur between 10 and 30 Gy.
- ***Hematopoietic (H)***. Can occur at 2–3 Gy, which affects the blood supply within the body along with white blood cells and other blood cellular elements depression.

rem	
1,000	Fatal within weeks
600	Chernobyl workers died within a month
500	LD$_{50}$ with a month single dose
100	Radiation sickness starts risk of serious illness later in life
40	Fukushima Japan (highest daily level
35	Chernobyl residents
2	Yearly limit for radiation workers
1	Full body CT scan
0.2	Typical radiation per year
0.3	X-Ray - Mamogram
0.01	X-Ray - Chest
0.001	X-Ray - Dental

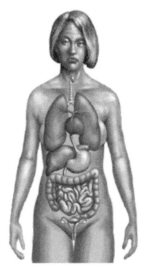

Typical signs and symptomology:

Eyes: High dose can cause cataracts.

Thyroid: hormone glans are very susceptible to damage.

Lungs: DNA damage with inhalation of radioactive isotopes.

Skin: Redness and burning

Stomach: Damage with ingestion of radioactive isotopes.

Bone marrow: White blood cell production, red blood cell mutations

Reproductive: DNA Damage which may produce mutations

Figure 3.22 This image demonstrates the types of internal and external injuries that take place based on the level of radiologic exposure.

- ***Cutaneous (C).*** Development of ulcerations and lesions within the epidermis, dermis, subcutaneous tissue, and musculature, and/or bone.
- ***Gastrointestinal (G).*** at lower doses, the 1.5 Gy breakdown of the mucosal barrier within the GI tract starts to occur.

Rescue and Emergency Treatment

If information is received, either through the dispatcher or visual evidence on the scene, indicating a radiation accident, appropriate precautions must be taken. Usually, the same general procedures as for other hazardous materials incidents hold true. Place the responding units at a safe distance from the incident and approach from uphill and upwind, placing emergency units out of any smoke, fire, dust, or gas emitted from the scene.

As with any hazardous materials incident, the following agencies must be contacted and advised of the accident.

1) Local law enforcement
2) Local fire department/hazmat team
3) Local medical facility
4) Any local individual or organization known to be trained in radioactive emergencies.
5) Regional Poison Control Center
6) CHEMTREC 1-800-424-9300

Heavy clothing such as fire turnout gear, coveralls, coats, and jackets will all assist in stopping the penetration of alpha and some beta particles; however, this clothing will not stop the exposure to penetrating gamma rays. The best protection from gamma rays is a dense metal such as lead, which has limited use in the field. Most radiation sources emit a combination of radiation that may include alpha, beta, and gamma rays. Using good judgment is the best protection.

If a rescue is needed, remember to use the principles of time, distance, and shielding. The best approach to provide a rescue is to divide the work to be done in the hazard zone between teams of workers, limiting the time spent in the hazard zone by each worker.

The use of self-contained breathing apparatus (SCBA) is selected based on the possibility of airborne radioactive particles found in gas, dust, or smoke. If it is determined that the rescuer needs an SCBA, then so does the victim. Arrangements should be made to provide one for the victim while being removed from the hazard zone.

The use of radiologic survey meters is recommended during initial entry into the hazard zone. These are not routinely kept on emergency medical services units. Stabilization of a victim should never be performed in the hazard zone. The patient should have a primary survey performed; any life-threatening procedures, such as opening the airway, control of hemorrhage, or placing the victim on a backboard, are quickly performed; and the patient should be removed from the hazard zone. External decontamination on the scene is recommended, and supportive care is started.

If extrication of a victim is needed, there is no effect from radiation on machinery used to extricate. Disposable gloves and other disposable assessment equipment should be used to reduce contamination to the rescuer. If the patient is not breathing, artificial ventilation should be accomplished with the use of a disposable bag-valve device. Any non-disposable equipment must be decontaminated and surveyed for residual radioactive particles after use.

As with other hazardous materials exposures, the medical responder must provide protection for themselves and their units. Once the stretcher is covered with a piece of plastic sheeting, then a blanket is draped over the stretcher. The victim should be placed on the blanket, and the blanket wrapped around them with only the head and one are extending through it. These areas are left exposed to assist in assessing the level of consciousness, vital signs, and respiratory status. The receiving hospital must be advised of the situation, so they can prepare and activate their own radiation policy and procedures. They need to be informed of:

1) Number of patients involved.
2) How many of these victims are known to be contaminated?
3) The medical condition of each victim.
4) What kind of decontamination was conducted on the scene?

Once the patient is transferred to the hospital staff, then the transporting unit should be placed out of service. A radiologic survey of the unit should be taken to determine the number of contaminants that still exist. If contamination is found, further decontamination efforts need to be taken to ensure that all radioactive and chemical contamination has been removed.

Treatment
Basic Life Support
- Decontaminate the victim(s) and meter with a radiological instrument
- Initially determine the history of the injury and the conditions the injury occurred
- Treat life-threatening conditions
- All care is supportive

Advanced Life Support
- Start IV/IO for fluid maintenance and drug administration.

Pediatric Considerations
1) Decontaminate the victim(s) and meter with a radiological instrument
2) Initially determine the history of the injury and the conditions the injury occurred
   a) May upgrade patient to a higher level of care due to age and size.
3) Treat life-threatening conditions
4) All care is supportive

Associated Toxic Conditions

Closed Space Fires

History
Both fire and EMS responders frequently encounter victims of closed-space fires. Conditions found in these fires represent probably the oldest hazardous materials emergency, and one of the most studied and analyzed. Since every fire condition is different, the toxicity of each fire is different. In more recent years, the use of synthetic materials in construction and furnishing has greatly increased the variety of chemicals found in smoke.

The lethal effects of smoke have been recognized and utilized as far back as the first century. Romans used the smoke of greenwood to execute prisoners. The United States Fire Administration state that, on average, more than 3000 people die as a result of fire each year, and more than 17,000 injuries result from fire and the toxicity of smoke. Many dramatic events in our recent history have illustrated the seriousness of the problem related to smoke inhalation. See Figure 3.23. In November of 1980, a fire at the MGM Grand Hotel in Las Vegas claimed the lives of 84 visitors and workers at the hotel. At least 68 of these victims died as a result of the inhalation of toxic gases; in January 1990, a fire in Zaragoza, Spain, claimed the lives of 43 who died from apparent exposure to toxic gases generated during the fire. These incidents are only examples of the devastation that can be caused by the inhalation of smoke gases generated during a fire. On a smaller scale, virtually every community has experienced deaths caused by fires.

Figure 3.23 Every fire situation is different and the toxicity of each fire is different. Toxicity is directly related to the materials burning or exposed to heat.

No professional is more aware of the effects of smoke than firefighters. Prior to 1940, most of the fire encountered by these professionals contained typical organic products used in construction. Exotic plastics were not yet invented, and synthetic materials were not widely used during this era and, therefore, not a concern. Firefighters believed that the smoke was not toxic, a belief that we now know is not true. For example, burning Douglas fir wood gives off in excess of 75 harmful chemicals. Other woods and cotton products give off harmful levels of carbon monoxide and carbon dioxide. When burning, wools and silks produce large amounts of both cyanide gas and hydrocyanic acid. Unfortunately, these beliefs led to the early deaths of most of the workers who participated in this profession over a long period of time.

Today, firefighters use equipment that protects them from many of the products of combustion that are found in typical structure fires. Self-contained breathing apparatus provides excellent protection while in the atmospheres containing the toxins. Unfortunately, compliance through the firefighting process including, attack, knockdown, and overall is still not 100%. Therefore, firefighting remains one of the most hazardous occupations in the United States. Today's structures contain a wide variety of exotic materials that, when exposed to heat and fire, release a myriad of chemicals that not only enter the respiratory system but also can penetrate the firefighters' protective clothing, making an entrance into the body through the skin.

The challenges that modern and future firefighters have are many, when it comes to typical fires. Wood, which has been the usual material used for construction, has become an expensive commodity. In an effort to make housing more affordable, new materials have been introduced into the construction industry. In recent years, aluminum wiring, laminated trusses, and lightweight roofs utilizing plastic materials have become more the standard. In each of these construction techniques, money, affordability, and ease of construction are the motivating factors. PVC (polyvinyl chloride), which has virtually replaced steel pipes and is used for other household items, has been identified to release 55 toxic products of combustion when involved in fire. Polystyrene and polyamide (high-temperature plastic), used in plastic polymer blends to make coffee makers, hairdryers, power tools, pump housings, and other household items. Both of these chemicals release more than 30 different hazardous products when exposed to fire.

Recent statistics on the life expectancy of a career firefighter range from 10 to 15 years less than the average population. This, at least, is in part due to the continued low-level exposures to the chemical in their workplace. Many states are now being forced to enact presumptive fault legislation for firefighters who develop cancer during their employment. This is a step in the right direction but does not address any liability for those firefighters who retire and develop cancer that probably was associated with exposures while on the job. Hydration, decontamination, and medical surveillance are discussed in this book with the hope that better education and compliance concerning the use of protective gear will change past practice.

Fire Toxicology

Fire is a somewhat unpredictable rapid chemical reaction and, depending on the fuel, amount of oxygen present, and the heat generated, an unstable reaction. Therefore, studies relating to this subject have been difficult. Many studies, usually utilizing a firefighter, involve the collection of the gases produced during structure fires. By these studies, scientists are able to determine what chemicals are typically found in fires of this type. Unfortunately, the number of synthetic materials and the variety of these materials used within structures are changing almost daily. The study of gases produced during fires involving modern structures looks like a hazmat practitioner's worse nightmare.

Most common toxins found in the fire environment

Carbon monoxide	Ammonia	Nitrogen oxides
Carbon dioxide	Hydrogen cyanide	Hydrogen chloride
Sulfur dioxide	Isocyanates	Halogenated acid gases
Hydrofluoric acid		

Other toxic gases that may be present

Acrolein	Methane	Ethane
Ethylene	Acetaldehyde	Benzene
Toluene	Chromium	

Examining the acute effects that lead to injury or death, these chemicals can be divided into two main groupings: asphyxiants and irritants. However, more chronic in nature, carcinogens and those chemicals that cause long-term effects on organ systems are also discussed.

Asphyxiants include simple asphyxiants or those chemicals that displace oxygen. The most common of these found in a smoke environment are carbon dioxide and nitrogen. Chemical asphyxiants that interfere with the transportation or use of oxygen within the body include carbon monoxide, cyanide, and nitrates, and nitrites. Asphyxiants incapacitate victims by causing sensory deficits and diminishing the victim's ability to reason and escape from harm. If a victim is inside a building involved in a fire, these symptoms, combined with decreased visibility and respiratory irritation, make escape almost impossible. Many victims never receive a burn injury but instead die from respiratory exposure to toxic gases.

Irritants are found in both solid particulate matter and gaseous form. The size of the particulate matter determines if the irritant affects the upper airways, lower airways, or alveoli. Particles less than 3 microns can gain entry into the lower airways, while particles of 1 micron or smaller can penetrate the alveolar space. The smoke itself is made up of carbon aerosol and soot, which is less than 1 micron and can coat the linings of the alveolar surface. Respiratory irritants found in smoke include acrolein, ammonia, chlorine, hydrochloric acid, nitrogen oxides, phosgene, sulfur dioxide, and formaldehyde. Irritants that reach the respiratory system injure the underlying tissues causing irritation, inflammation, and tissue destruction. If the irritants reach the alveoli, the surfactant can be destroyed or displaced. This condition, combined with tissue damage to the alveolar-capillary membrane, allows fluid from the intervascular space to easily penetrate the respiratory passageways and further hinder gas exchange.

The signs and symptoms associated with victims of smoke inhalation are related to either injury of the airways or hypoxia caused by some interruption of oxygen transportation to the cells. The signs and symptoms associated with irritation include coughing, shortness of breath, tachypnea, bronchial and/or laryngeal spasm, rales, chest pain, and tightness. Those signs and symptoms caused by hypoxia include headache, confusion, dizziness, and coma. Other signs include facial burning, singed nasal hairs, gross soot in sputum, and poor pulse oximetry and capnography readings.

Signs, such as singed nasal hairs, indicate that the person has inhaled active fire and not just hot air. But, even with the sign of singed nasal hairs, it does not mean that the lower airways are injured. As stated in the respiratory section, the lungs are very efficient when cooling hot air. In reality, unless the air is charged with superheated steam, the lower airways are usually untouched by the hot atmosphere. It must be emphasized that although a lower airway injury may not result from heat exposure, the most common cause of death during the early phases of burn treatment is an upper airway occlusion, secondary to an upper airway burn injury. Hot steam, to the contrary, has 4000 times the heat capacity of superheated air and can easily produce thermal injuries all of the way to the alveoli.

Victims of smoke inhalation should be held and monitored at the hospital for at least 24 hours, even if no symptoms originally existed. Changes in pulmonary tissue permeability can take place up to 24 hours after exposure. In these cases, the admission X-ray may be unremarkable only to show diffuse infiltrates many hours after the exposure. Long-term changes in the fine bronchioles and alveolar structure have been noted after continued exposures. These subtle changes may be noted on spirometry tests. For this reason, firefighters throughout the country are tested utilizing some form of pulmonary function test on their annual or biannual physicals.

Of course, asphyxiants and irritants are not the only dangers noted with exposure to smoke and combustion gases. Carcinogens are also prevalent in the smoke environment and are responsible for ending the careers of many professional firefighters. One study conducted at 24 fires, where firefighters wore air sampling devices, indicated that within all 24 fires, 6 known carcinogens were found. Among these were acrylonitrile, arsenic, benzene, benzopyrene, chromium, and vinyl chloride. These chemicals, in the levels found during a single fire, would probably leave no lasting effects on an exposed victim, but in the case of a firefighter who may experience repeated and combined exposure, the danger of cancer is greatly multiplied.

Danger to Firefighters

Advancements in personal protective gear have allowed firefighters to gain access deeper into structures involved in fire. These advancements have permitted firefighters to make dramatic rescues, and fire stops that in the past days would have been impossible. As a result, this protective gear is becoming even more contaminated with toxins generated during firefighting activities. Throughout the nation, there is a move to regular cleaning of firefighting gear to lessen the continued effects caused by wearing this contaminated gear. In fact, NFPA 1851: Standard on Selection, Care, and Maintenance of Structural Firefighting Protective Ensembles, sets specific requirements for cleaning of turnout gear. This is a consensus standard and is not enforceable, but represents best practice for the care and maintenance of firefighting gear. Unfortunately, many departments are not universally participating in this practice. Safe storage practices of contaminated gear have been addressed in standards throughout the fire service, yet, even today, during heightened awareness of contamination, many firefighters are storing their gear in clean clothing lockers, back seats of their autos, and even keeping items such as helmets and gloves in their food lockers.

If we are to realize a change in the life expectancy of professional firefighters, these practices must be altered. Many departments are still requiring firefighters to hang their gear in the same place as the apparatus is kept, where exhaust fumes penetrate the fabric day and day. This is another practice that should be remedied to lessen the long-term effects that firefighters have experienced throughout the history of the profession.

Treatment

Treatment of victims of smoke inhalation should be aimed at increasing oxygenation to the cells. This can be accomplished by providing 100% oxygen via a non-rebreather mask on the breathing patient. If the patient is not adequately breathing or an oximeter reading indicates saturation of less than 90%, CPAP or intubation should be considered with positive pressure ventilation. If auscultation of the lungs reveals wheezing or other signs of airway constriction, an updraft or aerosolized solution of Alupent or Proventil should be considered. In addition, consider the administration of Solu-Medrol, prednisone, or dexamethasone to reduce inflammation of the airways related to the inhalation of an irritant. Other bronchodilators, such as epinephrine or theophylline that are cardiac stimulators, should be avoided as some products found in smoke are myocardial sensitizers.

Some literature has suggested the use of a cyanide antidote (utilizing nitrites) kit because one of the toxic findings in smoke poisoning studies indicates cyanide as a significant toxicant. The Cyanide Antidote Kit primarily relies on the conversion of hemoglobin to a non-oxygen-carrying compound, methemoglobin. A smoke inhalation patient will already have a level of methemoglobin and carboxyhemoglobin. By adding up to 25% more methemoglobin, the patient's condition is likely to get worse, and this practice may even lead to death.

The CyanoKit that relies on the use of hydroxocobalamin to bind with the cyanide ion to form cyanocobalamin is the best bet. Cyanocobalamin is simply vitamin B_{12}. This kit has very few side effects and will directly remove the cyanide component without a detrimental effect on the patient's condition. Except for some raised red areas of skin, rash, and red urine, the overall side effects are very manageable.

Another consideration is that of hyperbaric oxygen (HBO). HBO has been proven to be a viable treatment for smoke inhalation. Because the primary toxins affecting the body are carbon monoxide and cyanide, it stands to reason that HBO would work. The emergency responder dealing with a smoke inhalation patient should consider the possibility of transporting a patient to a hospital that can provide this treatment instead of transporting to a hospital that does not have the capability.

Decontamination is another issue that must be considered. In the past, reference books did not recommend decontamination for firefighters after fighting a fire. But more recent research has identified that simple decontamination of firefighters after firefighting reduces the exposure to cancer-causing agents. Therefore, decontamination of fire victims is an important step both for the victim and for those who are treating the victim after exposure. At the minimum gross decontamination should always be done on a victim of fire.

This gross decontamination should involve the removal of all the contaminated clothing. The gross decontamination is performed for two reasons. First, it will lessen the exposure experienced by a patient by removing clothing that is holding contaminants against the skin. Many times, because of the firefighting efforts, this clothing is wet, increasing absorption of any chemical agents embedded in the clothing. Second, it lessens the secondary contamination suffered by those providing care, including the fire crews, the EMS transporter, and the hospital staff. The probability of secondary contamination from a fire victim may be minimal but does exist. It is better in this field to be safe than sorry.

Basic Life Support
- Immediately administer 100% oxygen if conscious; if unconscious secure an airway to deliver 100% oxygen.

Advanced Life Support
- If conscious, begin CPAP with the PEEP setting >10 cm of H_2O.
- If unconscious, provide an advanced airway and monitor end-tidal CO_2 ($ETCO_2$).
- Start IV of 1000 cc normal saline at a KVO rate.

- Treat unconscious patients per local protocol. Consider evaluating glucose levels and administering Naloxone (Narcan) 2 mg, IVP, IO, IM, or intranasal every 3–5 minutes up to 8 mg.
- Give CyanoKit; 5 g hydroxocobalamin IV infusion over 15 minutes.
- Start a dedicated IV line.
- Reconstitute each 5 g vial with 200 ml of 0.9% sodium chloride. (Lactated Ringers and D5W can also be used if Sodium Chloride is not available). Renders 25 mg/ml.
- Invert or rock the vial. Do not shake.
- Administer 5 g over 15 minutes (~15 ml/min).
- An additional 5 g dose may be administered if necessary.

Pediatric Considerations

- CyanoKit – Hydroxocobalamin
- The safety and effectiveness of the CyanoKit have not been established in the pediatric population (per package insert). The CyanoKit has been successfully used in the non-US market at a dose of 70 mg/kg to treat pediatric patients.
- The CyanoKit has been successfully used outside of the United States at a dose of 70 mg/kg to treat pediatric patients.

Wheezing Secondary to Toxic Inhalation

Overview

As discussed in other areas of this chapter, after exposure to specific chemicals, wheezing presents some concerns post-exposure. If the responding medical provider knows what chemical the patient is exposed to, it is important to follow those specific protocols. In the case of a patient complaining of post-exposure difficulty breathing without identification of the offending chemical, several procedures must be followed. First, the use of oximetry will provide oxygen hemoglobin saturation in the blood. But, keep in mind that some hazardous materials exposures can cause the oximeter to give false readings. In a hazardous materials situation, oximetry should always be backed up by capnography. Accurate reading in both of these instruments will give an indication of the patient's breathing status.

Wheezing is caused by two different pathologies. First is the involuntary constriction of the smooth muscle that makes up the bronchial walls. This occurs after an allergic reaction or a sensitivity to a chemical. The reaction stimulates the bronchial smooth muscle to constrict, narrowing the airways. The treatment for this type of condition is to provide a medication that relaxes the bronchial smooth muscles. These include Alupent, Albuterol, Brethine, and Epinephrine.

The second cause of wheezing is swelling of the bronchial walls. This occurs when there is exposure to a respiratory irritant such as chlorine or ammonia. Both of these chemicals are water-soluble and easily mix with the water-soluble respiratory mucous that coats the inside of the respiratory airways. Once the irritant is mixed into the mucous, the chemical is in constant contact with the tissue, causing injury and swelling. In treatment for swelling caused by this exposure, a steroidal anti-inflammatory drug must be used. These include drugs such as prednisone or methylprednisolone (Solu-Medrol).

Ideally, the best approach to treating wheezing from an unknown chemical is two-pronged. First with a bronchial dilator, followed next by an anti-inflammatory drug such as a steroid. Using both will cover any possible pathology causing the wheezing.

Advanced Life Support

- Initiate a nebulized updraft of either albuterol (Proventil or Ventolin) and/or ipratropium bromide (Atrovent).
  - Albuterol 2.5 mg (0.5 ml of 0.5% diluted to 3 ml with sterile normal saline) give via nebulizer. It may be repeated three times.
  - Ipratropium bromide 0.5 mg (500 mcg)/2.5 ml via nebulizer repeat at 20-minute intervals for a total of three doses (usually only one dose given in the field).
- Administer methylprednisolone (Solu-Medrol) 125 mg, IVP slowly.
  - Monitor heart rate. Contact medical control if HR is over 150 bpm.
  - Consider the administration of Brethine (terbutaline sulfate) 0.5 mg given a subcutaneous injection (0.5 ml of a 1 mg/ml solution).
- Maintain adequate ventilation and oxygenation
  - Assess:
    - o Oximetry
    - o Capnography
    - o EKG
- Provide
  - CPAP and set the PEEP setting at or above 10 cm of H_2O and 100% oxygen (or select the high setting).
  - Provide advanced airway if needed.
  - Control seizures or anxiety with Valium or Versed
    - o Diazepam (Valium) 2–10 mg or
    - o Midazolam (Versed) 2–2.5 mg IV/IO (Max 10 mg) or
    - o Lorazepam (Ativan) 2 mg IM/IV/IO. It can be repeated in 5–10 minutes.
  - Consult specific toxidrome protocols for additional medical treatment guidelines.
- If the wheezing is related to known exposure (hydrofluoric acid, chlorine, chloramine, ammonia, phosgene), access the specific protocols for further guidance.

Tachycardia Secondary to Chemical Exposure

There are a number of chemicals that can cause tachycardia post-exposure. These chemicals stimulate the CNS and, as such, cause the heart rate to increase. This is most often seen in nervous systems stimulation from recreational drugs such as methamphetamines or cocaine. In addition to these nervous system stimulators, there are botanicals that, when misused, also cause extreme tachycardia.

One in particular that has been trending across the United States on a periodic basis is the flowering plants called "Angel Trumpets." The flowers are found in the tropical states and grow a beautiful, large flower the faces down, at the ground, when fully grown. Teenagers have found that they can get high from this plant. They will pick the flower and several leaves, return home and boil them. The root, which contains the largest concentration of the drug, is often used as well. After boiling, the liquid is called "Angel Trumpet Tea," which contains belladonna alkaloids, including atropine, hyoscyamine, and scopolamine.

Patients present with hallucinations, restlessness, sobbing, sexual excitement, or aggressive behavior. In one case, the psychosis was so severe, the victim amputated his own tongue and penis. But one of the most severe issues is the toxically driven tachycardia that can be extreme. This tachycardia does not allow filling of the ventricles prior to contraction and ultimately results in poor cardiac output and a loss of blood pressure.

The toxically driven tachycardia does not usually respond to the typical treatments for supraventricular tachycardia. Valsalva maneuvers, adenosine, and overdrive pacing are generally unsuccessful, and prehospital care providers are left with a victim who is in critical danger and uncooperative, and sometimes aggressive.

Hypotension Caused by Exposure

There are a number of chemicals that can also cause hypotension. These do not include those chemicals that cause hypotension because of hypovolemic shock, as seen in organophosphate and carbamate poisoning or with arsenic and ricin poisoning. Specifically, when evaluating chemicals that cause hypotension, the evaluation involves those that change the hemodynamics of the body and not the changes in blood volume. The chemicals involved in this are any that cause a decrease in CNS function.

Those chemicals that affect the CNS function include most hydrocarbons. Hydrocarbons, and more specifically, halogenated hydrocarbons, are not only CNS depressants but also cause cardiac sensitivity.

Phenol, also called carbolic acid, is known for its ability to suppress the CNS. A victim who has been exposed to phenol should be watched for at least 24 hours as the onset of symptoms may be delayed and severe and will directly involve the depression of the CNS.

Depression of the CNS is characterized by a slow pulse, drop in blood pressure, slow breathing, and lower core body temperature. It can become so severe that death can follow a significant exposure.

The other chemicals that hemodynamically and physiologically change blood pressure are nitrates and nitrites. When nitrites/nitrates enter the body, much of the chemical is converted into nitric oxide. Nitric oxide relaxes the smooth muscle found around the blood vessels and essentially lowers blood pressure.

It is interesting to note that nitric oxide is the common agent used in erectile dysfunction drugs. Treating a patient who may be in need of nitro glycerine to treat chest pain, who is currently using an erectile dysfunction drug may cause a critical loss of blood pressure and death to the patient.

Fertilizers, crayons, paints, and dyes are commonly high in nitrates and nitrites. All of these have been known to cause nitrate poisoning. The vasodilating effects of nitrates and nitrites represent only the first part of the poisoning. The second issue with these chemicals is their ability to change hemoglobin to methemoglobin. This is discussed in detail in the nitrate and nitrite chemical asphyxiant section of the book.

Seizures Post-Exposure

Chemicals that cause seizures are called *seizurogenic* chemicals. They include a variety of toxic chemicals, from industry and chemical warfare agents to natural toxins. Chemicals cause seizures through two different means. Those that cause hyperstimulation of cholinergic receptors and increase neurotransmissions, such as organophosphate pesticides and military nerve agents. Others block the inhibitory regulation of neurotransmission through antagonism of inhibitory GABA (gamma-aminobutyric acid). GABA is an amino acid that inhibits neurotransmission through the CNS. In fact, it is these principles that regulate the "seizure threshold." The seizure threshold is the balance between the excitatory and inhibitory signals of the brain.

The seizure threshold is different in every person, so a seizure cannot be reasonably predicted after exposure. Antiepileptic drugs work on either the function of potentiating the inhibitory signals (GABA) or decreasing the excitatory signal (glutamate) conduction.

Seizures can be provoked by trauma, stroke, fever, medications, and chemical exposures. Medications that can provoke seizures include antidepressants, antipsychotics, and stimulants. Chemicals that can lower seizure thresholds and cause break-out seizures to include organophosphate pesticides, hydrocarbons, and possibly carbolic acid.

Opioids Overdose/Exposure

Over the past few years, there has been an explosive increase in the misuse of Fentanyl and its analogs (including Carfentanil) in the United States. These synthetic opioid-based drugs can be very dangerous to emergency responders. To add to the confusion of an emergency response to modern-day synthetic opioids, much of the information that is taught about these drugs is simply NOT true. In all honesty, this misinformation was taught to make the point that some of these agents are dangerous, very dangerous. But teaching fear in order to get respect has never been a good style of teaching and leads to other issues.

The technical information is based on fact, science, and a bit of common sense. It has always been better to have emergency responders who are well educated instead of making them fear what they do not understand. With many of these drugs/chemicals, the danger does not stop at the scene. Instead, having a lack of understanding about contamination and secondary exposure may lead to additional injuries down the line. Contaminating a fire truck, ambulance, or police car is a real concern. Just as bringing contaminants to the hospital and causing secondary exposure to hospital staff or uninvolved patients should also be a concern. Expected Audience Law Enforcement, Fire Fighters, EMS, HazMat, and Hospital Emergency Department Staff will all benefit from this program. The misuse of a chemical or drug always generates a law enforcement response. If there has not been a report of an injury, then law enforcement will just be dealing with the incident, and, more than likely, arrests will be made. If an injury is related to the misuse, then EMS and usually the Fire Department are dispatched.

In some places in the country, the fire department is only called if there is a danger of fire, and a third Service EMS agency is responded. Eventually, this will mandate the transport of an injured/ill person to the emergency department. It is easy to see how one misinformed decision early in the response can affect everyone down the line.

History

Although drugs have been misused for as long as recorded history, there are some landmark events that should be discussed in order to understand why it is such an issue today. In 1905 cocaine first became popular in the United States, but the use of cocaine exploded in the 1980s.

Opium Alkaloids

Opium was popular in the eighteenth and nineteenth centuries, especially in China, and it entered the United States in medications, elixirs, and smoked in opium dens. Opium was eventually controlled in the United States and only sold under physician orders, but misuse has continued to this day. The original opium was made from Poppy seeds and was used to make many other analgesics like morphine.

The actual poppy seed plant contains alkaloids that are known for their pain-relieving capabilities. Natural opiate drugs come from these alkaloids. These include:

- Morphine
- Codeine
- Heroin
- Thebaine
- Oripavine

Synthetic Opioids

Eventually, the opiate alkaloids were developed in the laboratory allowing for the creation of much stronger synthetic opiates. By using a laboratory to create these drugs, they could also be produced in much greater quantities. The most common pharmaceutical synthetic opiates include:

- Demerol
- Lortab
- Fentanyl
- Atarax
- Dilaudid
- Methadone
- Norco
- Buprenorphine

Fentanyl is one of those synthetic opioids. The first reported non-pharmaceutical manufacturing on Fentanyl was done in the early 1990s. A self-made chemist named George Marquardt (called a chemical genius) made Fentanyl that was sold on the street as "China White" or "Tango and Cash." These drugs were sold as extremely potent heroin. Heroin addicts heard about the strength of China White and would seek it out. If the addict used the drug in the same quantity as heroin, they would almost immediately die from an overdose. In fact, George Marquardt was believed to have caused the death of over 300 people. Some called him the most prolific serial killer of all times. George once said that drug deaths were good for business. As word spread about the strength of this new "heroin," more and more addicts wanted it and would pay a lot just to have it. Once George was arrested and sent to prison, the Fentanyl issue went away or went away until sometime after 2010. The primary producer of today's Fentanyl is China. The drug is smuggled into the country, where it is distributed to drug dealers.

Today, emergency responders and hospital personnel are just worried not about Fentanyl but also its analogs, including the much more powerful Carfentanil.

Semisynthetic Opioids

There are also a group of opiates called semi-synthetic opioids. These are made by mixing natural opium alkaloids and synthetic alkaloids. The name of these drugs depends on which natural opium alkaloid is mixed into it. These drugs are often prescribed for pain following surgery and have contributed to the opiate addiction crisis currently occurring in the United States. These include:

- Hydrocodone – contains the natural alkaloid, codeine
- Hydromorphone – contains the natural alkaloid, morphine
- Oxycodone – contains the natural alkaloid, thebaine
- Oxymorphone – contains the natural alkaloid, thebaine

Today's Fentanyl and Carfentanil

Fentanyl and its analogs (including Carfentanil) pose a real danger to both the emergency response community and the receiving hospital staff. An analog is a chemical compound that is very similar to another chemical compound, but the compound structure differs in a slight way. Analogs can be much more toxic or less toxic than the original drug. These illegally manufactured drugs are usually found in either powder, pill, or liquid form. There have been over 12 different illegally formulated fentanyl analogs, including Carfentanil. These drugs have poor skin absorption but can be ingested, the powder can be inhaled, or they can be injected.

From a secondary exposure hazard, touching it, holding it in your hand or, contacting with an arm or a leg will not be significant enough of an exposure to cause any symptoms as long as it is removed in a timely manner. It is important to remove any known contamination from the skin because of the possibility of secondary contamination to another person or rubbing it in the eye or brushing contaminated skin on the mouth.

Although some may not think of a Fentanyl or Carfentanil overdose/exposure to be a hazmat emergency, it certainly qualifies as such. Because these opiate drugs are so powerful that accidental exposure to the eyes, respiratory system, or mucous membranes can cause serious injury or even death, it is important to recognize these as toxic chemicals and the use of chemical personal protective equipment (PPE) is necessary.

There have been many stories and many more classes teaching both fire and law enforcement personnel about the dangers of touching fentanyl or its analogs. Some have claimed that death could result from touching these drugs. That is a great exaggeration, and because of what has been taught, many psychological responses have resulted. Emergency responders have become dizzy, short of breath, and even lost consciousness after just being in the general vicinity of a synthetic opioid.

Sometimes these psychological responses are treated with Narcan only to find that their later drug test (usually taken at the hospital) never showed an exposure. Just remember that these psychological responses produce real physiologic symptoms but are NOT caused by the opioid but instead by an over-reactive psychological response.

Making matters even more confusing for emergency responders is the common practice of mixing these drugs together. Mixing fentanyl with either cocaine or heroin is common and sold as more potent cocaine or stronger heroin. These mixtures are dangerous, as drug users who buy cocaine but instead get a mixture of cocaine and fentanyl may be unknowingly supplying a lethal dose resulting in death. The same is true for mixing heroin with fentanyl. These mixtures often lead to overdoses and death.

According to the American College of Medical Toxicology (ACMT) and American Academy of Clinical Toxicology (AACT), with their current understanding, significant toxicity cannot be obtained with small, unintentional skin exposure to tablets or powder, and if symptoms were to develop, they would occur at a very slow rate and only after extended exposure. Fentanyl and its analogs are by far the most dangerous of the common street drugs. When compared to methamphetamine, cocaine, and heroin, its toxicity is much greater, making fentanyl the most dangerous to emergency responders.

For emergency responders, all of these drugs look very similar, and it is difficult to identify their differences in the street.

Signs and Symptoms

Responders who may encounter fentanyl or fentanyl analogs must understand the signs and symptoms of opioid toxicity in order to provide appropriate treatment. These include:

- Drowsiness
- Nausea and vomiting
- Pinpoint (small) pupil
- Rapid onset of respiratory depression
- Slow/shallow breathing (hypoventilation)
- Decreased level of consciousness

These symptoms should be treated with Narcan (Naloxone).

NOTE: If the reported symptoms are vague or just dizziness and anxiety, these are probably not related to the exposure and should not be treated with the antidote (Narcan).

Experience: In East Liverpool, Ohio, A police officer nearly died during a traffic stop after coming in contact with fentanyl. The officer was attempting to pull over a driver who was allegedly conducting a drug deal. The driver had warrants for possession of Carfentanil; the driver rubbed a white powder into the floor of the car when the officer attempted to pull him over. The powder was later confirmed as fentanyl. After the arrest, the officer started feeling ill, and an ambulance was called. The officer fell to the floor, and the crew administered naloxone; he was treated at the hospital and later released.

Experience: In Winnipeg, Manitoba, a firefighter paramedic was given Narcan after allegedly being exposed to fentanyl during a medical call. Crews were responding to a possible fentanyl overdose when the firefighter-paramedic said he started experiencing respiratory distress. When crews made it back to the station, paramedics administered naloxone to the firefighter paramedic; he made a full recovery and has since returned to work.

In both of these cases, there were others in proximity to the drug, which never had any signs or symptoms, including the first case with the driver of the car in the confined cockpit of the vehicle. More than likely, both cases were psychological responses and not actual exposures to the drug.

Summary

A toxidrome is a group of findings, from a physical examination or from diagnostic testing. The term toxidrome was first coined by Doctors Joseph Greensher and Howard C. Mofenson. Both were pediatricians and had relationships with poison control centers. Part of the reason for developing common toxidromes was to assist in developing differential diagnosis' and to help narrow the possible causes of related symptoms into a singular diagnosis. This book identified toxidromes that are common in the world of chemical exposures.

The seven toxidromes discussed in this chapter contain materials from similar categories but not necessarily similar symptoms. They are (i) Corrosive and Irritant Toxidrome, (ii) Asphyxiant Toxidrome, (iii) Methemoglobin Generator Toxidrome, (iv) Cholinergic Toxidrome, (v) Hydrocarbon and Derivative Toxidrome, (vi) Biohazard and Infectious Disease Toxidrome, and (vii) Radiologic Toxidrome. A health care provider cannot be expected to know every possible offending agent. Instead, the intent of this chapter was to present some of the more common exposures then display the physiology associated with that exposure. Also, this chapter contains the proposed treatments for the toxidrome exposures. The treatments in this chapter have been vetted through three poison control center toxicologists and represent the current spectrum of treatment modalities.

It has always been the goal of the authors to provide enough detailed information to understand the physiological changes that take place after and exposure during the poisoning process. This understanding of the physiology is critical to interpret the diagnostic findings and to provide the most efficient and beneficial treatment, even if that treatment is only supportive in nature.

As noted in this chapter, hazardous materials are not just chemicals. Radioactive substances and biological materials are also discussed here. Over the past several years, especially after Ebola, SARS, MERS, and COVID, biological infectious materials have become part of the hazardous materials world and responding hazmat medics and emergency department staff must be familiar with the changing environment presented by contagious diseases.

The overall goal of this chapter is to provide detailed enough information to both emergency responders and emergency department staff to develop a differential diagnosis, identify critical signs and symptoms, and to provide rapid and beneficial treatment in an effort to save lives or reduce the occurrence of long-term injury.

Reference

Golsousidis, H. and Kokkas, V. (1985). POISINDEX: an emergency poison management system. *Drug Inf J.* 9: 2–3.

4

Event Conditions

Introduction

This section will provide an overview of scene assessment. Hazardous materials responders and other emergency response personnel should already be very familiar with the concept of scene evaluation, but for those whose responsibilities do not include emergency response, this will be a new concept. This process provides clues for identifying the offending chemical and the potential hazards.

Recognition of hazards and the ability to establish the degree of danger at an emergency incident is a basic skill for all emergency responders. Situational awareness, which is the ability to recognize changing situations and predicting future outcomes, is taught to every fire and law enforcement officer. Being aware of the dynamics of an emergency incident is important for survival and injury prevention.

It is important to, once again, stress that many chemical exposures take place at home and places of business and not necessarily on a large hazardous materials incident. Emergency responders must be in-tuned to look for clues while on the scene in those locations as well.

This example demonstrates the importance of recognizing clues and acting on those clues. On this emergency scene, the clues were initially missed allowing the emergency responders to enter a hazardous environment and suffer from illness as a result:

Emergency response fire department units, which included an engine company and rescue, were dispatched for a wellness check at a private residence. When they arrived, they met with the caller, who was the resident's girlfriend. She was concerned because the day before, her boyfriend had complained of a migraine headache and was going to lay down. Later that evening, she attempted to call him to see if he was feeling better and could not reach him.

The responders first knocked on the door with no answer. The firefighters then forced the locked door and made entry. Just inside the door was a large aquarium with a pet snake inside. The snake was lying upside down and was obviously dead. Each responder made a remark about the snake as they walked past it. Upon entering the bedroom, they found the boyfriend unconscious on the floor.

Once they moved the victim to the living room to begin providing care, one by one, the crew began to report that they were dizzy, nauseous, and unable to provide care. Eventually, the officer began to suspect that there was something in the house causing the group of events to take place. The dead snake, unconscious resident, and now several crew members

Hazardous Materials Medicine: Treating the Chemically Injured Patient, First Edition.
Richard Stilp and Armando Bevelacqua.
© 2023 John Wiley & Sons, Inc. Published 2023 by John Wiley & Sons, Inc.

feeling sick caused the officer to order everyone out of the house. They later found that the house contained very high concentrations of carbon monoxide.

In this example, the clues were subtle and initially missed by the responders, but as more clues became apparent, they recognized the danger and appropriately evacuated the structure.

Eventually, the hazmat team was dispatched, and the chemical was identified. Although identification of the chemical does not delay the treatment of symptoms, it does provide the information necessary to provide more specific treatment and even, in some cases, antidotal therapy. When the clues are recognized, the characteristics can be determined and the appropriate response actions put in place.

There are many clues that come into play when responding to an emergency event. Most fall into a category of the obvious, or what is referred to as a risk assessment. These are generally placed into one of the following recognition and identification categories;

1) Occupancy and Location,
2) Containers, Shape, and sizes,
3) Placards and labels,
4) Facility documents, and
5) Personal accounts.

The less obvious clues, but still can have a dramatic impact on the incident, are the operational hazards. These are subtle clues that, on the surface, may not have value but, when placed with other pieces of information, becomes significant. It is different than the typical process of hazard identification or risk assessment found on a fire scene or auto crash, where the clues are largely easy to see. Instead, it involves the assessment of potential threats by gathering groups of information and analyzing them together to assist in determining the hazards. Some of these include weather conditions, dispatch information, subtle suspicions (gut feelings), patient signs and symptoms, and witness accounts. (See the continuum on page 8)

Case Study – Fertilizer Explosion in West Texas

On April 17, 2013, a fire occurred at the West Fertilizer Company storage facility in West, Texas. The city of West is located about 18 mi north of Waco, Texas. The local volunteer fire department responded and attacked the fire with hose lines. Approximately 20 minutes after their arrival, the site exploded killing five volunteer firefighters from the city's fire department and four volunteer firefighters from three different neighboring fire departments who were attending an emergency medical services (EMS) class in the city and responded with West Fire Department to assist. In addition, one off-duty career fire captain from Dallas Fire Department, two civilians who were assisting the firefighters, and three civilians living nearby were also killed by the explosion. In total, there were 15 fatalities and more than 160 people injured; 150 buildings were damaged or destroyed.

Approximately 150 tons (300,000 lb) of ammonium nitrate fertilizer was on the site. It is estimated that about 30–45 tons detonated causing the explosion. The explosion created a crater 10 ft deep and 100 ft wide. When the blast occurred, it propelled flaming debris a half mile in all directions. One piece of debris was found 2.5 mi away.

The explosive force was comparable to 7.5–10 tons of TNT and was felt more than 30 mi away and registered as a 2.1 magnitude earthquake.

Ammonium nitrite has a significant history in creating disasters. In 1947, the largest non-military explosion in history took place in Texas City, Texas, when a ship's hold, filled with 4,600,000 lb of ammonium nitrite began to burn. The resulting explosion killed 581 people and all but one member of the Texas City Fire Department.

In 1995, 5000 lb of ammonium nitrite was used in a terrorist attack to blow up the Alfred P. Murrah Federal Building in Oklahoma City. The blast killed 168 people and injured hundreds of others.

The dangers of ammonium nitrite were well known in the fire service. The NIOSH investigation revealed a number of issues with the response and, as a result, made recommendations for every fire department to follow. First and foremost, every fire department should implement and enforce the use of the Incident Management System (IMS) during all emergency responses. In addition, fire departments should conduct pre-incident planning inspections to identify high hazard/high-risk structures and occupancies and have a written risk management plan and use risk management principles at all fires especially those involving high-risk hazards.

Operational Hazards

Dispatch Information

The initial dispatch provides critical information about the scene. The location, condition of the response, type of leak or spill, and information concerning the number and types of injuries are all critical initial information. Size-up involves analyzing the information about the scene. This process actually starts during preplanning with the records of Tier II reports. However, it should continue as information about the incident is received.

The gathered information should include the location of the incident so population densities can be predicted and the time of day to further assess life hazards. If the incident involves transportation of a material, it is important to know if it exists on an interstate, city/county road, rural setting, or railway system. Each of these settings presents a different set of considerations for the command staff and emergency response teams.

Building types and occupancies are important information for the responding units. Is the incident in an industrial park, or is a light industry within the business area? School and university chemistry laboratories contain large varieties of chemicals with varying hazards. Public and private sewer treatment plants, recreational areas, and waste sites also store or house large volumes of chemicals. Hospitals, clinics, and doctor's offices have a variety of hazards (biological, radiological, laser emissions, and chemical agents) that under the right condition may present negative consequences.

Note: Tier II reporting is the submission of reports to the local emergency management office and/or fire department and is required under EPCRA section 312 (Emergency Planning and Community Right to Know Act) of 1986. It provides the local authorities with the specific information on potential hazards that a facility or location may have. It includes types and names of chemicals on premise, both the highest and lowest amount that may be stored at the facility and where on the site the storage is located. These reports are provided on an annual basis.

As dispatch is gathering and assembling information, the responding units also have a variety of information available to consider. Responders who have extensive knowledge of the incident area may help assess the particular needs of that area. From the information given at the time of dispatch, several items can be evaluated for the initial scene setup. If the product is airborne, wind direction and speed will provide clues on the direction of travel.

When the location is identified, preplans should be reviewed and examined for the resources that may be required.

From the initial information, identification of the chemical and recognition of the hazards become an early consideration. After this is accomplished, isolation and evacuation may be considered. It would be ludicrous to say that you know or are able to read all of the information on hand about the chemical in question. However, if you give yourself a brief overview of the reference material, size-up will become a much easier task.

Note: Prepare chemical reference cards for specific chemicals that you know you have in your response area. Identifying the chemical and physical properties, toxicological data and supportive information such as medical treatment for both basic life support and advanced life support. These can be prepared and duplicates made as a one- or two-page scan sheet for quick referencing. These Quick Response Guides should be updated annually identifying the facilities where the chemical can be found along with quantities.

Preplanning of target hazards, including certain industrial occupancies, schools, shops, or other high hazard areas, can give exact evacuation and hazard zone parameters. By preplanning, a determination to defend in place may be considered in lieu of a full evacuation.

It is commonly thought that the two populations presenting the greatest challenge during an evacuation are children and the elderly. If these two groups are found in high numbers within the hazardous area, defending in place may be a real alternative to evacuation.

If there is a likelihood of exposed patients leaving the scene, timely notification to hospitals is critical. The hospital can then start to prepare its operation in order to meet the demands of the incident. This is discussed in detail when reviewing the HazMat Alert in this text (page 262).

Scene Safety

Emergency responders have been overstimulated with the words "establish scene safety." This has become a benchmark for every type of assessment, whether it is a practical paramedic exam or a lieutenant's promotional exam. It has been overused to the point that most responders never even look to establish true scene safety. When the act of assuring scene safety is conducted, everyone on the scene should be tapping into their skill of situation awareness. This involves evaluating everything that is located on the scene, including any active processes that are taking place, then determining the amount of danger that these findings create. After this quick evaluation, a responder or incident commander should then consider what they believe will happen next. Then plan for that event.

The danger of not performing scene safety or having situational awareness may lead a responder into a dangerous situation. Not being aware of a small fire in the same area of large quantities of flammable, combustible, or explosive chemicals on the scene can lead to significant injuries and deaths. Not long ago in Texas, there was an industrial fire at a fertilizer plant. Fertilizers are oxidizers and can cause combustibles to burn more rapidly and hotter because they produce oxygen as they are broken down by the fire. The responding fire crews placed hose lines into service to extinguish what they believed was a relatively routine warehouse fire. But no one recognized that the fertilizer was a particularly dangerous blend called "ammonium nitrate." This was not a simple fertilizer fire but a catastrophic event waiting to happen. When it did, the loss of life to fire department responders was quick and unforgiving.

Assessment of the scene for both your safety and the safety of the crew is vital. Missing important clues can cost lives. Additionally, firefighter intuition may play a vital role. If you are uncomfortable about the situation, then there is probably something to be uncomfortable about. In the end, the question is: do you notice the things happening around you, and do you question the clues that must be noticed to keep you and the crew safe?

Once responders arrive on the scene, the true work begins. Both physical and mental demands will continue until the termination of the incident. The first unit on the scene must continue a complete and accurate size-up using all of the available information. The on-scene information must be considered along with the information gained through preplanning and the dispatch process. The accuracy of the information gathered by the first arriving units is critical in planning a safe and efficient response.

Experience:

Odessa, Texas, October 26, 2019

As with most oil extraction processes, a variety of substances are removed from the earth. During this process, one of those chemicals removed is hydrogen sulfide. As oil is extracted from the well, a percentage of water is within the solution. Part of the process is to place this extracted oil into large bins for the oil and the water to separate. Once separated, the oil is then pumped to tankers or pipelines for further processing. The "produced" water is pumped to a cleaning station (water flood station). It is identified as such because oil and other chemical contaminates can still be within the solution. The produced water is placed into a storage tank and moved into the pump house. The produced water is pressurized and used to inject the oil field for further oil extraction. The injected water displaces the oil within the field allowing greater amounts of oil to be extracted.

At the site in Odessa, two pumps pressurize the water within the pump house. The pump houses are not typically occupied but an employee does periodically check the pump pressure. When there is a malfunction, an automated alarm is trigged alerting the employee to check the situation.

The station was outfitted with a hydrogen sulfide detection and alarm system as sulfides within oil extraction is common. On this date, both the pump pressure alarm and the hydrogen sulfide alarm were activated and the employee was notified via cell phone.

Each alarm was a separate alarm notifying the worker of the specific problems. In addition to the cell phone alarm, a separate visual alarm system was located on the outside of the pump house consisting of a rotating red-light beacon. The visual alert beacon was not operable that day.

On October 26, 2019 at 6:38 p.m., the automated alert system was activated for a pump oil level malfunction, five minutes later the worker/operator was notified of a pump problem. The operator arrived some minutes later to investigate. The hydrogen sulfide notification beacon was not operating causing the worker to leave his personal hydrogen sulfide monitor in the truck. His investigation took him to the control panel which indicated that the problem was at pump #1.

During an alarm, the pumps are operating but not flowing any product. Once a normal level is detected, the pump starts flowing again. On this evening, the worker failed to de-energize the pump, closed the discharge valve, and partially closed the intake. This allowed the produced water, containing hydrogen sulfide to escape and release the hydrogen sulfide gas into the room. Within seconds the worked was fatally injured.

After a few hours, because the husband had not returned home, the spouse, along with their two children, elected to go search for their father. Upon arrival at the pump house, the wife soon found her husband on the floor of the pump house and was overcome by the hydrogen sulfide and was also fatally injured. Emergency services was called, and upon arrival, a strong odor was detected at the gate of the facility. Crews made haste to enter the facility with air packs and found the worker and his wife deceased at pump #1. The children, who were still in the car, were rescued without injury.

The investigation of this event showed several distinct failures:

1) The alerting system through the workers cell phone and visible alerting system that were inoperable.
2) The worker was assigned a personal air monitoring device which was left in the truck and found to be in full H_2S alarm. Failure to use safety equipment, understanding how it reads, and the training associated with the device is critical.
3) The pump was not isolated, potentially allowing the pressures to become reduced created the movement of water releasing more H_2S into the space. A lock-out, tag-out procedure was not performed.
4) There were inadequate security measures preventing nonemployees from entering potentially dangerous areas. A locked gate or some form of an entrance protocol along with maintenance to ensure visual notification of a hazardous environment.

Upon Arrival – Stop, Look, and Listen

Stop and stage far enough away to be out of possible danger, yet close enough to assess the situation. This distance may not be the true safe distance; rather, it is a distance that will give you time to react if the situation deteriorates (Staging for the incident should be initially set up at a distance initially determined using the Emergency Response Guide). Next, *look* for smoke, vapor clouds, or visible gases, and most important, indications of fire or fire potential. Finally, *listen* for high-pitched noises, an indication that fluid is creating a gas or a gas is under pressure and is being released. Remember to *stop, look,* and *listen* during the initial size-up.

Information gathering is a key component when answering an alarm. This assessment process is an ongoing progression, constantly observing the incident, the patient, and the situation as a whole. This must be continued until the termination of the incident.

Weather Conditions

The initial actions and decisions made in the first few minutes of any incident will often decide the direction of the incident. A hazardous materials emergency is no different, and the medical approach to a chemically contaminated patient makes these actions even more crucial. This is compounded by the complexity of call, multiple patients affected, and changing conditions such as weather.

Weather conditions, as with many scene conditions, can add to the difficulty a responder may encounter. Evaluating how weather can affect the chemical cannot be ignored. For example, if the ambient temperature is well below the boiling point of the chemical involved, the evolution of gas or vapor is extremely low; thus, the potential exposure is greatly reduced. However, take into consideration an evening that the temperature dew point will be reached, and the product involved is a vapor-producing water-soluble chemical. Either you have to stabilize the scene before you reach the temperature dew point, or your resources will have to be used to evacuate an area. (Remember, plume modeling does not include fog creation and cannot give you accurate predictability of how great an area may be affected.)

Responders have trained on mass decontamination efforts, and numerous videos are demonstrating different techniques for providing decontamination to large groups of contaminated victims. However, what is not discussed frequently is the weather in relation to a decontamination event. Many responders feel that a mild 75° day is warm enough for decontamination when actually given the right set of circumstances 80° outside when wet may cause hypothermia if the contaminated population is elderly. Evaluating the weather conditions upon arrival and projecting the weather conditions in the future must be done to manage the incident safely. Some considerations with weather conditions:

- Temperatures below 80 °F can cause hypothermia during and after decontamination (wind will accelerate the hypothermic response).
- Temperatures above 80 °F can cause heat stress (this is accelerated with increase in humidity, level of activity, poor heat acclimation, and working in direct sunlight).
- Temperatures above the boiling point of your chemical – gases or vapors will be produced.
- Temperature dew point is the point in which the saturated water in the air comes out and produces fog. At this point, any gas or vapor that is soluble can mix with the fog, or the fog acts as a cap and holds the vapor or gas closer to the ground maintaining higher concentrations of the chemical.
- Weather fronts that will bring rain, lightning, wind, sleet, or snow will make the incident more difficult to manage due to the increased demand for resources.

Witnesses Accounts

Often, victims/patients/responders have described a smell or color or even a sign or symptom that may not be described exactly the way we were taught. For example, on one incident that the authors were on, a female hazmat technician described an odor as being musty. She did not describe it as a rotten egg smell but rather musty. At that incident, it was a reaction that produced hydrogen sulfide. Yet, no one identified the hydrogen sulfide. The important point to take from this statement is; what is learned in books is often different than what is expressed in personal accounts. It is important to listen to personal accounts and evaluate what is being said. In the Sarin attack in Tokyo, Japan, on March 20, 1995, many of the patients described ocular pain and headache around the front of the forehead (this has been described at many organophosphate emergencies over the years).

Evaluating the most common complaints of the many victims in Tokyo, it was initially odd that the most common signs and symptoms were headaches and eye pain. But, evaluating the physiology of the exposure, these symptoms are understandable. One of the determining signs of sarin exposure is myosis or pinpoint pupils. The iris will dilate in low light situations and constrict in bright light situations through the use of muscles in the iris. Pinpoint pupils are caused by an extreme constriction of the pupillary muscles. Over time these muscles become fatigued and begin to generate pain as a result. This is perceived by the patient as eye pain. The point here is that emergency responders have been taught to look for myosis but have not been taught that continued myosis will cause eye pain and headache. It is important to look at the surroundings, be alert to the symptomology of the chemical, understand basic physiology, be alert to what the patients or witnesses have to say; they may provide critical clues.

Risk Awareness

On-scene indicators are those clues that all responders should be aware of and the risks that are associated with each. Most fire department responders are very familiar with these clues. They are listed here as a review and for basic information for the nonresponders studying this subject matter.

Occupancy and Location   As an emergency responder, you should be aware of those target hazards found in your city or response district. This is especially important for extremely hazardous substances (EHS). There are also target hazards that may contain smaller quantities of a variety of hazardous chemicals that can cause injury. For example, hardware stores, gas stations, auto repair shops, chemical plants, or transportation hubs all provide clues. Responders should know what type of products or processes are used on these sites. In many of these cases, pre-incident planning is required to develop a list of potential hazards. However, if you are responding to the chemical manufacturing plant containing numerous hazardous chemicals, one of the first questions you will need to have answered is: what are the chemicals that the victim was working around, or what is the offending chemical?

Other occupancies may not have obvious risk or are so commonplace that we have grown complacent to the potential threats. It is our experience that these places often generate single exposed patients who are suffering symptoms of chemical toxicity. Auto repair shops have solvents of many different kinds, including hydrocarbon fuels and ethers, jewelry stores that have chemical cleaning agents; some containing cyanide, big box stores containing fertilizers, chlorine-generating chemicals, acids, and solvents. All have the potential to cause an exposure.

Containers, Shape, and Sizes   The container that a chemical is stored in, the size, and the shape all give clues to the potential risks that can be found (see Figure 4.1). Containers are used to transport and store chemicals in a variety of quantities. The design of these containers can give clues to the state of matter, whether it is a gas, liquid, or solid. Each container and its shape are designed to maintain that material under desired conditions, and many of these containers are regulated in some way by federal law.

Containers are placed into very general categories, bulk and non-bulk. These are regulated by the Department of Transportation 49 Code of Federal Regulations, 171.8 (DOT 49 CFR 171.8). Bulk containers include highway cargo tanks, train cars, and intermodal

Figure 4.1   Container shape and design are intended to safely maintain the material it stores or transports in its desired state of matter.

tanks. They usually have a maximum capacity of more than 119 gal, or 450 l for liquids, or a water capacity of greater than 1000 gal (454 kg) for liquefied gases, or a maximum net capacity of more than 882 lb (401 kg) for solids. If the container contains less than this, it is called non-bulk. But do not be mistaken; a non-bulk container can potentially create a greater hazard depending on the toxicity of the chemical being stored in it and the conditions that the container is under. The DOT Emergency Response Guidebook provides a pictogram page to assist emergency responders in the identification of highway tanker truck (see Figure 4.2).

Placards and Label   Emergency responders must have a basic understanding of the rules and regulations that are involved in placarding, labeling, and identifying fixed storage facilities. Federal law mandates placarding of vehicles that transport hazardous substances (DOT 49 CFR). Reviewing federal laws is beyond the scope of this text, but it is worth mentioning that not all who transport or store hazardous materials clearly state the substances they have. It is estimated 60–70% of all placarding is correct. In some cases, one may also notice placarding of warehouses and businesses, referred to as *fixed-site facilities*. In many jurisdictions, providing fixed-site placards is a voluntary standard. Some localities have adopted the standards as part of the local code, but they are not mandated by federal law.

It would be nice if all transporters and occupancies had a warning sign, but they do not. Rules and regulations are only as good as user compliance which, as noted, is not always correct.

There are three basic marking systems that have some form of a placard, label, or stenciled numbers and/or letters. These are Classification placards, Identification

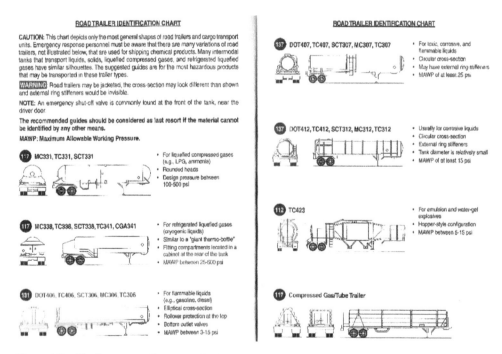

Figure 4.2   The Emergency Response Guidebook provides pictograms to assist responders in identifying types of hazardous materials trucks.

numbers or labels, and stenciled markings for dedicated containers. Some of the marking systems may be required by local ordinances or derived from the federal register. Others are loosely based upon a variety of systems. It is strongly recommended that you become familiar with these systems and their intended meanings. The following is the most commonly found systems:

The Department of Transportation (DOT) has defined nine hazard classes. These hazard classes are used in packaging, containers, and vehicles. Depending on the chemical properties and the volume, these chemicals are labeled and/or placarded in an effort to notify emergency responders of the hazard should an emergency situation occur. Information concerning labels and placards are found in the Emergency Response Guide (see Figure 4.3). By identifying the placards on a transport vehicle or a label located on a package the responder can begin to identify the hazards and determine their initial actions.

The most commonly seen hazardous materials markings are the DOT placards used on vehicles. Two concepts of regulations come into play when transporting hazardous material, and all placard regulations are based on these two concepts.

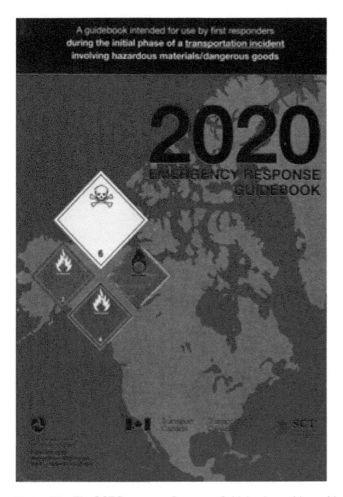

Figure 4.3 The DOT Emergency Response Guidebook provides guidance to all emergency responders in the initial stages of an incident.

1) A hazardous material is defined as a substance, material, or compound that has been determined by the DOT to have the capacity to do harm to the environment and human and animal populations. It is or has an unreasonable risk to the safety, health, and property upon contact.
2) Quantities of generally over 1001 lb require placards except for Table 1 materials that require placarding even in very small quantities (see Figure 4.4 for a general description of placard placement).

Those materials determined to be the most hazardous according to the Code of Federal Regulations (CFR) are called **TABLE 1 materials** and must be placarded any time they are transported regardless of the weight. In this category are:

Explosives, hazard classes 1.1, 1.2, and 1.3
Poison Gas, hazard class 2.3
Dangerous When Wet, hazard class 4.3
Organic Peroxide, hazard class 5.2
Poison Inhalation Hazard, hazard class 6.1
Radioactive 7, Radioactive Yellow III

All other hazardous materials are found in TABLE 2. Placards for TABLE 2 materials are only required when the amount of the chemical exceeds 1001 lb. In other words, 990 lb of a hazardous material may not require a placard at all, but at 1001 lb, it will (Figure 4.5 displays both the yellow pages and blue pages where an emergency responder can locate the chemical and access the guide that provides the initial response actions). From a health hazard point of view, does 11 lb create a safer situation?

Placards are signs made of durable material so that they will not deteriorate under normal environmental conditions. This does not mean that when the vehicle is on fire or if the

Figure 4.4 When placards are required on a vehicle, they must be placed on the front, the rear, and each side as outlined by DOT requirements.

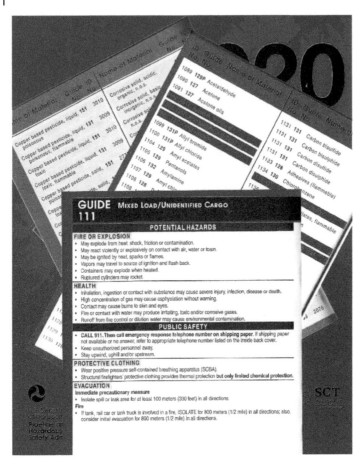

Figure 4.5 Responders can access information by chemical name or United Nations (UN) number which will lead them to an initial response guide.

chemical is leaking, the placard will not be destroyed. All vehicles that require placarding must display the placard in four locations – the front, the rear, and each side as outlined under the DOT guidelines (See Figure 4.6). Placards are square, 10 3/4 in. by 10 3/4 in., and placed on a point giving them the look of a diamond. Each placard has four general characteristics to assist in identifying a substance:

Color. Different colored backgrounds identify the classifications (1–9) of the hazardous material.

Symbol. Specific symbols identify the type of hazard. (Each classification has its own symbol.)

United Nations (U.N.) Identification Number. A four-digit identification number found in the middle of the placard or in an orange panel located *above* or below the placard must be displayed. The number identifies the chemical or group of chemicals contained within the vessel. The North *American Emergency Response Guidebook* contains a list of chemicals that corresponds to these numbers and can be referenced quickly to give the emergency responder information. (Some placards will include the hazard name.)

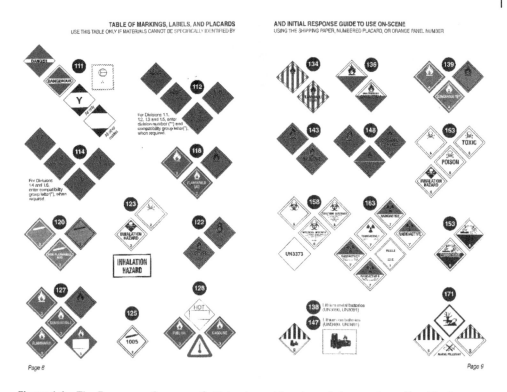

Figure 4.6 The Emergency Response Guidebook provides placard clusters that will guide the responder towards response guidelines.

Classification Number. A subclassification number to identify the subcategory of the substance and compatibility codes.

Labels look very similar to placards and contain much of the same information. Labels are placed on two sides of the container and are normally 4 in. by 4 in. Under the DOT transport placard and labeling system, chemicals are categorized under nine hazard classes and two-word classifications.

Class 1: Explosives. This classification placard shows a symbol of an exploding object (bursting ball) in the top corner and has an orange background. There are six subdivisions within this hazard class:

1.1 Materials that *have* a mass explosion hazard
1.2 Materials that have distinction through projection hazard
1.3 Fire hazard materials
1.4 Materials with a minor explosive hazard
1.5 Materials that are considered insensitive
1.6 Materials that are dangerous but are considered extremely insensitive

Class 2: Gases. This classification includes a variety of hazards, with the hazard denoted by a symbol representative of the danger. A symbol of fire against a red background is used for the flammable gases (2.1). A cylinder against a green background represents the nonflammable gases (2.2). For a poisonous gas, the skull and crossbones symbol is used on a white background.

2.1 Flammable Gas
2.2 Nonflammable Gas (lower flammable limit of no higher than 13% and a flammable range of greater than 10%)
2.3 Poisonous Gas
2.4 Corrosive Gases

Class 3: Flammable Liquids. In this classification, a fire symbol against a red background is used. The liquid can either be combustible (those liquids with flashpoints above 140 °F) or flammable, which is rated against the material's flashpoint.

3.1 Materials that are flammable with a flash point less than 0 °F.
3.2 Materials that are flammable with a flashpoint between 0° and 72° (equal to 0° but less than 73 °F).
3.3 Materials that are flammable with a flashpoint between 73° and 141 °F.

Class 4: Flammable Solids. This class includes a variety of placards, each designed to identify the hazard classification. The red vertical-striped background with a flame symbol at the top denotes the flammable solid (4.1). The spontaneously combustible solids (4.2) are denoted by the white top and red bottom placard with the symbol of a flame at the top. The Dangerous When Wet placard (4.3) is blue in color with a flame symbol at the top.

4.1 Flammable Solids
4.2 Spontaneously Combustible
4.3 Dangerous When Wet

Class 5: Oxidizing Substances. Those substances that support combustion are symbolized by a ball on fire with a yellow background. The words oxidizer (5.1) or organic peroxide (5.2) is across the middle. The subdivision numbers 5.1 and 5.2 must also be placed on the placard.

5.1 Oxidizers
5.2 Organic Peroxides

Class 6: Poisonous* and Infectious Substances. These placards have the symbol of the skull and crossbones with a white background for the poison and a diagram of the biohazard symbol for the etiologic materials.

6.1 Poisons*
6.2 Infectious Material

Note: Poison, poisonous, and toxic are all synonymous.

Class 7: Radioactive Materials. This class of materials emits ionizing radiation. There are three subdivisions to this classification, each denoting the level of ionization potential. Each placard shows the radioactive propeller on a yellow background and the word radioactive on the bottom half on a white background. Class 1 is all white and is the lowest in hazard ($5\,\mu Sv/h$). The middle hazard is class two with a yellow top half and a white bottom ($5\,\mu Sv$ to 5 mSv/h). Class three is the most hazardous and has emissions above the 5 mSv/h range.

Class 8: Corrosive Material. This class contains those materials that are extremely basic or acidic. The placard is white on top with a black bottom. The top diagram area shows a hand and an object being destroyed by a liquid that is being poured out of a test tube, denoting the corrosivity.

Class 9: Miscellaneous Materials. This classification denotes hazards that do not fit into any of the above definitions, but they present a significant danger. This placard has vertical black lines on the top half and white on the bottom half.

9.1   Miscellaneous Dangerous Materials**

9.2   Environmentally Hazardous**

9.3   Dangerous Waste**

*Note: ** Canada's classification.*

ORM-D: Consumer Commodity. These are mostly labeling with ORM-D (other regulated material, classification D) in the middle. They are for household commodities such as bug spray, drain cleaner, etc. The letter D is from an old classification. Class nine was the ORM classification. There were five subdivisions: A, B, C, D, and E. The present-day miscellaneous materials are the old ORM A, B, C, and E. The new class of ORM-D are those commodities from the old class nine ORM-D. ORM-D was phased out by DOT on January 1, 2021; however, it may still be seen on packages.

DANGEROUS: Mixed Loads. When the total weight of two or more Table II materials is 1001 lb or more, they are identified by this sign.

NFPA 704 Placard   Warehouses and businesses, referred to as fixed-site or fixed-storage facilities, also may be placarded. The NFPA 704 fixed storage placarding system is designed specifically for storage facilities as shown in Figure 4.7. Many communities have adopted the standard as a local ordinance to alert emergency responders to the dangers inside, but it is voluntary, not mandated by Federal Law (See Figure 4.7).

The Building Officials and Code Administrators (better known as the BOCA code) adopted by many localities requires the use of the NFPA 704 system. The 704 system consists of a large diamond-type placard that is colored in four quadrants: blue, red, yellow, and white. Each color represents a different type of hazard: health, flammability, reactivity, and special

Figure 4.7   The NFPA 704 Marking System for alerting emergency responders to the hazards inside a structure is a voluntary standard in most communities and not mandated by Federal Law.

information, respectively. Each of these quadrants contains numbers on a scale from 0 to 4. Zero indicates no hazard for that color, and 4 indicates the worst possible hazard. In the special information (white) quadrant, two symbols, W for do not use water and OXY for oxidizers, may indicate additional hazards.

Although this system provides much-needed information, it is still limited as it does not identify the chemical found at the structure, only the potential hazard. The occupancies that actively use the marking system provide important information to those responders who may be faced with entering such a facility during an emergency.

Hazardous Materials Identification System (HMIS) Another type of identification label is the Hazardous Materials Identification System label. This label utilizes the same coloring scheme as the NFPA 704 system, but the numbering may not be the same. The HMIS system was not designed for the emergency responder, so it is rarely seen during transport. It is primarily used in storage and manufacturing facilities so that employees can quickly reference the hazards associated with the chemicals they are handling. This label helped the industry meet portions of the hazard communication standard. Most HMIS labels also contain information on personal protective equipment needs associated with the use of that chemical.

Shipping Papers and Facility Documents Another source of information available on the scene is the shipping papers or the facility documents. These papers/documents can provide an excellent starting point for your chemical research. These documents have the chemical name and the quantity being shipped or stored, and they identify the manufacturer. Most companies that produce chemicals have an emergency contact number, and some even have response teams or technicians to assist with the technical questions about the chemical that you are dealing with.

For over-the-road transportation accidents, the shipping papers are found with the driver in the driver's side door panel. Due to incident dynamics, they may be left in the hot zone or contaminated prior to removal from the vehicle. In these cases, the papers should be placed in a large plastic food storage bag. If needed, the storage bag can be decontaminated for safe handling in the cold zone. The bag will prevent the possibility of secondary contamination and allow the papers to be safely read in the cold zone. Most trucking companies have these documents in an electronic format, which can be transmitted to the dispatch center or command vehicle.

For over-the-rail accidents, the consist is found with the engineer of the train. This consist is a document listing every rail car contained in the train by the car number and contents. These documents are also maintained in an electronic format and can be easily transmitted from the rail line to the command staff. The railroads have a very active hazardous materials training and supportive division. These are very good contacts to have if you have a major rail in your response jurisdiction.

All facilities and businesses that manufacture, store, or use any quantity of hazardous materials are required to maintain Safety Data Sheets (SDS). These used to be called MSDS – Materials Safety Data Sheets on the premises so employees can be kept informed about the hazards of their use. This is part of the Right-to-Know standard. They contain a wealth of information, including the chemical's name, synonyms, chemical properties, physical properties, health and safety issues, fire potential, flammability, and reactivity, which are but a few items addressed on these documents.

Reference Materials   The shape of the container may provide responders with clues to the general type of hazard. Cylinders that have a pressure cap, like those found on an oxygen cylinder, may indicate a compressed gas or poison. Larger containers that have an arrangement of valves and input gates should alert the responder to the possibilities of compressed gas and/ or a cryogenic product (a product usually existing in a gas form but placed in liquefied form through pressure and refrigeration). Large cylindrical objects stacked lengthwise on flatcars or semitractor-trailers may indicate high-pressure compressed gas.

Emergency responders have a limited time to gain information on an emergency incident. The primary goal is to quickly identify the product and to gather the appropriate information for mitigation. The *North American Emergency Response Guidebook,* along with the *NIOSH Pocket Guide to chemical hazards,* provides a good start.

The North American Emergency Response Guidebook (ERG)

The ERG is the quickest reference book available for the emergency responder as seen in Figure 4.8. These reference guides are available for free and should be located on every emergency response vehicle in North America. To date, and as of the writing of this book, more than 16 million copies have been circulated to first responders and emergency managers. In fact, every state requires an ERG in every emergency response vehicle (police, fire, or EMS). Another useful book for emergency responders is the *NIOSH Pocket Guide to Chemical Hazards*. This book is a bit more technical; however, once you are familiar with the basic organization, it can give you a wealth of information. These two books should be the minimum library carried aboard any medical unit that responds to a hazardous materials situation. Other, more advanced references and texts are listed at the end of this discussion.

The *North American Emergency Response Guidebook (DOT ERG)* is the one resource that was developed (and has been maintained) for the emergency responder. Its primary goal is to identify the product and some general properties by identification numbers. From the

Figure 4.8   The Emergency Response Guide is a quick reference that can provide almost immediate information concerning the initial response to a hazardous materials incident.

U.N. identification number, specific materials, groups of materials, and action plans can be researched. This book is excellent for the police officer, first response fire unit, and EMS unit in the field. It is even the first book used by professional hazardous materials response teams. In general, it describes initial action that should be taken in order to protect oneself, other emergency responders, and the public.

The *Emergency Response Guide* is divided into five sections, which are indicated by different colors. The white pages designate the general information and introduction to the book. This area assists the reader in how to use the book, with an explanation of its limitations. In this area, the user is also introduced to the CHEMical Transportation Emergency Response Center (CHEMTREC) in Washington, D.C., and the National Response Center (NRC), both of which are manned 24 hours daily. The introduction also displays pictures of placard types, shipping paper organization, and information for Canadian shipments and shipments to and from the Republic of Mexico. The color representations of the placards are given so that the user can compare what they see with what is pictured in the book.

At the back of the book are the protective action guides, which describe general actions to be taken. One point to remember is if there is no information available (placards may not be apparent due to dusk or dawn light, fire, or degradation of the placard) about the product, use guide 111.

Note: General Response Guidelines:

General DOT Guide 111
Explosive DOT Guide 112
Flammable DOT Guide 115/116
Radiological DOT Guide 164
Etiological DOT Guide 158

The yellow pages are a numeric listing (U.N. number of chemicals in numerical order) of specific materials or chemical families, each one followed by a guide number. The ID numbers start from 1001 to 9269 as of the last printing (2020). The guide numbers listed in this section refer to the action guides found on the orange pages. The material names identified in this section may be specific, such as chlorine with the U.N. number of 1017, or the number may identify a general commodity such as 2478 as four types of isocyanates (all use guide 155). PIH (Poison Inhalation Hazard, also as TIH – Toxic Inhalation Hazard) and LSA (Low Specific Activity, a radiation qualifier) are but a few abbreviations in the book. (LSA refers to packages of radioactive material and must be transported under special conditions.) When the product is highlighted, the responder should go to the green pages for immediate action. These are initial evacuation and isolation distances and are only good for the first 30 minutes!

The blue pages list chemicals alphabetically by their name. (Chemicals have synonyms. If not found, research for the synonym must be done.) Each is followed by a guide number and the U.N. identification number. After each material name is an action guide number, and again, this refers to the action guides found in the orange section. The corresponding ID number is listed in the last column. The highlighted areas indicate additional information located in the green section for immediate evacuation and isolation.

The orange section is the action guides that are referred to by the yellow and blue sections of the *Emergency Response Guide*. Each guide number is found at the top of the page. The numbers run from 111 to 174, and they list potential hazards as well as emergency response and public safety. These are general procedures that can and more than likely will be expanded upon. They are a starting point for the first responder. The potential hazards, such as fire probability and explosion possibilities, as well as the health hazards, are enumerated. Potential hazards identify health, fire, or explosion hazards with a public safety section identifying general safety procedures, protective clothing, and evacuation. Emergency response procedures identify fire control procedures, the size of the spill or leak, along the first aid to be taken. Decontamination is not always mentioned but is always considered. The green section gives the isolation and protective action distances, including a table of initial isolation and protection distances for a large and small spill. The chemicals identified in this section are those that were highlighted in the yellow and blue sections. The primary goal of this section is to identify materials that, because of their vapor production or the possibility of vapor production, can, under certain conditions, produce either explosive or poisonous effects. This table is only useful within the first 30 minutes of the incident. (Remember, there could have been an EMS notification delay!) The materials are listed numerically by their identification number and the name as it appears in the yellow and blue sections. Both identification numbers correspond. The very end of this section has a listing of dangerous water-reactive materials with a list of toxic vapors produced. Once the vapor is identified, then the new chemical can be researched in the yellow and corresponding orange sections. One primary rule about this section is if the substance is on fire or you are past the 30-minute time frame, this table should not be used. It is only good for vapors and vapor potential. This is not a medically oriented guidebook but rather a reference for the responder who arrives on the scene first. It provides information concerning what actions need to be taken to mitigate the incident.

NIOSH Pocket Guide

Produced by the National Institute for Occupational Safety and Health, this guide is more technical than the DOT guidebook. It contains more specific chemical information than does the DOT ERG but is quick and easy to use. The authoring services are the U.S. Department of Health and Human Services, the Public Health Service, and the Centers for Disease Control. Figure 4.9 demonstrates the use of the NIOSH Pocket Guide.

Figure 4.9   The NIOSH Pocket Guide gives more detailed information than the Emergency Response Guide.

The introduction gives information about the service and how the data was gathered. The reader must become familiar with the variety of abbreviations used within each chemical section. Each chemical entry is very specific (unlike a few entries in DOT ERG). The book is limited to information on commonly found chemicals in the workplace and does not contain many of the unusual compounds you may come in contact with.

The front matter describes the use of the pocket guide with an explanation of each chemical category. Each section in the body of the book is explained in this section. The listing of chemical data begins with Acetaldehyde and ends with Zirconium compounds. Each chemical listing is read from the left-hand page through the right-hand page with a variety of information on each chemical.

A DOT number, CAS (Chemical Abstract Service) number, chemical formula, RTECS (Registry of Toxic Effects of Chemical Substances) number, U.N. number, and guide number are found in the first column.

Headings of information include Synonyms, Exposure Limits, Physical Description, Chemical and Physical Properties, Incompatibilities, Personal Protection, and Health Hazards.

The limitations to the toxicological information are that it is very general, not specific toward any level of treatment. General signs and symptoms are given, but no true medical direction.

CAMEO

Computer-Aided Management of Emergency Operations (CAMEO) is a computer software application that provides information for hazardous materials incidents. Developed and maintained by EPA and the National Oceanic and Atmospheric Administration, it provides the first responder with tools to research and evaluate a chemical incident. This suite of applications comes with four distinct modules: (i) CAMEO Data Manager, (ii) CAMEO Chemicals, (iii) MARPLOT, and (iv) ALOHA.

The CAMEO Data Manager allows the operator to manage large amounts of information related to the Tier II reports stated earlier in this chapter. As stated, the EPCRA act states that chemical information must be given to local authorities, and this database is a convenient and useful way for emergency services within a local to use this information. CAMEO Data Manager stores the Tier II data, which can be used within MARPLOT.

CAMEO Chemicals is a searchable chemical database that has over 5000 and also contains the 1000 plus United States Coast Guards data sheets called CHRIS (Chemical Hazards Response Information System) codes, the DOT ERG, and links to the online NIOSH Pocket Guide and International Chemical Safety Cards. Each datasheet has common identifiers such as Hazard Labels, NFPA diamond (also referred to as 704, see above), along with a general description of the chemical. Hazards, response recommendations, physical properties, and regulatory information, along with synonyms, can be found.

Here you will find toxicological data in the rawest form with application toward community planning such as acute exposure guideline levels (AEGL's), emergency response planning guidelines (ERPGS), and Protective Action Criteria (PACS). PACS are based on AEGLs, ERPGs, and temporary emergency exposure limits (TEELs) (see Chapter 5). Additionally, chemical reactivity can be suggested when two or more chemicals are mixed together. These reactions are based upon hazard categories but can give clues toward potential issues with mixtures such as compatibility.

MARPLOT

Mapping Application for Response, Planning, and Local Operational Tasks (MARPLOT) is the mapping application of the suite. Maps with street and satellite views are available. Layers can be imported from common GIS (Graphic Information Systems) to give you locations of critical facilities such as schools, universities, industrial facilities. These can be linked to the CAMEO Data Manager, giving the ability to have contact numbers and site pre-hazard plans all loaded within the system. Maps can be generated to evaluate potential area impact to a certain location or critical infrastructure. It is here which detailed toxicological data can be imported with medical treatments and or assessments.

ALOHA

Areal Locations of Hazardous Atmospheres (ALOHA) estimates threat zones associated with hazardous chemical releases, including toxic gas clouds, fires, and explosions. A threat zone is an area where a hazard (such as toxicity) has exceeded a user-specified Level of Concern (LOC).

These threat zones can be imported into MARPLOT to gain geospatial information with reasonable assessment capabilities.

As a whole, this suite of applications can give the emergency manager and responder detailed information. However, the user must import the details. For example, regional plans and SOPs can be loaded in where appropriate but must be reviewed on a regular basis. Linked reports to facilities along with treatment protocols must be individually placed within the program.

WISER

Wireless Information System for Emergency Responders (WISER), maintained by the United States National Library of Medicine is a database designed to assists first responders at the scene of a hazardous materials incident. It gives quick and concise information. However, it is limited to a shortlist of commonly found chemicals. Under the chemical file, you will see a medical folder that will give you a list of choices. Here in this list, you have a treatment overview, health effects, toxicological values, along with a summary. The medical treatment area can give you a detailed outline of medical procedures, differential diagnosis, along toxicokinetics, to name a few.

At the bottom of the screen, you will see a link to PubChem (see next discussion area) and PubMed, two additional databases for your reference. These are robust databases, and you can get lost in the amount of information, so pre-incident planning and practice using these is a must.

WISER is very user-friendly with a toolbox of referencing; searching a known and unknown material is also available. It is an application that you can download to your smartphone and use on the fly. The unknown chemical search does have a sign and symptom picklist that will start to give you direction toward a chemical. However, sometimes it will mislead you as your interpretation of a sign and symptom may be slightly different than what the database has, overall a very powerful component of this application.

ToxNet and the Hazardous Substance Database

These programs merged and have been placed within PubChem as an online database. It is a very detailed database of chemical information maintained by the National Library of

Medicine. With 98 million chemicals within the database, you should be able to find most chemicals of interest.

Searching can be a little cumbersome; however, when using the PubChem website, enter the chemicals name, several choices may come up. Here a small degree of chemical nomenclature knowledge may be necessary to navigate toward the chemical that is being researched. IUPAC names are used as a descriptor; however, synonyms are also presented. Once you have the entry you are looking for on the right-hand side, you will see a contents table to jump to that section of the document.

Here you will see Toxicity; once selected, the panel will open with several more selections "antidote for emergency treatment" can be found. This detailed list of content is not laid out in a step-by-step process but rather science papers entries, several pages in some cases. This will take some time to read through and gain the information that is required. However, a robust database for detailed information can be a little frustrating and laborious to read through under stressful conditions.

If under toxicity, you cannot find what you need in terms of medical data at the bottom of the contents, you will see information sources this gives additional information from a variety of sources. At the top of the page, the menu down arrow will give you a list of sources, and you can select HSDB (Hazardous Substance Database), and a new screen will appear. This information is more succinct toward the medical application.

Resources

In any community, there will be a variety of resources that a hazmat team, especially the medical component can draw from. As an example, the local emergency department should be an integral part of the medical hazmat system. Poison control centers are also a great resource. These are all part of a national Poison Control Network throughout the United States. Every state has at least one poison control center and many states have two or three.

The poison control center will assist on-scene medical responder with researched information and forwarding that information to the hospital receiving the patient to ensure a continuum of care. Additionally, the involvement of these centers within training sessions helps in building that relationship between the field and the medical staff.

Veterinary clinics that work with local law enforcement K9's also can be an informational resource. Although these individuals do not practice medicine on humans, they do have contact with other toxicologists and other veterinarians.

CHEMTREC, a 24-hour service offered by the Chemical Manufacturers Association, can give general recommendations and contacts. The National Response Center (NRC) is manned by the United States Coast Guard and can provide you with a variety of informational contacts. Each of these emergency sources gives general information and contacts, but as a rule, they do not tell you what to do.

Summary

Any time you are dealing with the research of chemicals, and especially when you have an unidentified compound(s), several research modes and sources should be employed. The simple fact is that databases, books, and research articles are written by someone.

During the process of publication, mistakes and mistranscription can and does occur. Sometimes the literature is not precise in its description of a sign or symptom. Many times, these issues are just copied from one author to the next without deep-dive research. With that in mind, in order to have true documentation, the information on the chemical must come from a minimum of three sources. This applies to initial identification and referencing of the material's chemical and physical properties along with the toxicology of the material(s).

Clues and cues are sometimes hard to connect. This is the reason that the confidence cycle was described at the beginning of the book. To have confidence that the offending chemical is identified and to determine the level of exposure, three areas of the response must be identified. This includes Patient presentation (Exposure and toxidromes), Event conditions (Scene evaluation), and Scene assessment (Hazard identification). If these are methodically followed, most incidents will be handled safely and efficiently, and any person exposed and injured by the chemical will receive the appropriate and efficient treatment.

5

Hazard Identification

Introduction

The emergency agencies, which include fire, emergency medical services, law enforcement, emergency managers, and hospitals, have become the responsible entities when a disastrous events occurs.

These are the organizations that develop teams, plan, and train for a variety of emergencies yet, very little effort is given to chemically exposed patients. Because of this gap, this text presents a systematic approach toward chemically injured patient management. Part of managing a chemically injured patient is having an understanding of how the toxic properties of the chemical affect the victim. Therefore, it is the hazmat medical technicians and the hospital's emergency department staff's responsibility to understand the basics of chemistry and toxicology in order to treat these complex patients safely and efficiently.

Many times, within a framework of an unknown chemical, the emergency services must mitigate these incidents that involve uncontrolled situations. In most areas of the country, chemical emergency response is handled by the local or regional hazardous materials teams. Often these teams fall under the fire department or law enforcement agencies and are supported on the fringe by private or public emergency medical services. These emergency professionals must understand command and control, incident management, detailed mitigation efforts, technical, and at times, high-level didactic information.

Incidents involving small amounts of chemicals make up the majority of the situations requiring emergency intervention. Because these minor incidents often lead to complacency, safe practices must always be stressed to avoid injury to the responder.

For example, emergency units are dispatched to a patient having difficulty breathing. The dispatcher fails to acquire the pertinent information surrounding the incident and therefore is unable to inform the emergency responders that the patient's difficulty was caused by a mixture of chemicals that led to a respiratory injury. As a result, the incident is dispatched as a medical emergency rather than a suspicious chemical emergency and without the support of a hazardous materials team. It is for this reason that all emergency responders must have an understanding of basic chemical and physical properties and the application of these properties to patient exposure and the resulting injury.

In Chapter 1, the Situational Assessment Continuum (SAC) was presented. That principle included the concept that to evaluate any scene/patient; three areas should be considered. Each of these areas will provide a clue during the patient/scene assessment process.

Hazardous Materials Medicine: Treating the Chemically Injured Patient, First Edition.
Richard Stilp and Armando Bevelacqua.
© 2023 John Wiley & Sons, Inc. Published 2023 by John Wiley & Sons, Inc.

The ring of the SAC that will be emphasized in this section is the Scene Assessment (Hazard Identification) ring. In this part of the continuum, the chemical and physical properties are evaluated concerning how they may affect the patient and potentially the emergency responders. Additionally, how evaluation instruments are used to identify these potentially hazardous threats will be discussed.

There are a lot of chemistry and hazardous materials response books available that discuss the chemical and physical properties of dangerous substances. Although these properties must be defined and each component evaluated, this chapter will focus on the level of concern that is expected when an exposure has occurred. In doing so, first, the properties will be defined in detail, then identify how these properties affect the health and well-being of the exposed person.

Case Study – Phosgene Exposure

January 23, 2010: Belle, West Virginia
   DuPont De Nemours & Co., Inc
   While checking the weight of a 1-ton phosgene cylinder, the transfer hose failed and spraying a worker in the face. The worker was checking the weight of the cylinder. The employee, who was working alone, immediately went to the emergency rinse area located on the far wall of the transfer facility and called for help using the in-house emergency phone. Coworkers assisted him to the plant's occupational health clinic, where the patient showed no symptoms and was transferred to the local hospital for observation and treatment. He deteriorated over the next 24 hours and died the following day.

The critical mode of exposure for phosgene gas is through the respiratory system. Directly after exposure, phosgene causes a mild dry cough, skin and mucous membrane irritation. Within a few hours after inhalation the respiratory tract irritation becomes worse with coughing, chest tightness, and wheezing. If the inhalation injury was significant the irritation advances to pulmonary edema, difficulty breathing, and hypoxia followed by adult respiratory distress syndrome. There is no definitive treatment for phosgene and no antidote. Medical care is supportive in nature and may include mechanical ventilation with positive end expiratory pressure (PEEP). The use of extracorporeal membrane oxygenation (ECMO) has shown some positive results and allow the lungs time to heal but do not completely cure the damage created by the phosgene. This is the reason that the mortality rate of phosgene poisoning remains very high. See Figure 5.1 for a quick reference to the properties of Phosgene.

States of Matter

All elements can be found in one of three states: solid, liquid, or gas. Within these states of matter exist tiny discrete particles called atoms. All atoms are made up of neutrons, protons, and electrons, which are arranged in a particular configuration. The atom is the smallest unit of an element, and elements comprise all matter at the atomic level and above. Each element can combine in a variety of configurations leading to a diversity of compounds and states of matter.

Each state is not less harmful than the other but must be evaluated under different considerations and analyzed under the situation in the way they are presented on the scene. As the states of matter are presented, exposures related to these states will be reviewed.

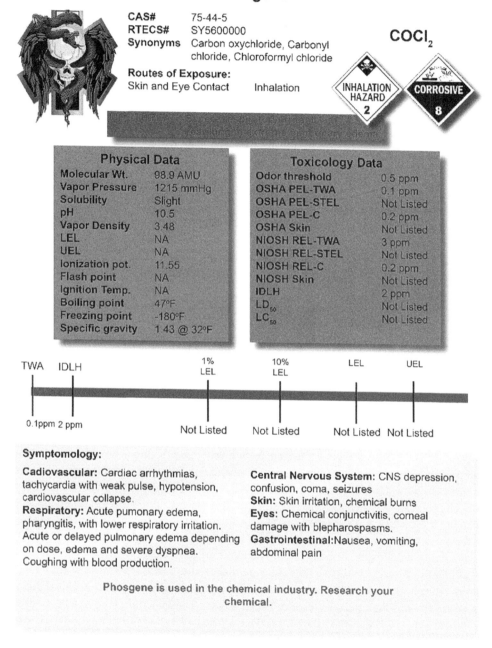

Corrosives & Irritants **Phosgene**

CAS#	75-44-5
RTECS#	SY5600000
Synonyms	Carbon oxychloride, Carbonyl chloride, Chloroformyl chloride

Routes of Exposure:
Skin and Eye Contact   Inhalation

$COCl_2$

INHALATION HAZARD 2

CORROSIVE 8

Physical Data

Molecular Wt.	98.9 AMU
Vapor Pressure	1215 mmHg
Solubility	Slight
pH	10.5
Vapor Density	3.48
LEL	NA
UEL	NA
Ionization pot.	11.55
Flash point	NA
Ignition Temp.	NA
Boiling point	47°F
Freezing point	-180°F
Specific gravity	1.43 @ 32°F

Toxicology Data

Odor threshold	0.5 ppm
OSHA PEL-TWA	0.1 ppm
OSHA PEL-STEL	Not Listed
OSHA PEL-C	0.2 ppm
OSHA Skin	Not Listed
NIOSH REL-TWA	3 ppm
NIOSH REL-STEL	Not Listed
NIOSH REL-C	0.2 ppm
NIOSH Skin	Not Listed
IDLH	2 ppm
LD_{50}	Not Listed
LC_{50}	Not Listed

TWA	IDLH		1% LEL	10% LEL	LEL	UEL
0.1ppm	2 ppm		Not Listed	Not Listed	Not Listed	Not Listed

Symptomology:

Cadiovascular: Cardiac arrhythmias, tachycardia with weak pulse, hypotension, cardiovascular collapse.
Respiratory: Acute pumonary edema, pharyngitis, with lower respiratory irritation. Acute or delayed pulmonary edema depending on dose, edema and severe dyspnea. Coughing with blood production.

Central Nervous System: CNS depression, confusion, coma, seizures
Skin: Skin irritation, chemical burns
Eyes: Chemical conjunctivitis, corneal damage with blepharospasms.
Gastrointestinal: Nausea, vomiting, abdominal pain

Phosgene is used in the chemical industry. Research your chemical.

Figure 5.1 Phosgene quick reference card.

Note: In a solid, the molecules are in proximity to each other. Although the atoms are vibrating ever so slightly, the movement cannot be seen. In the liquid, the molecules are a bit farther apart, slipping by one another. This can be observed when a liquid flow out of its container. Breaking from the liquid environment to the vapor state is easy with enough molecular

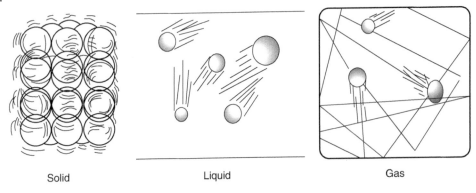

Figure 5.2 Molecule activity in a solid, liquid, and gas.

momentum. In the gaseous state, the molecules are the farthest apart, colliding with other molecules and the container itself. In each state of matter, an increase in temperature increases the molecular movement; likewise, decreasing the temperature slows the molecular movement. See Figure 5.2 for a graphic representation of these properties.

Solid

A solid, which is thought to be on the lower end of the hazard spectrum, can present with concentrations that have the potential of severe injury. First, the solid is that state of matter that all the molecules are packed together so tightly that each constituent of the material cannot move. For most solids, when the temperature reaches a certain point, the substance will start to melt and becomes a liquid. As a liquid, and as the temperature continues to increase, vapors are released that can cause injury.

From an exposure perspective, the real danger of a solid is when the physical size is transformed into dust or fine particulate (smoke) and inhaled. In addition, a solid, exposed to temperature changes, can melt into a liquid, then vaporize or can go through a process of sublimation. Sublimation is a solid's ability to change directly into a vapor without going through the liquid state. Only certain chemicals have the ability to go through sublimation. Paradichlorobenzene is a fairly common fumigant for insects (mothballs). This chemical has the ability to change directly from its solid form into the vapor form at normal temperatures. See the quick reference card in Figure 5.3.

Some forms of solids can become airborne and can be influenced by temperature and airflow, similar to gases and vapors. These include fibers, particulates, and dust, which do not react to typical monitoring efforts performed for vapors and gases. Fibers are those solids produced from raw materials during a chemical process or production. They can be natural or synthetic and are longer than they are wide, and can be less than 1 micron in size. These fibers can make their way down deep into the lungs resting in the bronchioles or alveoli. Asbestos fibers are excellent examples of these fine fibers that gain access into the lungs.

Particulates can be suspended in air and include dust, smoke, soot, and liquid droplets. This fine particulate matter is generally the byproduct of manufacturing and is found in all kinds of industries, from pharmaceuticals to furniture. They are also produced in mining and construction. Generally, in industries where fine particulate is produced, engineering measures are used to reduce the exposure. This may include the use of particulate masks and sophisticated ventilation systems.

Hydrocarbons

Parachlorobenzene

CAS# 108-90-7
RTECS# Not Listed
Synonyms Benzene chloride, Chlorbenzol, CP-27, Monochlorbenzene, MCB

C_6H_5Cl

Routes of Exposure:
Skin and Eye Contact   Inhalation
Skin absorption   Ingestion (solution)

FLAMMABLE
3

THREAT: Cardia Arrythmias, Respiratory failure, Pulmonary edema, CNS depression

Physical Data		Toxicology Data	
Molecular Wt.	112.56 AMU	Odor threshold	0.21 ppm
Vapor Pressure	8.8 mm Hg	OSHA PEL-TWA	75 ppm
Solubility	< 1 mg/ml	OSHA PEL-STEL	Not Listed
pH	Not Listed	OSHA PEL-C	Not Listed
Vapor Density	3.88	OSHA Skin	Not Listed
LEL	1.3%	NIOSH REL-TWA	10 ppm
UEL	7.1%	NIOSH REL-STEL	Not Listed
Ionization pot.	9.07	NIOSH REL-C	Not Listed
Flash point	74°F	NIOSH Skin	Not Listed
Ignition Temp.	Not Listed	IDLH	1000 ppm
Boiling point	270°F	LD_{50}	1.66 g/kg
Freezing point	-50°F	LC_{50}	Not Listed
Specific gravity	1.11		

TWA	1% LEL	IDLH	0.13% 10%LEL	1.3% LEL	7.1% UEL
10 ppm	130 ppm	1000 ppm	1,300 ppm	13,000 ppm	71,0000 ppm

Symptomology:

Cadiovascular: Cardiovascular collapse, Tachycardia, R on T leading to venctricular fibrillation.

Respiratory: Upper respiratory irritation, cough. Bronchospasm, dyspnea, respiratory failure.

Central Nervous System: Headache, seizures, CNS depression, Disturbances in hearing with tinnitus.

Skin: Dry and cracking of the skin.

Eyes: Chemical conjunctivitis, corneal burns.

Gastrointestinal: Nausea and vomiting, excessive salivation.

Research your chemical.

Figure 5.3 Quick reference card for paradichlorobenzene.

Coal miners have long been victims of working in an environment filled with ultra-small particles. The small particles of coal dust are less than 1 micron in size and gain access to the fine bronchioles and alveoli, where disease forms years later. Coal workers' pneumoconiosis (CWP) is commonly called black lung. But coal mine dust also causes other lung diseases.

This spectrum of diseases is called "coal mine dust lung disease" (CMDLD). This spectrum of diseases includes CWP, silicosis, mixed dust pneumoconiosis, dust-related diffuse fibrosis, and chronic obstructive pulmonary disease (COPD).

Fuming is technically a condensation of solid particles that are produced after heating/melting a solid. However, it is also used to describe the volatilization of liquids when acids are present. Fume is commonly used to describe the vapors that come off of concentrated acids.

Liquids

In liquids, the molecules are less concentrated than the solid but more concentrated than the gas form. They are free-flowing and have a defined volume but not a defined shape, and must be contained. Although different liquids may have the same volume, at the same time, they have different weights. Liquids, when subjected to heat, will start to vaporize; under pressure, they mist and aerosolize; and when the liquid is an acid, they fume.

Vapors and Gases
Vapors should not be confused with gases. Gas is a natural state of matter under normal temperatures and pressures. Vapors are produced when a liquid is evaporating. As an example, diethyl ether is defined as a liquid. When diethyl ether is at a temperature of 94 °F or higher, it boils and rapidly becomes a vapor.

Mists and Aerosols
A mist or aerosol is when a liquid is condensed, or droplets are suspended in the air. Either are classified as the suspension of liquids. These can be caused by the movement of a liquid through a nozzle or a pinhole leak in a liquid line, or in some cases, evaporation within a closed space and mist condensation occurs.

Gases

Gas is a state of matter consisting of particles that have neither a defined volume nor a defined shape. In order to store and transport gas as a commodity, it must be contained in a sealed vessel. It may be contained as a pressurized gas, a liquified gas under pressure, or a liquified gas under a very low temperature (cryogenic). In all of these conditions, when the gas is released, it will go back to its natural state. All gases will expand at different ratios (compressed volume to uncompressed volume) called "expansion ratios."

Compressed Gas
A pressurized or compressed gas is gases that are simply put under pressure within a cylinder to increase the usable volume. The oxygen tank that is used for healthcare or the SCBA air cylinders on the back of a firefighter are examples of compressed pressurized gases.

Liquified Gas
Liquified gases are gas that has been placed in a liquid form and maintained in a compressed gas cylinder. These liquified gases are much more concentrated than a pressurized compressed gas and have much more volume when expanded than the compressed gas cylinder. This is an efficient way to transport and store gases, as large quantities can be stored in

smaller cylinders as compared to just compressing a gas. The propane cylinder used at home gas grills is a great example of stored liquified gas. Other gases that are currently stored in a liquified form include anhydrous ammonia, chlorine, natural gas, and carbon dioxide.

Cryogenic Gas

A cryogenic gas is a gas that has had pressure and temperature reduction applied to liquefy the gas. Then, the temperature and pressure are maintained to keep the gas in a liquified form. They exist at temperatures at or below −243 °F (−153 °C). Typically, these are gases that are not easily managed the same way as compressed or liquified gases. Or they are placed in a cryogenic state because larger quantities can be placed in even smaller cylinders (larger expansion ratios). Some examples are liquified natural gas, nitrogen, hydrogen, and oxygen, to name a few.

Chemical and Physical Properties

Every substance is made up of atoms, but it is how they are organized that is important. The atomic building blocks properties are due to the type of atoms and how these atoms are arranged within space. Each element and molecule also have its own characteristic chemical and physical properties. A molecule is a substance that is prepared from two or more atoms of an element. The properties of these combined elements are different from the properties of the elements separately, thus behaving differently during an exposure (see Figure 5.4). Compounds are a type of molecule in which the atoms (elements) are different within the structure. And again, the properties can behave differently during exposure. Table salt is a compound; drinking alcohol is a compound within a mixture; oxygen, nitrogen, and ammonia are molecules.

These substances have their own set of chemical and physical properties, which are useful when describing that chemical and ultimately the detection of it. Every substance has different qualitative numbers (like fingerprints) that are peculiar to that chemical. For example, a sulfur dioxide molecule, in any state, always possesses an individual set of chemical and physical properties. So, if a sulfur dioxide molecule melts, boils, looks, and freezes like a sulfur dioxide molecule, it must be sulfur dioxide.

When testing different chemicals and determining the chemical and physical properties, the testing is done under "standard temperature and pressure (STP)." These are a standard set of conditions for experimental measurement to keep laboratory testing consistent. STP involves a standard temperature of 32 °F (0 °C or 273 K) and a standard pressure of 1 atm (760 torr Hg or 14.7 psi) that the reaction during a laboratory has taken place in. If the testing varies from the established STP, it will be described in the data and made clear in the publication of findings.

When two or more elements combine to make a compound, the individual elements and the resultant compound have different properties.

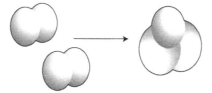

Figure 5.4   The elements combining to make a compound with different chemical/physical properties.

Chemical Properties

All chemicals try to reach equilibrium, and in their quest for equilibrium, chemicals undergo change. These changes give chemicals their properties that include the pH, flammable range, ignition temperature, flash point, and transfer of heat.

pH (Corrosivity)

pH is a measurement of how acidic or how alkaline a substance is. This measurement is reflected in qualitative numbers and ranges from 1 to 14, with 7 representing neutral. Acids are measured with any number less than 7, and alkaline is measured with any number greater than 7. The numbers are calculated from the chemical's ability to produce the hydronium ion, H_3O^+, or the production of a hydroxide ion, OH^-.

The acid is a compound that is capable of transferring a hydrogen ion in a solution, or it can donate a proton. A base is a group of atoms that contain one or more hydroxyl groups. The base is referred to as the proton acceptor. This acceptance and donation produce the acid–base reaction. Acids and bases exist on their ability to dissociate. An equilibrium between the hydronium ion and the hydroxide ion creates a neutral solution.

Ignition Temperature (IT)

Ignition temperature is the minimum temperature to which material must be heated to be ignited (by an outside source) and sustain combustion. Materials that have a low ignition temperature will have a relatively high flash point. This concept is most important when dealing with volatile organic compounds. Heated to the point of ignition without an outside source is the auto-ignition temperature.

Flashpoint (FP)

Flashpoint is a component of vapor pressure and is the minimum temperature under which a liquid will give off vapors to form an ignitable mixture in air, but not enough vapors to sustain combustion.

Heat Transfer

Heat transfer is the process of thermal energy that can cross a well-defined boundary and is further defined as conductive heat, convected heat, or radiant heat. With conductive heat or *conduction* (sometimes called diffusion), heat flow from an area of high temperature to the area of lower temperature through a solid object (second law of thermodynamics). Convected heat or *convection* occurs with a movement of liquids, or gases, or a flow. This flow can have external effects such as wind (airflow) or continued heat production, which increases the buoyancy of the gas. Radiant heat transfer *radiation* occurs as electromagnetic waves.

All chemicals are subject to change, each exhibiting its own chemical and physical properties. They interact to produce a chemical conversion. The transformation into a stable compound is the result of a chemical reaction. These reactions take place because of the specific architecture of the elements or compounds within the surrounding environment.

Physical Properties

Appearance, viscosity, melting point, freezing point, boiling point, lower/upper flammable range, density, specific gravity, vapor density, solubility, vapor pressure, and sublimation are all examples of physical properties, which are characteristic of a particular element or molecule. They are qualities measured without changing the chemical makeup of the compound or molecule.

Appearance

Appearance is the form of the chemical. The form may be reported as solid, liquid, or gas. Appearance may also be the size of a particle such as a powder, dust, or fume and can even be the color. The form of the hazardous material dictates the management strategies toward incident stabilization. For example, if the material is a liquid, then leak and spill control may be the tactics of choice. When a gas is involved, limiting the release or changing the physical form may be the tactical procedure. If the material is in solid form, confinement of the material may be the action chosen to minimize exposures.

Viscosity

Viscosity is a measure of flow. It is a determination of the thickness of a liquid. A low viscous liquid will flow like water. The lower the viscosity, the higher the tendency for the liquid to spread (and emit a vapor). On the other hand, high viscosity liquids flow more slowly (like molasses).

When dealing with combustible and flammable hydrocarbons, viscosity can be related to the production of static electricity. Moving hydrocarbons within piping such as PVC can create static electricity. These molecules move over one another, and against the sides of the PVC, a static charge can build. Once this electricity finds a ground, a spark can be created, and if the conditions are right, vapors ignite.

Melting Point (MP)

Melting point is the temperature at which a material changes from a solid to a liquid. From an emergency response point of view, generally speaking, a solid is easier to manage than a liquid. If a material has a low melting point, it may be expected to become a liquid in an emergency. Liquid materials maintain the shape of their container but have no form of their own. When they are not in a container, liquids present responders with challenges such as containment or confinement and contamination due to the state the chemical is in. Check the melting point of materials to see when they change states.

Freezing Point (FrPt)

Freezing point and melting point can be thought of as synonyms, depending on the context of the chemical reaction. If, for example, the chemical is moving from a solid to a liquid state, then the term *melting point* is used. If the product is going from a liquid to a solid state, then the term *freezing point* is used. The amount of heat that is required to move the chemical from a solid to a liquid or the amount of heat that must be removed to move the liquid to a solid is dependent on the chemical itself. Each chemical and the state of matter that it is being contained in will have an impact on these two properties.

Boiling Point (BP)

Boiling point is the temperature at which a liquid's vapor pressure equals atmospheric pressure. At this point, the material changes from the liquid into the gaseous state. For example, a pot of water placed on a hot stove to boil in Florida (sea level) will boil more slowly than the same pot of water placed on a hot stove in Denver (mile-high elevation) because the pressure being applied to the surface of the water in Denver is much less than the pressure in

Florida. Therefore, it takes less time for the vapor pressure to equalize with the atmospheric pressure, so the water boils faster. This same principle that applies in the kitchen also applies to hazardous materials incidents. Hazardous substances that are in a liquid state and possessing very low boiling points must be kept under pressure, or they will boil and change form. Most of these chemicals present potential fire, reactivity, or health hazards. Conversely, the high-boiling-point liquids have relatively low vapor pressures. These liquids need an active energy source (fire) in order to convert from the liquid state to the vapor state. Materials having high boiling points are usually much safer than those with low boiling points.

Flammable Range

Flammable range describes the minimum and maximum concentration of a hydrocarbon and air at which the mixture will ignite or combust. The minimum concentration that will burn is referred to as the lower explosive limit or lower flammable limit (LEL or LFL). The maximum concentration is referred to as the upper explosive limit or the upper flammable limit (UEL or UFL).

LEL/UEL is sometimes called an explosive range (i.e. LEL and UEL) or called the flammable range (LFL and UFL), which is defined as the percent of vapor that has the proper stoichiometric relationship of oxygen to fuel. We use a 0–100% measurement scale to identify this area of potential flammability.

Two additional conditions are also present in that the liquid vapors are above the flashpoint and that the temperature of the ignition source is above the ignition temperature of the substance. This relationship is used to perform a hazard risk assessment. The area between these two points represents a flammable and potentially explosive atmosphere. With the area above the UEL/UFL, the fuel to air ratio is too rich (too much fuel in the stoichiometric relationship) and below the LEL/UEL too lean (not enough fuel in the stoichiometric relationship). However, below the LEL toxic values start, and by the time you have an LEL and beyond, the level of toxicity is high enough to ensure an exposure.

Density

Density is thought of as how heavy a substance is. Although not a bad way of thinking about density, it is actually a relationship between weight and quantity or volume of a material and can be applied to all three states of matter. Chemically speaking, density is a ratio between mass and volume. For solids and liquids, this ratio is expressed as grams per centimeter cubed, and for gases, grams per liter.

Density = Mass / Volume

Both specific gravity and vapor density are applications of this principle for liquids and vapors, respectively. They both use the mass per unit volume.

Specific Gravity (SG)

Specific gravity is the weight of a liquid or solid compared to an equal volume of water. Water has a value of 1.0 in relation to the compared material. If the tested material has a specific gravity of less than one, the material will float. If it is greater than one, the material will sink. At hazardous materials emergencies, it is often imperative to know if the liquid material will float or sink in order to decide the type of hazard control techniques to be used. If the material has a specific gravity of greater than one, then a viable tactic may be to contain vapors by floating water over the top of the material. If the value is less than one and water is used, the material will flow over a larger area carried on the surface of the water. Given the specific gravity of a liquid, the weight per gallon of the substance can be calculated. For example, the specific gravity of sulfuric: acid is 1.84. This number is based on the weight of a chemical as compared to the weight of water. Water weighs 8.35-lb/gal, so to calculate the weight per gallon of sulfuric acid, multiply the weight of water per gallon times 1.84 specific gravity. The calculated weight of sulfuric acid is 15.36 lb/gal.

Vapor Density (VD)

Vapor density is the weight of a vapor or gas as compared to an equal volume of air. Air is assigned the density of one. When a vapor or gas has a vapor density value of greater than one, the vapor or gas will settle below the air. If the vapor density of the gas or vapor is less than one, it will generally rise above the air. There are factors that may keep those gases and vapors closer to the ground. These include humidity and wind.

Under certain conditions, especially when dealing with the lighter-than-air gases in high humidity environments, the water vapor in the air must also be considered. For example, methane, which is lighter than air, has shown qualities very much like those of liquid petroleum gas (LPG), which is heavier than air. This can be explained if we look at the vapor density of methane and water vapor as a combined effect. Vapor density is equal to the molecular weight of gas over the molecular weight of air. See Figure 5.5 for a description of this principle.

In North Carolina a natural gas line was fractured during road construction, causing a serious leak. The repair crew attempted to place a jacket over the broken line that was under about 3 ft of water. The gas bubbled up through the water, becoming humidified. When the gas escaped the hole, instead of rising and dissipating, it traveled the ground and found an ignition source a distance from the leak, and exploded.

NIOSH uses rVD (relative vapor density) or relative gas density, which is defined as the mass of gas relative to the air, the same as our definition above.

Solubility (SOL) is the ability of a material to blend uniformly within another material. The material that is in the greatest amount is called the *solvent,* whereas the material that is in the lesser amount (usually the additive) is called the *solute.* The blend is called a *solution.* Certain materials are soluble in any proportion, while others are not. Some of the dependent factors in solubility are the polarity, concentration, and temperature of the solvent of the materials involved. Whether the solute (the stuff you are placing into the solution) is a liquid or a solid also has a bearing on the ability to form a mixture. *Miscibility* is often used synonymously with solubility. When a compound is said to be miscible in water, it means that the substance is infinitely dissolvable in water.

Polarity has a lot to do with solubility. Polar substances have a positive end and a negative end, and non-polar substances do not. Generally speaking, like will dissolve in like, polar with polar and nonpolar within nonpolar.

$$\text{Vapor Density} = \frac{\text{Molecular weight of chemical}}{\text{Molecular weight of air}} = \frac{\text{Methane (16) + Water vapor (18)}}{\text{Air (29)}}$$

$$= 1.17$$

$$\text{Molecular without water vapor VD} = 0.55$$

Figure 5.5 Calculating the vapor density of a gas when exposed to high humidity or water vapor.

If the solute is solid, the polar/nonpolar dissolvability still holds true; however, there is a limited capacity of the solid to dissolve. Each solvent has a saturation point. Above this point, the added solute will not dissolve.

Solutions represent a homogeneous mixture where all the parts of the end mixture are composed of the same material. For example, if water is used to dilute a hydrochloric acid spill, then the solution running off the road is a less concentrated mixture of hydrochloric acid and water. Although the solution is less hazardous in terms of concentration, the molecular structure has not changed; thus, the strength has not changed.

On the other hand, if a spill of carbon tetrachloride occurred and the intention was to dilute it with water, the outcome would not be the same. Carbon tetrachloride is not soluble in water; therefore, the water would only displace the carbon tetrachloride causing it to move around and contaminate other areas. If a patient is contaminated with carbon tetrachloride and is decontaminated only with water, will the procedure be successful? Probably not!

Reference materials list solubility as a percentage rather than listing chemicals a soluble or non-soluble in water. The higher the percent found in the reference literature, the higher the potential for that chemical to become soluble in water. To the contrary, the lower the number in percent, the less soluble in water.

When looking at reference material, it is assumed that the chemical has a certain degree or percentage of miscibility with water. If the chemical is not soluble in water to any degree, then the subscript of hexane will be noted, meaning that the chemical is only soluble in a hydrocarbon. If the reference material listed a chemical as 0.7% soluble, then the chemical is poorly soluble in water.

To better explain this concept of 0.7% solubility of a chemical convert 0.7% into ppm simply multiply $0.7 \times 10{,}000$ ppm $= 7000$ ppm. So, for every one million parts, 7000 will be soluble in water.

If the vapor pressure of this chemical is low, contact is limited but, if the vapor pressure is high, and in this case it is, the contact hazard becomes extremely high. In this example chlorine gas has a vapor pressure of 6.8 atm (atmospheres) at Standard Temperature and Pressure (STP). One atmosphere is equal to 760 mm/Hg. If 760 mm/Hg is multiplied by 6.8 atmospheres, the vapor pressure can be displayed in the real pressure of 5,168 mm/Hg.

Vapor Pressure (VP)

Vapor pressure is the pressure a material exerts against the sides of an enclosed container as it tries to evaporate or boil. Each atom within the liquid material is bouncing about until it reaches enough velocity to escape the liquid. Once this molecule has escaped the liquid form, it has been changed into a molecule traveling through the air space, giving it a vapor state. This movement of the atom in the gaseous state is measured as vapor pressure. All liquids have vapor pressure. Vapor pressure is measured in millimeters of mercury (mm/Hg).

For example, water has a VP of 21–25 mm/Hg.

Those materials having a higher vapor pressure will maintain a higher pressure within a closed container. The same materials that have high vapor pressure also have low boiling points. These materials have a greater potential for container breach, especially if heated.

Vapor pressure is measured in mm/Hg (millimeters of mercury) but also can be expressed in Kpa (kilopascals where 1 Kpa = 7.5 mm Hg). Torr, millibars, and psi (1 torr = 1 mm/Hg used in medicine, 1 millibar = 0.75 mm/Hg used in weather prediction, 1 psi = 51.71 used in hydraulics) are all used to describe the pressure a vapor exerts. Each is used in a different field of study. In this book, we will use mm/Hg at STP.

Expansion Ratios

All gases want to remain as gas. For transportation and storage efficiency, gases are pressurized, refrigerated, and liquified. In these processes, gas is changed to a liquid state of matter. When these products escape their containment, they revert back to their natural state, which is a gas. When this occurs, the volume of liquid gas or the volume of compressed gas expands back to its natural state. We call this the expansion ratio. The expansion ratio is expressed as one volume of liquid/pressurized content to volume of gas in a natural state as seen in Figure 5.6.

Properties and Their Medical Implications

Each state of matter and physical property has an influence on an injury; some are obvious, like heat transfer, while others are more insidious such as viscosity. Looking at these properties, one must remember that it is the circumstance and the conditions that make one property more important or hazardous than another. It is important to evaluate these properties as they relate to chemical exposure, as each has a different impact on the person exposed.

Solids can produce dust or particulates in the air that can affect the skin, eyes, and respiratory system. The size of the particle will determine how far in the respiratory system it will travel.

When a liquid, vapor, or gas is involved, the depth of travel in the respiratory system is greatly affected by the solubility. Poorly soluble gases and vapors are able to penetrate deep

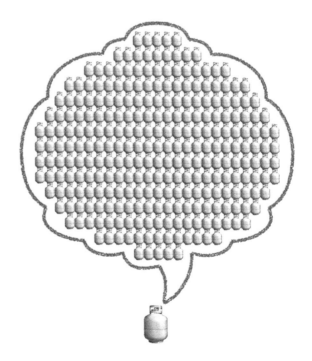

Figure 5.6 When a chemical goes from a compressed (or cryogenic) liquified gas into a gaseous state, it will expand multiple times. For example 1 liquid container of propane is equal to 270 times the size of the container in vapor form.

into the respiratory system and even reach the alveoli. Those chemicals that do reach the alveoli have the ability to create damage there or, through diffusion, enter the bloodstream. Once in the bloodstream, injury can take place far from the respiratory system affecting other organs.

Note: Differences between a gas and a vapor:

The terms gases and vapors are at times used synonymously. However, these two forms of chemicals in the air are different and can produce different effects of toxicity. Both are related to the vapor pressure of a chemical. The gas has, within its natural state, a vapor pressure above 760 mm/Hg where a vapor has a vapor pressure below 760 mm/Hg. The advantage that the gas has is its energy or pressure that can move within a room or environment, where the liquid vapor has less energy and thus has less potential to move throughout a room. For example, a gallon of acetone nail polish solution in a closet vs an average sized living room will develop different concentrations. The concentration is also dependent on the volume of the acetone spill and the temperature with each room. This is in contrast to a BBQ propane cylinder released in either room. Propane being a true gas has the ability to fill the space and displace oxygen in the room.

Two issues arise when a liquid chemical is released, the hazard of the chemical spill and the hazard created from the vapors being produced. Take, for example, an employee at a hardware store who accidentally spills a solvent. There now exists two problems: the chemical contact on the skin and the vapors being produced by the spilled chemical that the employee may also be breathing in.

Vapor Pressure, Medical Implications

The vapor pressures of a chemical can greatly influence the injury. Vapor pressure above 40 mm/Hg is considered to be an immediate inhalation hazard. Vapor pressure is the outward force of a liquid chemical to become a vapor. The higher the vapor pressure, the more vapors are being forced into the air. For comparison, water has a vapor pressure of 21 mm/Hg, considered to be relatively low. Ethanol, on the other hand, has a vapor pressure of 44 mm/Hg. One is below the 40 mm/Hg reference point, while the other is above. The ethanol is vaporizing much faster than water and becoming an immediate respiratory hazard.

To the greater extreme, acetone has a vapor pressure of 180 mm/Hg. When entering a room where acetone is being used to remove nail polish, the immediate odor can be overwhelming because of the concentration of vapors readily entering the air. In fact, just a few minutes in that environment may leave a person light-headed.

Depending on the conditions and circumstances of an incident, both acute and chronic injuries may occur. Chemicals possessing lower vapor pressures, those around 40 mm/Hg, may not cause immediate symptoms, but exposure over time can cause symptoms and injury that appear hours to days later. When the vapor pressure is above 300, the symptoms may be immediately overwhelming. Once the vapor pressure of a chemical is 760 mm/Hg or higher, the chemical is not a liquid but a gas at normal atmospheric pressure. Gases present a different hazard in that they displace oxygen and create an asphyxiating atmosphere. See Figure 5.7.

When researching the chemical and physical properties of a substance involved in an incident, finding the vapor pressure is important to both determine the immediate danger to responders and evaluate the level of injury suffered by the victim.

Vapor pressure is not the only chemical/physical property that must be evaluated when a chemical is released or spilled into an environment. Additionally, the chemical should be evaluated to determine if it is an acid or base, its solubility in water, and the flammable range.

Figure 5.7 Depending on the molecular weight of a chemical, the vapor pressure of the liquid can exhibit as a respiratory hazard.

Vapor Density, Medical Implications

Vapor density defines whether gas or vapor is heavier or lighter than air. As an example of the dangers of a chemical's vapor density, hydrogen sulfide will be evaluated. Hydrogen sulfide is a gas that is heavier than air. It naturally occurs when organic material decays. This includes plant and animal materials; when this occurs in below-grade confined spaces, the gas concentrates in the space and presents a very dangerous environment.

There are numerous reports of workers entering these below-grade areas unaware of the danger. After a short period of time, the worker drops a tool and, when bending down to retrieve the tool, suddenly lose consciousness and, if not immediately rescued, die. There are numerous reports of the initial entry person falling unconscious; then, the co-worker enters the space to perform a rescue only to succumb to gas as well. It is this scenario that has played out time after time that caused the propagation of the OSHA confined space regulation (29 CFR 1910.146).

Specific Gravity, Medical Implications

Specific gravity is similar to vapor density. Vapor density measures the weight of a gas or vapor to the weight of air where air is given the value of 1. Specific gravity compares the weight of a liquid or solid to the weight of water where water is given the value of 1. Any chemical with a specific gravity of less than one will float on water, and any chemical with a specific gravity of greater than one will sink.

This physical characteristic may create a dangerous situation when the specific gravity is less than one. In this case, the chemical will float on top of the water and has the ability to continue to vaporize while moving in any direction that the water moves. In the case that the spilled chemical has a specific gravity of greater than one, the chemical will be found under the water and, as such, unable to vaporize.

Solubility, Medical Implications

From a medical perspective, solubility plays an important role in absorption through the skin and respiratory system. When a substance is determined to be miscible, this means that it is completely soluble in water (also called polar). Non-water-soluble substances are typically nonpolar and do not mix in water. In the body, those chemicals that are non-water-soluble are typically lipid-soluble. This is important for the responder as some level of injury can be predicted if the solubility of the chemical is known.

For example, the inhalation of anhydrous ammonia, which is very water-soluble, will cause its initial injury in the upper respiratory system, where the water-soluble mucous coats the airways. The ammonia will rapidly combine with the water-soluble mucous and cause alkali injuries in the upper respiratory system. Phosgene, on the other hand, is not water-soluble (it is lipid-soluble), so when a respiratory exposure takes place, the phosgene moves past the water-soluble mucous and directly into the alveoli. The chemical in the alveoli responsible for decreasing surface tension and keep the alveoli open is called surfactant. Surfactant is a lipid-soluble chemical and is easily disrupted and destroyed by the non-water-soluble phosgene.

Not only is the respiratory system of concern with solubility so are the skin and eyes. Water-soluble gases and vapors rapidly affect those areas of skin and mucous membranes that stay moist. Anhydrous ammonia, as stated above, quickly affects the groin, under the arms, and the folds under the chin, where sweat keeps the skin moist. Eyes, because of the team film containing both lipids and water, are susceptible to either lipid or water-soluble.

In the liver, water-soluble toxins are quickly detoxified during phase I, and the water-soluble chemicals are returned to be bloodstream for filtering out at the kidneys. Lipid soluble chemicals are sent to phase II of liver detoxification, and either made water-soluble and sent to the bloodstream or remain lipid-soluble and stored in fat tissue.

The solubility of a chemical plays an important role when evaluating the level of injury a patient will have after exposure. Some signs and symptoms can be quickly treated on the scene, while others will be undetectable on the scene but cause chronic and late-onset disease processes.

History of Toxicology

It would be foolish to think that all of the information that falls under the heading of toxicology could be covered in one short chapter. The discipline is vast and quite complex. The true understanding of how chemicals and toxins affect the human body is still quite young, relatively speaking. The effects that occur are primarily the results of cause and effect and that creates challenges in diagnosing the patient. What cause (influence) gave the effect (response); in many cases, assumptions must be made. In other cases, where a lot of case history exists, we can see the correlation of cause and effect.

Therefore, emergency responders have to look at the basic principles of toxicology to understand potential causes and effects. The only way this can be done is to evaluate the chemical and physical properties along with some basic principles of toxicology to build a toxicological picture. Evaluating chemical families and the exposure history may give rise to exposure events. In essence, this becomes a process of enemy identification. It is essentially identifying the chemicals and properties that can harm those exposed to it. By applying some basic principles and a general understanding of properties, the danger presented by the "enemy" can be identified and used to base the decisions of treatment on.

Ancient writings describe "chemical compounds" as being used for medicinal purposes, poisoning animals for their easy capture, and as an assassin's tool. Even the great scholar Socrates fell prey to the ingested poison hemlock (which, incidentally, was the Grecian government's "standard" poison). However, it was not until the sixteenth century that development within the discipline of pharmacology (toxicology) was seen. A Swiss alchemist and physician, Philippus Aureolus Theophrastus Bombastus von Hohenheim, better known as Philippus Paracelsus, was the first to recognize the dose–response relationship that it is the quantity of a poison that is harmful. He stated that "it is the dosage that makes it either a poison or a remedy." Paracelsus realized the difference between the therapeutic levels of a substance and the toxic levels of the same substance is a fine line over time.

For years the subjects of toxicology and pharmacology followed the same curriculum. Both were taught at the same time, with pharmacology receiving the highest concentration of time and educational intellect. Toxicology became the offshoot of pharmacology. Only in 1961, with the formation of the Society of Toxicology, did the science of toxicology begin to develop. One reason for this change in focus was the idea of poison detection rather than analyzing possible antidotes. It is interesting to note that while this philosophy was changing, another important development was also taking place: A new discipline, emergency medicine.

The establishment of emergency medical systems across the country also provided a reduction in the time between the poisoning and the response of medically trained technicians. This response of medical technicians lowered the mortality and morbidity in some cases, while in other situations, secondary exposure occurred, with the medical technicians becoming patients and some suffering long-term effects.

Today, there is a vast diversity of disciplines under the heading of toxicology-environmental toxicology, industrial toxicology, and industrial hygiene, clinical toxicology, biological toxicology, to name a few examining chemical effects from food, drugs, the workplace, and the environment. All provide one goal: understanding how chemicals affect our environment and the living organisms within their surroundings.

There can be more than one response produced by a drug or chemical. The extent of injury may be one target organ, or it may be several. The organism as an individual directs the outcome of the toxicity of a chemical or chemical family. There are no single solutions to the technical problem of health risk assessment; rather, the solution of any analysis is weighing all the information, then gathering viable solutions. This solution process is not static but dynamic in nature.

Several interrelated topics are discussed as they relate to the chemically injured patient. Each topic is discussed in order to gain a further understanding of the material presented. As in any discipline, the terminology of the science helps to establish the building blocks for clarification. Toxicology is no different.

Exposure vs. Contamination

It is not always easy to discern the difference between exposure and contamination. But, in reality, they are significantly different. For the sake of this book, when contamination is discussed, contamination means that an unwanted and possibly dangerous chemical is on or in a person or exists on a piece of equipment that may eventually cause harm to a person. For a person to be contaminated with a chemical, they first must be exposed. Exposure, therefore, means that a person is in the presence of a chemical substance.

Exposure is a bit more difficult to understand as it can cause significant injury and not cause contamination. An analogy might be walking into a cow pasture. As soon as you enter, you can smell the cow patties. You have not touched them in any way, so you are just exposed at this point. If you were to leave the pasture, you would take none of the cow patties with you. But as you continue into the pasture and suddenly step into a fresh patty, you are now contaminated. If you leave the pasture at this point, you will carry part of that patty with you. You are now contaminated. It is as simple as that.

Toxin vs. Poison

Another set of terms that becomes somewhat confusing is toxicity, poison, toxin. Many use these terms interchangeably, but, in reality, they are very different. For clarification, below are the official definitions of these terms.

Toxicity is the degree to which a chemical or substance can cause injury, cause damage to, or harm a person (or animal).

A **poison** is simply a substance or chemical that is capable of causing an illness or death of a person (or animal) when introduced or absorbed.

A **toxin** is a specific poisonous substance that is a product of the metabolic activities of a living organism. These are notably toxic when introduced into the tissue.

The term poison in this book will be used in the general sense. Specifically, it is a hazardous chemical that has the potential to be toxic if it affects the organism, usually in a negative context. For example, some compounds are reported to have toxic properties in the pure or diluted form. Combined with another "activating" compound or impurity, the combination can be lethal. In this presentation, we will consider both.

Toxicity of a Poison or Toxin

When evaluating the toxicity of a poison or toxin, the route of entry the substance takes to gain access into the body must also be evaluated. Depending on the route of entry and the toxicity, the symptoms can be immediate or delayed. There are other external factors that can affect the toxicity of a chemical as well. Factors such as temperature, humidity, trauma, previous exposure, sensitivity, and type of personal protection all may affect the severity of the toxic event.

So, when exposure and dose relationships are discussed, what is actually being discussed are the biochemical reactions, the risk of influence, and the individual's body response to the chemical stimuli. This stimulus is described as a biochemical reaction and is expressed by a toxicological numerical term. Biochemical reactions are basically chemical reactions that take place inside the cells of living organisms. During normal biochemical reactions, those unaffected by an invading chemical, the sum of these reactions is called metabolism. Therefore, when chemicals invade the body and cause an untoward reaction, it changes the normal metabolism of the body.

When a chemical exposure takes place, and there is an effect from that exposure, the healthcare provider must ask the following questions:

- What is the chemical(s) involved?
- What is the amount (or concentration) of the chemical(s)?
- What are the defined harmful effects?

These answers will usually lead to the appropriate actions that need to be taken to successfully treat the patient. However, when an individual is contaminated internally or externally, and decontamination is not performed or not possible, the exposure may lead to chronic effects. Asbestos has a history of creating internal contamination on the alveolar level, then 20 years later, lung disease is the result. The reality is that chemicals are harmful. Chemicals can and will cause harm at some level.

It is interesting to see that people who work in a setting where chemicals are present and exposure takes place on a daily basis, their response to a toxic level produces fewer symptoms than they do to a person who is not exposed regularly. For example, evaluate a person who smokes two packs of cigarettes a day. These smokers do not suffer any acute symptoms to their daily intake of smoke and the chemicals carried in the smoke. But if a non-smoker was to attempt to smoke two packs of cigarettes in a day, they would be very ill to the point of hospitalization.

Some books refer to this effect as tolerance or hyposensitivity. Others state that a person becomes acclimated to routine and regular exposure. In reality, the body has developed processes to detoxify more efficiently the chemicals it is exposed to on a regular basis. This does not imply that either person (the smoker and the nonsmoker) will not have chronic effects. Firefighters who experience all different kinds of smoke and chemical exposures during their career rarely display acute symptoms, but the long-term effects of these exposures are well known and documented. Many times, the result of these exposures is cancer or long-term lung disease and a documented lower life expectancy.

Although laws have been propagated from a public outcry to control chemicals within our society, society has also become more complacent to those chemicals used in the household.

Society today is chemically dependent. Every day there are vast amounts of solid, liquid, vapor, and gas chemicals used in the household, including cleaning agents, insecticides, fertilizers, cosmetics, dyes, plastics, fuels, synthetics, etc. These chemicals improve the quality of life and make life easier for all of society. But a person cannot lose sight of the fact that chemicals used improperly can cause devastating effects.

Chemical materials of all types are dangerous, possessing the potential for toxic effects. Emergency responders must become fully aware of the health hazards these materials pose. As stated in the introduction to this section, the idea is to learn enemy identification. The nomenclature and testing procedures used to define the toxicity of chemicals must be known before the referenced materials at an incident are understood. It is up to emergency workers to understand and apply this information so that decisions based on a risk versus benefit analysis can be accurately made.

In all of chemistry, toxicology (for our purposes) must be thought of as a biochemical reaction. In other words, a chemical compound that is foreign to the human body is reacting with the bio-metabolism and causing an effect. On occasion, that effect may be very slight, or it may be catastrophic.

Thousands and thousands of chemicals are present in our environment, millions of chemicals within our society. Some are naturally occurring (organic), and some are man-made (synthetic). Because of this vast quantity of chemicals, it is sometimes hard to link the cause and effect to an exposure. Did the exposure (the cause) lead to the medical problem (the effect)? The fundamental reason this relationship is so hard to establish is that it is difficult to understand the human (and animal) physiological functioning on the biochemical (cellular/tissue) level. Our overall understanding of biochemical reactions and functions is very limited in terms of the influencing factors.

Exposure can be tracked, given the chemical compound and the route of entry. However, the intricacy of the human body is deeper than the function or structure of the tissue. In many instances, chemical reactions between the toxin and the organism take place at the cellular level. So, the real problem is the simple fact of whether we can detect an adverse reaction. Is it truly a reaction to the toxin, or is this reaction from a defensive mechanism within the body? In either case, it is difficult, maybe impossible, to determine if the reaction will affect the organism in any way.

Standards, Guidelines, and Acts Regulating Hazardous Materials

NFPA and OSHA

In the hazardous materials response community, there are both standards and guidelines that responders follow. For example, the documents produced by OSHA are standards that are backed by law. Those produced by NFPA are called consensus standards and essentially serve as guidelines unless adopted by a state or community. Both provide real guidance concerning how to handle a hazardous material when an emergency incident takes place.

The idea is to have a measurable level of potential harm and develop protective measures such that the activities surrounding the use of chemicals can be managed. Guidelines and exposure standards are largely determined by a risk management process. These assessments give estimates of potential exposures given a set of well-defined circumstances. It is based upon the fact of reducing exposure through the use of engineering and administrative controls. These controls are documented in publications that are taught prior to an incident and referenced during an incident.

Environmental Protection Agency (EPA)

As an example, pesticides, hazardous waste, air, and water pollutants all fall under the standards and guidelines of the Environmental Protection Agency (EPA), while workplace exposure guidelines and standards are based under the Occupational Safety and Health Administration (OSHA).

Federal regulation for the distribution, sale, and use of pesticides are referenced under the Federal Insecticide, fungicide, and rodenticide act or FIFRA (7 U.S.C.), which started in 1910 to ensure the quality of pesticide from fraudulent products. In 1947 and again in 1972, the act was amended to reflect issues within the environment due to the tremendous use of insecticide after World War II. Since then, it has been amended several times to address the risks and benefits of pesticides.

Hazardous waste is regulated under Resource Conservation and Recovery Act (RCRA), which is a law that provides a mechanism to properly manage hazardous and non-hazardous waste. Along with CERCLA, which is commonly referred to as the Superfund act. The main purpose of CERCLA was to clean up hazardous waste disposal sites and to set standards for reportable quantities. As a part of a congressional mandate, the Agency for Toxic Substances and Disease Registry (ATSDR), which is to look at the health effects of hazardous chemicals within the environment.

The Clean water act (CWA) and the Safe Drink water act (SDWA) under the EPA conduct risk assessments of water quality and issues the MCL's for water (Maximum Contaminate Levels) contaminates. These contaminate can be man-made or naturally occurring episodes. EPA then provides health advisories for drinking water contamination. Data is based upon NOAEL (No Observed Adverse Effect Level- highest dose which the toxicological studies have shown no adverse effects) or LOAEL (Lowest Observed Adverse Effect Level – which is the lowest dose, which the toxicological studies have shown no observed toxic or adverse effects). Air emission standards under the clean air act (CAA) and the National Quality Standards (NAAQS) as they relate to atmospheric pollutants requires EPA to regulate toxic emissions.

This brings us to the Occupational Safety and Health Administration (OSHA). OSHA publishes exposure limits used in industry to determine if a work environment is safe or not. The exposure limits focus on the health and welfare of the employees and mandates that the employer provides a safe work environment. There are many different exposure limits. Each of these will be reviewed below.

Time Weighted Average

The Permissible Exposure Limits or PELs are for airborne contaminants. This is the level of a chemical in the air that is safe to breathe for an 8-hour day, five days a week, to establish a 40-hour workweek. If an employee works at the level of a chemical that is at or below the PEL, they will not suffer damage from that chemical exposure. Since this level is determined safe by evaluating it over a specific timeframe, this is called a time weighted average (TWA). Documents refer to this as a Permissible Exposure Limit – Time Weighted Average or PEL-TWA.

Other regulating agencies also publish these TWAs but call them something different. For example, the National Institute for Occupational Safety and Health (NIOSH) has named this level the, Recommend Exposure Limit (REL) or REL-TWA. The American Conference of Governmental Industrial Hygienists (ACGIH) denotes it as threshold limit value (TLV) and is identified as TLV-TWA. Working in an environment above the PEL, REL, and TLV will generate a toxic effect. See Figure 5.8. While working under these published levels, a worker will not suffer from adverse effects.

Short-Term Exposure Limits (STELs)

Another limit is the STEL or Short-Term Exposure Limit, which is an exposure that only occurs for 15 minutes and is not repeated more than four times a day. Each 15-minute exposure event is interrupted by a 60-minute non-exposure environment so that in the course of an 8-hour day, the individual can only be exposed to a chemical for 15 minutes with an hour break in between exposures not to exceed four times within a day. The STEL is sometimes referred to as the emergency exposure limit (EEL is an administrative control to guide worker safety). The ACGIH has recently recommended the use of excursion limits (EL). The EL is a weighted average with an exposure time that can exceed three times the published 8-hour TWA. This can only occur for 30 minutes within any one workday that is 8 hours in duration and cannot in any circumstance exceed five times the TLV-TWA.

Immediately Dangerous to Life and Health (IDLH)

IDLH was developed as a cooperative guideline between OSHA and the National Institute for Occupational Safety and Health (NIOSH), which is the parent agency of the Center for Disease Control (CDC). So, we have two definitions, one from the OSHA perspective and the other as NIOSH, and how they see the problem:

OSHA. "an atmosphere that poses an immediate threat to life, would cause irreversible health effects, or would impair an individual's ability to escape for a dangerous atmosphere."

NIOSH. "an airborne exposure likely to cause death or immediate or delayed permanent adverse health effects or prevent escape from such an environment."

From OSHA's perspective, it is an all-hazards approach to the "toxic" environment, while the NIOSH definition is looking at respirator selection. It is worth noting that the NIOSH definition does not include oxygen deficiency below 19.5, whereas the OSHA defines all hazards by using the word impair rather than prevent.

Exposure limit comparison					
OSHA		NIOSH		ACGIH	
PEL-TWA	Maximum concentration over a 40-hour workweek without adverse effects	REL-TWA	Maximum concentration over a 40-hour workweek without adverse effects	TLV-TWA	Maximum concentration over a 40-hour workweek without adverse effects
PEL-STEL	Maximum concentration not to exceed 15 minutes occurs 4 times in a day with an hour between exposures	REL-STEL	Maximum concentration not to exceed 15 minutes occurs 4 times in a day with an hour between exposures	TLV-STEL	Maximum concentration not to exceed 15 minutes occurs 4 times in a day with an hour between exposures – TWA is not exceeded
PEL-c	Maximum concentration exposure not to be exceeded	REL-c	Maximum concentration exposure not to be exceeded	TLV-c	Maximum concentration exposure not to be exceeded
IDLH	an atmosphere that poses an immediate threat to life would cause irreversible health effects or would **impair** an individual's ability to escape from a dangerous atmosphere		an airborne exposure likely to cause death or immediate or delayed permanent adverse health effects or **prevent** escape from such an environment		
Example chemical - acetone					
PEL-TWA	1000 ppm	REL-TWA	250 ppm	TLV-TWA	250 ppm
PEL-STEL	N/A	REL-STEL	N/A	TLV-STEL	500 ppm
PEL-c	N/A	REL-c	N/A	TLV-c	750 ppm
IDLH	2500 ppm				

Figure 5.8 Exposure limits comparison.

The American Conference of Governmental Industrial Hygienists (ACGIH) has developed a ceiling level that pertains to airborne concentrations that should not be exceeded under any circumstance. In situations in which the ceiling level is not identified, in other words, the toxicological data does not support a ceiling level conclusion five times the TWA-TLV can be used in the TLV-c's place and only three short-term exposures to that identified level not to exceed 30 minutes/day.

Control Banding

Control banding is a technique that is used to limit exposure to an employee. It may include engineering controls such as the addition of extra ventilation, containment, and/or PPE. It may also include administrative controls such as policies and procedures written to prevent unnecessary exposures. Typically, these written procedures are found in the company's health and safety plan. These are based upon a range or "band" of hazards that may be present with the chemical. However, control banding must be used along with other health and safety practices.

When large populations are effected or when decisions on community evacuation is considered. or protect in place issues present themselves at an incident, several protective action guidelines (PAG) have been developed. This criterion is based upon exposure limits, evidence-based methods, and evaluation. The AEGL's and ERPGs are derived from human and animal studies, where the TEELs are derived from levels of concern. Public Action Criteria (PAC) have a hierarchy system of use AEGL > ERPG > TEEL.

Note: AEGL's (Acute Exposure Guideline Levels) were developed by a collaborative effort from the U.S. Army, EPA, ATSDR (Agency for Toxic Substances and Disease Registry), NRC (National Research Council), DOE (Department of Energy and others). There are three levels of concern. They are used by emergency planners to identify and give guidance to responders when there is an accidental airborne release into the environment. These values are meant to protect the old, young, compromised, and sensitive populations. These are calculated in short-time cycles, 10, 30 minutes, 1 hour, 4 and 8 hours.

AEGL – 1. Discomfort is noted, possible irritation, transient and reversible effects.

AEGL – 2. Potentially an impaired ability to escape, with irreversible adverse health effects.

AEGL – 3. Life-threatening adverse health effects or deadly consequences.

The American Conference of Governmental Industrial Hygienist (ACGIH) has Emergency Response Planning Guidelines (ERPGs). They are an estimate of the point that effects from the airborne chemical can be felt by the general population for an hour. This guideline does not take into consideration the old, young, or compromised victim in that these sensitive populations may sustain health effects well below the ERPG (see Figure 5.9).

These are placed into three tiers:

ERPG – 1. The maximum airborne concentration that a person, after exposure for one hour, would show only mild or the absence of transient health effects or complaints of an objectionable odor. This is the lowest and least concentrated environment.

ERPG – 2. The maximum airborne concentration to which persons exposed for an hour will be able to take protective actions without developing serious health effects or symptomology. Most would only experience temporary effects from the chemical. This is the medium level of concentration.

ERPG – 3. The maximum airborne concentration in which persons exposed for an hour do not experience any life-threatening health effects. People may experience significant health effects but are not life-threatening. This is the highest of the concentrations.

These guidelines should be used to protect the public when Acute Exposure Guideline Levels (AEGLs) are not available. These are usually used when there is a chemical release into the environment, which is short-term in duration.

Temporary Emergency Exposure Limits (TEELs) are guidelines to protect the public during a chemical incident when AEGLs or RPGs are not available. They are estimations of the

Emergency planning guidance comparison		
AEGL	**ERPG**	**TEEL**
AEGL 1- Airborne concentration above which Discomfort is noted, possible irritation, transient and reversible effects.	**ERPG 1** – the maximum airborne concentration which persons exposure to for 1 hour would show not or mild, transient health effects, or an objectionable odor. This is the lowest and least concentrated environment.	**TEEL – 1** airborne concentration which persons exposed for more the one hour could experience discomfort, irritation which are transient and reversible
AEGL 2- Airborne concentration above which Potentially an impaired ability to escape, with irreversible adverse health effects	**ERPG 2** – the maximum airborne concentration to which persons exposed for an hour will be able to take protective actions without developing serious health effects or symptomology. Most would only experience temporary effects from the chemical. This is the medium level concentration.	**TEEL – 2** airborne concentration which persons exposed for more the one hour could experience serious adverse health effects which could be irreversible or the inability to self-rescue.
AEGL 2- Airborne concentration above which Life-threatening adverse health effects or deadly consequences.	**ERPG 3** – the maximum airborne concentration in which persons exposed for an hour do not experience any life-threatening health effects. People may experience significant health effects but are not life-threatening. This is the highest of the concentrations.	**TEEL – 3** airborne concentration in which persons exposed for more the one hour could experience life-threatening adverse health effects or deadly consequences.

Figure 5.9 Comparison between actue exposure guideline levels, emergency response planning guide, and temporary emergency exposure limits.

concentrations which, when exposed, individuals will begin to experience health effects to the airborne contaminants. As with the other two (AEGLs and ERPGs), there are three tiers corresponding to a specific level of health effects. This exposure limit takes into consideration susceptible populations.

TEEL – 1. Airborne concentration in which persons exposed for more than one hour could experience discomfort, irritation which are transient and reversible.

TEEL – 2. Airborne concentration in which persons exposed for more than one hour could experience serious adverse health effects, which could be irreversible or the inability to self-rescue.

TEEL – 3. Airborne concentrations in which persons exposed for more than one hour could experience life-threatening adverse health effects or deadly consequences.

Emergency Exposure Guidance Levels (EEGLs) were developed by the U.S. Navy to protect submariners from the effects of air contaminates for very specific tasks during emergency conditions lasting from 1 to 24 hours. They are meant to allow task performance of an objective during emergency operations which would allow self-rescue. These levels do not impair judgment or cause irreversible harm; however, they may cause some transient effects. They are used during incidents in which a greater risk poses a threat greater than discomfort.

Short-Term Public Emergency Guidance Levels (SPEGLs) are short-term exposures in emergency situations toward the general public from airborne contaminants. All sensitive populations such as old, young, disease consequences are considered. Produced for the Department of Defense by the National Academy of Science (NAS), they are based upon a 60-minute exposure. SPEGL is set at a fraction (one-tenth to one-half) of the EEGL with a safety factor of 2 to provide adequate protection to susceptible groups.

Within industrial hygiene, guidance is given to control health hazards to supplement their occupational safety and health program. Developed by the ACGIH, they do not endorse their use; however, in certain instances, they make use of these indexes. Biological Exposure Indices (BEI) are meant to assist in potential workplace health hazards.

Basic Toxicology Definitions

There are several shortcomings to the testing designs used within the toxicology field. Despite these problems, the testing procedures employed are the most accurate. The main problem is so many body systems interact, along with the chemistry involved, that it is very difficult to set testing parameters.

For emergency responders to make the decisions that they will be faced with in the field, several problems with the testing procedure must be mentioned. The reference material unfortunately does not identify any of these possible problems. These tests only reference a single point in time, which under testing conditions was used as an effect parameter. Basically, the effect a chemical has on an organism has to do with the rate of absorption, detoxification, and excretion. The amount of the chemical is also a factor; however, some of the testing procedures give such a high dose that any animal would show an adverse reaction.

To consider the possibility of intermediates having a profound effect on the organism, then another parameter must be added within the testing procedure. Hypersensitivity, hyposensitivity, synergistic, additive, potentiation, and antagonist reactions are not truly considered within some of the testing methodologies.

Statistically, these factors are deliberately looked at from a mathematical model standpoint. In other words, a certain percent of a population will be hypersensitive, and a certain percent will be hypersensitive. By utilizing statistics, it is assumed that randomness in the population will equal itself out, using the above two factors as an example. However, humans are not statistical points on a graph. Each point observed on such a graph represents a life. As seen in the acute and in sub-chronic testing, the chance of having a statistically biased test is quite high. This occurs when the testing group is too small or limited to only a few species.

Some testing laboratories have been accused of misrepresentation on the basis of the dose rates that have been utilized. In some of the test procedures, the dose rates are so high that any animal, whether sensitive to the chemical or not, will produce an effect. If the group did not include sensitive individuals, then the results are skewed, giving an unrealistic result.

Gender differences, race differences, age, and physical activity are not usually considered. In terms of race, chemicals are not tested on humans. See Figure 5.10. To that end, there is uncertainty concerning how the genetic makeup will be affected from one race to another. In the testing process, some of the standard tests that are applied are administered to "genetically pure" groups. These groups have been fed a certain diet, bred to limit genetic anomalies, and controlled to a certain

Figure 5.10 Race and Gender as well as age and size can alter the intensity of signs and symptoms from a chemical exposure.

"clean" environment. This does not occur in human existence: humans do not live in a totally "clean" environment. Breeding does not take place in order to have genetically perfect offspring. Additionally, how many people, for the entire length of a life span, eat food that is nutritionally healthy?

However, blunt this may sound, the reality is that scientists are trying to block out as many factors as possible that may affect the toxic event. There does not exist a better practice in order to evaluate the toxicity of chemicals. When the health effects are being considered at the scene of a hazardous materials incident, all the concepts that have been mentioned must be considered in the decision-making process.

Acute Exposure

The term acute is used to describe a sudden onset of an exclusive episode. It relates to the exposure of the hazardous material as a single event that causes an injury. This single exposure is usually a short-duration type of exposure. It is sometimes classified as nonpredictable. This single dose occurs within a 24-hour period, or it could be a constant exposure for 24 hours or less. In certain cases, it could also be multiple exposures within the 24-hour time frame.

There is a testing routine that, for the most part, is a standard. However, any person or organization testing a chemical has the levity to determine what they believe is a sound testing procedure. A chemical testing process includes:

- Reference all information concerning the previous testing of the chemical.
- Define the testing technique that will be used.
- Publish the findings as it relates to previous testing and findings.

In terms of difficulty, the oral toxicity tests are relatively easy. The scientist uses rats or mice to study the chemical or drug in question. These tests are done initially to rate the toxic levels of the chemical. This initial test involves providing the test animals with an ever-increasing amount of the chemical until a lethal dose is attained. From there, they can evaluate the chemical in relation to other chemicals in terms of toxicity.

Then they take a group of animals and place them into four or five (sometimes six) groups. In group one, the rats and/or mice are given a set amount of the chemical that may or may not be at the threshold limit value. It is, in general, a no-death dose. All the animals are anticipated to live. However, there may be an observable toxicity event. From group one, increasing to group four, five, or six (depending on the chemical and the testing technique), the animals are given increasing dosages. The last group is given a dose that is known to completely destroy that group. From this observation, a further sub-ranking of the chemical is done. They observe for the next 14 days the death rate, sickness, fur loss, and other effects. The dose that produced a 50% death toll is determined to be the LD/LD_{50}

Each group produces a statistical picture of the chemical dose versus the death rate, fur loss, sickness, and so forth. This is compared to effective doses that are known for the chemical or like chemicals (chemical families). Both graphs are statistically "flattened" out into a straight line or what is called the *dose–response curve*.

Through statistical modeling, the effective dose and the lethal dose are calculated, and the published LD_{50} is now established. Both oral toxicity and inhalation toxicity are measured in this way. The oral toxicity levels are usually a one-time dose. Where the inhalation exposure is an exposure that occurs for one hour to each test group, each test group receives the appropriate dose. In each case, the groups of animals are observed for a period of 14 days. At the end of the 14-day cycle, the 50% death rate must be observed. If the percent is higher or lower, then the doses are reevaluated until the end result is 50% the standard baseline.

With dermal exposure, the test animals are usually pigs; however, rats and mice are also used. The pig's skin represents a very close facsimile to human skin. The animal is exposed for a period of 24 hours on the bare skin (if the hair is present, it is shaved off). Only 10% of the body surface area is used. The animals are again observed, and the reporting is matched with the known level of that chemical or chemical family.

If the chemical needs further study or the results are such that the human element needs to be established, then the procedure is repeated using different species. In general, if during the testing process all the test animal species respond in a similar manner and the statistical slope of the dose-response is steep, then it is considered to be an accurate LD_{50} or LC_{50}. On the contrary, if the testing battery showed that there was a diversity of slopes across the animal species spectra while shallow slopes were observed statistically, then the accuracy of the outcome is questioned. The problem is that the numbers that are seen, represent a point in time in which a toxic dose was received. It displays only a small window of time into the dose–response and does not demonstrate the angle of the testing slope. Was the slope steep, thus, a good correlation to the human event, or was it a shallow slope, which does not represent the true toxicity of the chemical in a man?

Lethal concentration or dose testing is not performed on a human. If a document states that a reported level of a chemical caused death in human, then this information was achieved through an autopsy after the death occurred from a suicide, homicide, or accidental release.

In general, the smaller the amount reported as a lethal concentration, the more toxic the chemical is. If, for example, the number is large, then the toxicity of the chemical is relatively low. It is hypothesized that it is the variation of exposure that gives the true toxic picture. In simpler terms, if the chemical in question has been demonstrated to be toxic in most plant and animal life without the documentation of a human experience, then it is considered to be a health hazard to humans. If the literature states that most plant and animal life is evenly destroyed when an exposure exists and is associated with a human event, then again, it is a human health risk and considered toxic.

Lethal concentrations and doses are only an estimate of the potential health problems that may exist. There is by no means an all-or-none limit. As with the TLVs, PELs, And RELs. They should not be taken as a set limit that will identify an exposure as a health effect or as a limit that no exposure will occur.

Most chemicals are weight-dependent. In other words, the more physical mass of the animal in relation to the quantity of chemical exposure, the higher the likelihood of the animal combating the exposure. To the contrary, the lighter the weight of the animal, the lower the potential of combating the exposure, or more accurately, the toxic event.

Looking at the LD_{50} of some chemicals, in order to estimate the lethal dose, we could take the weight of the person and multiply it by the quantity of the material. For example, in a small child weighing 40 lb (18.18 kg), the LD_{50} is multiplied by 18.18 kg. So, if this child ingested Sevin (a moderately toxic: carbamate pesticide, LD_{50} of 500 mg/kg), we would multiply 18.18 kg by 500 mg/kg or 9090 mg, or on a more convenient measuring system 9.090 g or 0.02 lb. If the patient was a 220-lb man, for example ($220/2.2 = 100$ kg; $100 \times 500 = 50,000$ mg or 50 g which is 0.11 lb), it would take a little over a tenth of a pound to reach the LD_{50}. Remember that metabolism, sensitivity, humidity, temperature, and vapor pressure are a few of the factors that may influence toxicity and must be considered. This discussion is only considering body weight as it relates to the LD_{50} of the chemical.

Generally, there are more LDs than there are LCs because most chemicals are in a solid or liquid state and do not change states of matter. Very few become a vapor problem (this is true as long as there is not flame impingement on the product). There are not many chemicals, relatively speaking, that are airborne contaminates. However, this brings up yet another issue related to the testing procedures.

Emergency responders, specifically those in the fire service and in hazardous materials response, have to deal with flammability, and reactivity issues that a chemical may possess. These research methods do not reflect the true health issues when chemicals are under fire conditions. Responders must consider what are the synergistic, additive, antagonistic, or potentiate reactions presented when fire is involved. How chemicals react, combust, and further jeopardize the health integrity of responders are all significant management problems.

Sub-Chronic/Sub-Acute

Sub-chronic and sub-acute are two terms that, at times, are used interchangeably. Both have been used to describe the same type of exposure in printed literature. However, in order to be correct, the appropriate term is sub-chronic. This type of exposure involves an acute exposure that is repetitive. It is, by definition, a recurring event. In total, it is an exposure that happened during approximately 10% of the organism's life span.

The testing procedures are based upon the LD_{50} or LC_{50}, which established death in 50% of the test group and are conducted on two species (two species as a minimum), with each having a control group. Three to four groups are tested simultaneously, as was discussed in the acute testing procedure. At the top end, the dose that is given is under the established LD or LC curve. The dose is high enough to show signs of injury, however, not death. At the low end of the dose range, there should not be any noticeable effects. Depending on the test chemical, a middle point is chosen. If the curve in the acute testing was shallow, then two or more middle points are picked. If the curve is steep, one point is selected.

The exposure is given for a period of 90 days. At the end of the time frame, all the animals are autopsied for any histological evidence of an effect. It is this effect, observed during the autopsy, that is compared to the control groups and histological evidence documented and reported in the literature.

Chronic Exposure

Chronic exposure is a long-term effect a chemical may have on an organism. Technically speaking, it is the length of time that the animal was subjected to exposure to a chemical, usually 80% of the total life span. Chronic effects are much harder to establish than acute effects. Many factors come into play when discussing the chronic effects of chemicals. The current understanding of toxicological responses and the knowledge of biochemistry is limited. Chronic exposure is deeper than just exposure that occurs for 80% of a life span. It also can influence some, any, or all the offspring. Accumulation, or acclamation, may lead to hyposensitivity or hypersensitivity.

As in the acute testing process, a certain degree of inaccuracy is also seen in the chronic exposure figures. Most of what we see and read about is an after the fact chronic event. In other words, an individual or a group have become ill or have died. What has to be considered is that not all injuries and deaths will have a defined cause and effect after a chemical exposure.

It is the acute oral study that is investigated first for the chronic toxicity of most chemicals. There is a good reason for this particular starting point. The first studies were to isolate those chemicals that could potentially become a problem if introduced into the food. For example, preservatives and food production additives were studied. This method of exposure study is easy to do. Only in recent years has this chronic exposure study been expanded to include chronic inhalation and topical exposure studies. These topical skin studies were propagated from the September 11 attacks in 2001. Many questioned the skin absorption studies concerning the work being done at the site. Additionally, new protective gear was proposed, adding to the argument of some study methodology that would validate the exposure protection garments.

The experiment is similar to the sub-chronic (sub-acute) method of evaluation. The experiment starts out by finding the toxicity range of the chemical in question. Most of the information is gathered from the acute studies. The chemicals are introduced into the animal by placing them in the food. This process lasts for 90 days. During this time, a variety of dose ranges are used. This testing establishes the high, medium, and low toxic ranges. In the high toxic range, the effect is severe. In the low, no desirable effect is noted. In the moderate range, the effect is mild. In addition to these three groups under study, there is also a control group.

Typically, many animals are used. There are also two or more different species utilized. The oral ingestion is started from birth and continues to approximately 80% of the animal's life expectancy. Every day the animals are observed for negative effects, and weekly, a battery of tests are performed. At the end of the experiment, all animals are autopsied, and histological tests are analyzed. From the data, a statistical model is used to give the dose-effect response.

As demonstrated, this type of testing can become extremely technical and expensive. Also, the length of time that is required to observe some effect is years in the making. If new concepts were introduced within the experiment during the testing period, the results would be skewed.

NOEL, NOAEL, LOAEL

No Observed Effects Level (NOEL), No Observed Adverse Effect Level (NOAEL), and Lowest Observed Adverse Effect Level (LOAEL) (see Figure 5.11).

Typically, these are toxicity tests that are done for pharmaceuticals and are used within clinical trials; however, there has been some discussion on the practicality and application to the emergency response community within medical surveillance programs. Most of this has

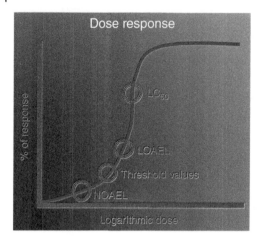

Figure 5.11 This logarithmic curve represents the comparison between the NOAEL, threshold limit values, LOAEL, and lethal concentration for 50% of the test population.

to do with the heightened attention to one's health and the health programs that are now in place within the emergency services.

This type of testing is evaluating the risk assessment in response to the question of what type of exposure can produce what effect. In other words, a given chemical, when exposed, may give rise to a variety of effects or potential effects of health concerns. In these definitions, exposures are evaluated a little differently, and compare epidemiological studies with laboratory studies.

Basically, the NOEL is the highest exposure that has no adverse effects when compared to the control. The NOAEL displays no statistically or biologically significant increases in the frequency or severity of adverse effects between the exposed population and its control. LOAEL is the lowest level that there is significant severity of effects either statistically or biologically.

Levels of Concern

In many of the plume modeling programs and toxicological literature, there is a common term and acronym used to describe the very edge or beginning of a dangerous level. The term is Level of Concern expressed as the acronym LOC. When plume modeling is discussed in CAEMO's ALOHA, the LOC is described as the threshold concentration that may injure individuals if they have the opportunity to inhale the specific chemical. Under this pretense, the lower the number, the more toxic the material is towards humans.

In the toxicological world, these are considered safe levels of exposure. The concept is that below these levels, there is no applicable risk to human health; therefore, at low levels of exposure, there is a negligible risk to harm below the NOEL.

In general, the LOCs are in essence exposure limits and used for planning out a potential event. These are considered as public exposure guidelines (see AGELs, ERPGs, TEELS).

Dose Response and Exposure

Two principles are used when describing the toxic levels of a chemical. Each, in itself, explains the pharmacological response to the chemical in question. The first is the concentration–response relationship, and the second is the dose–response relationship.

When a chemical is viewed at the physiological level, the concentration of that chemical plays a part. The concentration determines the ability of the chemical to bind to a receptor site, combine or otherwise attach with a biological target. Chemicals bind to normal receptor sites in order to inhibit, excite, or control the site.

The chemical must "fit" into the site as if it belongs there for one of these responses to take place. The fit, in turn, causes a response. The chemical may simply replace a different chemical that would normally fit into that receptor site, or a reaction can result in the enzyme production phase. The change in the molecular "signal" from the cellular level is what is observed as the signs and symptoms, all depend on the receptor sites affected. Logically this reaction must occur more than several times and often enough to produce an effect.

Graded Response

So, the true question may be, how much of a chemical will produce this observation? What concentration of the chemical will bind enough receptor sites, or disrupt enough enzyme pathways, to cause an effect? Two types of responses are seen in terms of concentration. The first is a *graded response,* which is represented by a gradual increase of receptor sites binding with the chemical. It is a gradually more pronounced response that is seen as the concentration increases – the greater the concentration, the greater the number of sites that are occupied. The increase of signs and symptoms displayed by the individual is proportional to the bound receptor sites.

Quantal Response

In the *quantal response,* an all or none response is observed. In other words, the increase in concentration does not necessarily mean an increase in observable effects. It may take a large concentration to produce an effect. Up until that time, no effect may be noted. The concentration producing the effect is thought to have a direct relationship with body weight and metabolism. The point at which the chemical produces an effect is called the effective concentration, abbreviated as EC. If a number is associated with the EC like, EC_{50}, this means that 50% of the tested population responded with an observable effect at this concentration.

Many models have been developed to understand the effects of drugs and chemicals. The body of any organism is a complex machine in which complicated biochemical reactions take place. Most physiological models look at the body as having five groups:

1) Lung group
2) Vessel-rich group
3) Blood-rich compartment
4) Muscle group
5) Fat group

The lung group consists of the respiratory tree, partial pressures within the respiratory system, and how these pressures relate to the other four groups. When the effects of vapors or gases are discussed, the lung group is isolated for injury study. The vessel-rich group looks at the site in which metabolism may occur, such as in the liver, kidneys, heart, and gastrointestinal tract. Some scientists place hormonal activity within this group. It is observed that this area is where most of the damage from a variety of chemicals occurs. The blood-rich compartment is the brain. An inability of the brain to metabolize some chemicals is observed as a central nervous system disorder. The muscle group is composed of muscle and skin. When studies are done with topical agents, this group is isolated for the experiment. The fifth and last group is the fat group, which is composed of adipose tissue and bone marrow. In this group, the adipose tissue may respond by absorbing the material, and toxic events can take place when the body chooses to metabolize fat.

All of the compartments have one thing in common – blood perfusion. It is assumed that as an individual assimilates the chemical and equally distributes it throughout the bloodstream. Through this equal distribution, the chemical's physical and chemical properties target a particular organ. For example, a fat-soluble chemical will target the fat group, hiding in the adipose tissue until metabolism starts to utilize that particular tissue.

The liver and the kidneys are very well perfused (this is also where phase one and two reactions take place), therefore there will be a higher degree of organ damage within these tissues. With some chemicals, once a reaction takes place (the intermediate or end product of the reaction process), the chemical can precipitate out of the solution causing clots that interfere with normal perfusion. Usually, it is within the brain and kidney that this occurs (see Chapter 3, Hydrocarbons and Derivatives Toxidromes, Toxic Alcohols). Most of the chemicals target the liver, lungs, and kidneys.

Response Curve

The dose–response is an overall observable reaction to a chemical. It takes into account the difference in tissue function by looking for a response as it relates to the particular compartment. The problem here is that there is a wide variety of responses that a particular chemical may elicit. When looking at dose–response, the total group of responses must be evaluated. Not only are the issues of time versus dose examined, but also the effect experienced by the total animal population. For example, two theoretical chemicals will be examined: one that affects a small percentage of the population and one that insults a large percentage. These chemicals will be referred to as chemical A and chemical B. When chemical A is given to a population of 100 individuals, an effect within a small population base is observed. With chemical B, a large population within a total of 100 was affected.

From these statistical curves (as seen in Figure 5.12), we could assume that a small concentration of chemical A is relatively safer than a small concentration of chemical B. At a hazardous material event where two chemicals are involved, reference may indicate that one is not as toxic as the other but may not be able to give the exact toxicity of a particular chemical, highlighting the fact that there may not be a safe level of exposure. For chemical A we could say that a one-tenth concentration is safer than the one-tenth concentration of chemical B, but that is all. The large range of individuals affected by chemical B shows us that a larger safety factor would be needed. It is for this reason that we as emergency responders should not predicate our decisions based solely upon the (lethal concentration of 50% of the populations airborne) LC_{50} or (lethal dose of 50% of the populations liquid or solid) LD_{50}.

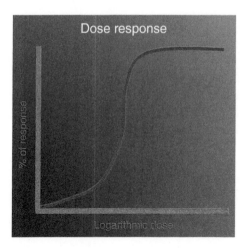

Figure 5.12 Dose response curve.

Lethal Concentrations and Lethal Doses

The next segment of terminology to be discussed comes from the dose–response curve. At the beginning of this section, the concept of graded response, quantal response, and effective concentration was discussed. Each by their own quality showed the exact moment in time for which the population showed an adverse reaction. These are called the threshold limit, effective concentration, or dose.

The qualifier here is that 50% of the total population under study died. This is based on the understanding that there will be some individuals within the population that are hyposensitive and some that are hypersensitive to the chemical agent. This qualifier refers to a finite time frame (usually 14 days), in which a total of 50% of the population died. If the chemical in question is airborne and is primarily an inhalation hazard, it is termed as the lethal concentration (LC). If the chemical poses a threat other than an inhalation injury, it is termed the lethal dose (LD). In both cases, it is the percentage of the population that is being described. A 50% death rate after the introduction of the chemical is denoted as lethal concentration fifty (LC_{50}) and is the calculation of an expected 50% death toll after the exposure of the chemical to test animals as an inhalation hazard. The lethal dose fifty (LD_{50}) is the calculation of an expected 50% death toll after exposure to a chemical, which may or may not include an inhalation injury.

For example, an LC_{25} or LD_{25} denotes a 25% death toll within the tested population. The number that is given with the associated LC or LD is relative. In general, the smaller the number, the more toxic the chemical. Likewise, the greater the number, the less toxic it is. For the most part, LD_{50} and LC_{50} are an average of lethal doses (ALD). These mean values are observation and calculation based on researched suicides or homicides.

One must remember that these are values observed in the laboratory under controlled conditions. These tests are not done on human beings. However, there are a few cases in which the human experience has been documented. In these cases, the reference texts usually denote this by placing the chemical exposure animal in parenthesis. At times this value is referred to as the (TDL) toxic dose low or (TLC) the toxic concentration low. As we noted before, the concentration enumerates the inhalation injury, and the dose is a liquid/solid state of matter. This could be absorbed either through the skin or by ingestion.

The EPA utilizes the values as presented from ACGIH, NIOSH, and OSHA. They are mostly concerned with the impact that the chemical may have on the environment and the organisms that make up that environment.

The reference books that enumerate the LD and LC for humans are actually statistical extrapolations. They are derived from the mean lethal dose or the average lethal dose or concentration. From there, they are calculated to the human experience.

LC_{lo}. This represents the lowest concentration of airborne contaminants that caused an injury.
LD_{lo}. The lowest dose (solid/liquid) that caused an injury.
LC_{50}. 50% of the test population died from the introduction of this airborne contaminate.
LD_{50}. 50% of the tested population died from the introduction of this chemical, which may be a solid, liquid, or gas.
LCT_{50}. A statistically derived LC/LD (LC_{50} statistically derived lethal concentration/dose).
RD_{50}. A 50% calculated concentration of respiratory depression (RD = respiratory depression of 50% of the observed population), toward an irritant, over a 10–15-minute time frame.

Toxicity is not a tangible, measurable quantity like chemical and physical properties, which can be measured and reproduced. When chemical properties are measured, the same points should be reached each time, given that the tests were performed under the same temperature and pressure. When measuring toxicity, this is not the case. Too many factors (metabolism, species differences, size, and weight, to name a few) propagate the quality of toxicity.

The toxic event can be displayed as a mathematical representation of what is observed. It is a graph that represents the observations pictorially. The numbers cited in the reference material are a single point on this graph. This single point only makes reference to a particular point in time with a very specific exposure and animal study group or, in some cases, groups of animals. The numbers that are found in the reference material do not give the standard deviation of the graph. Did the testing procedure account for 95% of the population (two standard deviations), and if so, what about the other 5%? As shown in the foregoing example, the differences between chemical A and chemical B had to do with the width of the graph, a greater or smaller standard deviation. These particular facts are not given in the referencing literature.

Before moving on, one more general concept must be discussed, that of flammable limits. As an example, flammable limits were defined as a range of concentration of a chemical that would burn, expressed in a percent mixed in the air. When discussing toxicity, these ranges also apply. Generally speaking, the toxic levels of a chemical lie just below the lower flammable limit (lower explosive limit, LEL). Just like these flammable limits, the toxic limits or levels have a graduation of toxic effects. For this reason, percentages are taken for "safe" levels when we are dealing with confined space (usually lower than 10% of our LEL). In general, under the LEL, we shall say it is a graduated toxic atmosphere.

Chemical Time lines

The easiest way to look at toxicology and get a visual representation of toxic events is to look at a chemical timeline or ChemLine. Here we are going to build a progressive line from left to right that describes the chemical and toxic values within context and compares them on a scale.

This line represents the spectrum of harm to the left: we have zero chemical (hence the circle at the beginning of the ChemLine), and to the right, we have some level of harm (As seen in Figure 5.13). Placing the LEL and UEL starts our building of how safe, unsafe, or dangerous the environment may be. In this case acetone will be used as the example chemical:

The LEL and the UEL is placed on the ChemLine, to display how the explosive limits fall on the line. The percentage of vapor in air is multiplied by 10,000 (1% = 10,000 ppm) to gain the respective ppm of the LEL and the UEL. In other words, below the 26,000 ppm, the mixture in air too lean to burn given an ignition source and above UEL too rich to burn. The ChemLine now shows where flashpoint occurs given the appropriate conditions and also ignition. (See Figure 5.14).

Evaluating Acetone's Flash Point of 0°F (−17.8°C), it is logical to see that at normal ambient temperature, there is enough vapor that will produce a flash. If there is enough vapors to produce a flash, then the acetone is well into the toxic range (Figure 5.15).

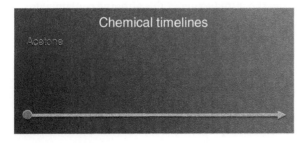

Figure 5.13 A ChemLine is established with the left side, at the dot indicating no harm. As the line progresses to the right, the level of harm is increased.

Figure 5.14 This ChemLine for acetone indicates the location of known chemical properties which demonstrates the level of harm.

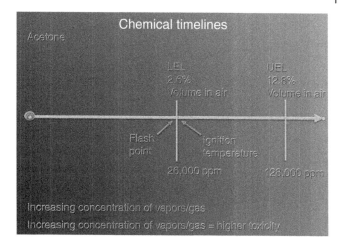

Figure 5.15 Adding more chemical property components the division between safe, unsafe, and dangerous.

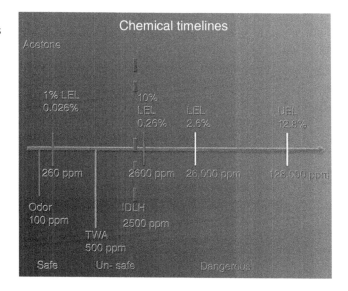

The toxic range starts somewhere above "zero" and extends to and beyond the LEL. It is a gradual accumulation of the chemical in the environment that creates an exposure. Reading this from left to right, odor appears at 100 ppm; for some individuals, this level may be enough to trigger some sort of response (quantal). Figure 5.16 demonstrates a completed ChemLine and clearly demonstrates the area defined as safe, unsafe, and dangerous.

For others, it may take more of the chemical. Continuing to the right, next is the TLV-TWA, the normal amount that is suggested as a safe working environment, and as before, there may be some individuals that are sensitive to, in this case, acetone and present with a reaction. But beyond the TWA, this area that can cause a toxicological response as the IDLH is approached. At this point, the chemical in the environment is strong enough to represent a significant exposure.

Redrawing the graph, the areas of safe, unsafe, and dangerous can be seen. Additionally, these numbers can be used to identify and isolate zones for work within the incident.

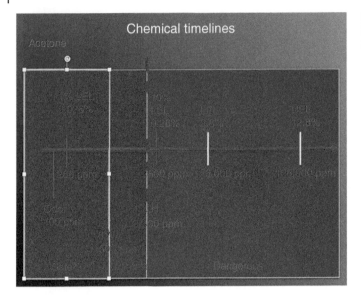

Figure 5.16 The areas of harm can be easily seen by color coding the ChemLine.

Safe as being defined as no chemical in the environment to the Threshold Limit time-weighted average/ceiling. Unsafe being from the TLV-TWA to the IDLH and above the IDLH is the dangerous area (IDLH or 10% of the LEL, whichever comes first, becomes the start of the danger area).

Additional Toxic Effects

So far, several concepts have been discussed to explain the degree of toxic exposure. Each exposure entails certain degrees of toxification. It must be realized that most often, the exposure is to more than one chemical. At a clandestine lab, for example, the exposure would include a number of chemicals, each possessing its own inherent physical and chemical properties. At what level does the chemical in question harm us? Does the chemical become enhanced, or does it cancel out the effects of toxicity? These chain reactions are defined as synergism, antagonism, additive, and potentiation.

Synergism occurs when the combined effects are more severe than the individual chemicals. In this case, there exists a chemical that by itself is moderately toxic; however, in combination with another, the toxic qualities are enhanced $(1+1 = 3)$. For example, cigarette smoke and asbestos each will contribute to lung cancer separately; the risk is high, together the risk is much greater.

Antagonism occurs when the combined effects cancel the effects of each other, decreasing the toxic event. What actually happens is that one of the chemicals acts to decrease the effects of the other chemical $(1+0 = 0.5)$. Example: giving oxygen to a patient that is suffering from carboxyhemoglobin due to carbon monoxide exposure.

Additive occurs when some chemicals that are different in chemical structure (shape and polarity) have the same physiological response in the organism. Thus, the effect is a twofold (or more) enhancement $(1+1 = 2)$. Example: ingesting a sedative with alcohol.

Potentiation occurs when a chemically inactive species acts upon another chemical, which enhances the chemically active one $(1+0 = 2)$. Example: combined effects of multiple exposures.

Toxic Influences

Responders are subjected to a wide range of chemicals from the rescue environment during their course of response duty. These are, but are not limited to, chemicals at the scene of an incident or even naturally found substances in the environment. Occupational exposures, including lifestyle factors, can all contribute to an exposure event.

You may have heard that the physical health of an individual can determine the outcome of a catastrophic disease process. This idea of holistic health is a primary goal in many medical philosophies. One must have a good emotional and physical health status in order to ward off disease. This same idea is also present when it comes to exposure to a chemical. One's state of health, both emotionally and physically, must be intact. It has been known that sequential exposures to multiple events can increase the potential for injury. Combining health issues with mixed chemical exposures can have even more detrimental effects on the responder. There are seven areas that can impact the severity of a chemical exposure. These areas are:

1) General health
2) Diet
3) Previous exposure
4) Age
5) Gender
6) Genetics
7) Sleep

General Health

To a point, people can control the physical condition of the "body temple" with exercise, limited alcohol ingestion, no smoking, and a well-balanced diet. However, the emotional side of our health may not be as controllable within the emergency response field. It can be hard to maintain or control the emotional ups and downs everyone experiences in life. Not to mention the type of emergencies that a responder is involved in. Yet, several studies have indicated that the emotional well-being of an individual can greatly influence the state of health of the responder. Additionally, with the growing concern over posttraumatic stress disorder (PTSD) within the emergency services, maintenance of general health may become harder for an individual to sustain.

The medical community is also not sure why some individuals are more susceptible to a chemical (hypersensitivity), and others are less susceptible (hyposensitivity). There have been several experiments wherein rats were used as the test animal against organophosphates, phosphorous-containing pesticides that overstimulate the synaptic response. It was found during these experiments that, at times, the female is more sensitive to the chemical than the male.

During the research for this book, several studies that looked at human exposure to organophosphates identified that the female can be more sensitive to the acetylcholinesterase inhibitors than a male. The male was identified as being hyposensitive to this category of poisons. However, research further identified that males are more sensitive than females to the chemicals that affect the liver. Within this same experiment, the affected systematic (average) exposure for the female was not appreciably affected. In other words, on average, the female did not show a sensitivity to the chemical.

Diet

It is easily seen that part of this concept is diet-dependent. From the discussion on metabolism, carbohydrates were identified as providing the fuel to produce energy. Beyond this, the body will find other substances, for example, fats and proteins, in order to maintain the energy output required of the body. This same concept of diet having an influence on toxicity has been demonstrated in laboratory animals. During the performance of toxicity testing in several documented cases, the caloric qualities were limited prior to the exposure. This change in diet alone made drastic changes in the outcome of the experiments.

During metabolism, energy is produced for the body to complete many tasks that it must perform. These tasks include interior functions such as digestion and breathing and outside tasks such as carrying heavy objects or running. The energy is dependent on the foods placed in the body. It is easy to see that the diet consumed by a person has a direct effect on one's physical health.

The other aspect of physical health is the current state of health a person maintains. Take, for example, a healthy individual applying additional table salt to food. This would probably not trigger any dramatic effect in this person. On the other hand, giving the same amount of salt to a person with reduced cardiac output because of a previous infarction could be devastating. This particular victim could suffer from pulmonary edema, increased blood pressure, and an additional workload on the kidneys, which, in effect, could be interpreted as a toxic event. It is reasonable to expect less of a toxic response from a healthy individual than would be seen with an unhealthy or compromised one.

When laboratory animals are tested for chemical toxicity, one factor that is in the experiments is the animal's diet. This concept of metabolism versus diet is further enhanced when experiments on cancer-causing agents are performed. In one experiment, two sets of animals were fed two different diets, one a nutritionally correct diet and one that was less than nutritious but tasted good (we don't know how they established that the food tasted good to a bunch of rats). The nutrient-enriched diet animals produced less than the expected response (cancer), while the nutrient-starved animals produced a greater frequency of tumors as an associated disease response.

Previous Exposure

In some individuals, a single previous exposure can potentiate the effects of later chemical exposure. In recent years there have been many exposures that correspond to this single event exposure and create a hypersensitive state. In addition, it is hypothesized that the genetic makeup of the individual may increase the occurrence of this single-dose response.

One group of individuals that seem to form hypersensitivity are those that have had repetitive small dose encounters with the organophosphates and carbamates (synthetic organic insecticides). This is especially true if the same individuals were exposed to DDT (an organochloride) in the early 1950s to the middle 1960s.

A chronic exposure to a chemical may also produce an "adaptation" to the chemical, and a hyposensitive state exists. This case is rare for most chemicals yet is often seen in low-level dose exposure over very long periods of time. Examples of this would be alcoholics (low level to ethyl alcohol) and COPD (low-level exposure to carbon dioxide/monoxide). The time element between each event is long enough for the chemical to detoxify, and a possible genetic rearrangement has taken place.

Age

Usually, when we refer to age, there is an assumption that the discussion is about middle-aged individuals or those who are in an advanced stage of life. For the most part, there exists very little information on these two-age groups. Most of the information pertains to the newborn and children. This is not to say the geriatric patient may not suffer under the same circumstances, but there is not supportive documentation to see what happens to the advanced aged individuals.

When trauma is taught in EMT and paramedic school, it is taught that two high-risk groups exist when an injury occurs: the young and the old. The reason for this statement is related to the inability of the young and old to compensate during shock.

This same idea holds true for poisonings in that the young do not have the body mass to compensate for the ingestion of a chemical. The older individual may be unable to compensate, possibly due to a disease process that is already present or other drugs that they take for other illnesses. The elderly, for the most part, do not have the health status conducive to ward off the effects of exposure. This is related to their overall health and dietary status. In either case, the chemical can invade the body and start to manipulate metabolism easier in these high-risk individuals than it can in a generally healthy middle age population.

It has been seen in the laboratory that newborn rats are extremely sensitive to organophosphates, but the effect of exposure to DDT was non-discernible. However, with the adult rat, the effects of both chemicals were apparent. This shows that the concept that young individuals are more sensitive to a chemical than an adult is not an honest statement.

Take, for example, an elderly individual who has angina. The angina is controlled through the use of nitroglycerin PRN. Other than an occasional bout with chest pain, this individual leads an active "normal" life. One day this person runs the car in the garage to warm it up before heading to the store on a cold winter day. The garage is left closed to keep out the cold, and a dangerous level of carbon monoxide builds up. Within just a few breaths, the individual starts to experience crushing chest pain that is not relieved with nitroglycerin. The reduced oxygen-carrying capacity of his blood has led to a serious event and may cause myocardial ischemia. If a young, healthy individual had experienced the same levels of carbon monoxide poisoning, the outcome would be much less severe and outcome more positive.

Gender

We can easily see that there are physical and physiological differences between a male and a female. In the past, the male was predominant in the workforce; therefore, most exposures occurred to the male. In more recent history, females have entered the same workforce exposing them to the same consequences as that of the male. The impact of this has not yet been fully studied, yet it could have a profound influence on genetic changes caused by exposure.

Today, women are working around the same chemicals as men but, the major difference is that women are capable of becoming pregnant. In some cases, women are working as long as they can during pregnancy. This exposure of chemicals to the fetus during the developmental first trimester is an area that has had limited research. The ramifications and ethical implications of this issue are well beyond the scope of this text. Nevertheless, the influence on reproductive functions, chemical exposure, and genetic modifications are all factors that influence the toxic event.

Enzyme production, lung capacities, and muscle mass to body frame ratios can all influence the toxic event. For example, consider lung capacity. In the male, the average air volume movement is 0.4–$0.5\,m^3/h$; with the female, lung volume movement is an average of 0.3–$0.4\,m^3/h$. It might stand to reason that a male would experience greater effect from a respiratory exposure than a female.

Bodyweight, on average, is less in the female as compared to the male. Heart rate and enzyme production is different between males and females. Fat stores are higher in females then in males. All may influence differing effects of exposure between a male and female. There is a difference in dose–response between the male and female, especially when dealing with chemicals that affect the liver as the target organ. Some research has implied that this is because males are more apt to have that "single" beer with the guys after work. Repetitive consumption of alcohol may depress the liver's ability to create the enzymes needed to detoxify chemicals. It will be interesting to see if females in the future become just as susceptible to these chemicals as males. Now that females are in the workforce, they too are going out with the gals for that beer. Could this also have an effect on these individuals? Only time will tell.

Genetics

The subject of chemicals causing mutations on genetic material is beyond the scope of this text, but a simple discussion of this subject is warranted in order to give the reader an overall picture of this issue. We do know that some chemicals can affect and change the DNA within cells. In the field of toxicology, these rearrangements and chromosomal breaks have been studied. Ethylene oxide, hydrazine, heavy metals, and ionizing radiation all have histories of causing genetic mutations. These mutations could have a role in causing cancers and birth defects as noted in some lab experiments.

The fire service provides a means of looking at this subject in a more practical and realistic manner. There have been many studies in an attempt to determine the exact cause of the much higher occurrences of cancer in firefighters. In some cancer categories, the occurrences in firefighters are twice as high as the general population. Many of the studies focus on the products of combustion. During a fire many chemicals and chemical compounds are released exposing the firefighter and their personal protective gear to these unknown hazards created by smoke. In the past firefighting gear was used repetitively without being washed. When the firefighter would don the contaminated gear and begin to sweat, the exposure would continue over and over again.

In many cases this contaminated gear was placed next to the firefighter's bed in the fire station and carried home in the back seat of the family car, not only exposing the firefighter but their families as well.

The result of these studies has caused these old practices to change. The development of NFPA 1500 and 1581 has guided fire departments from around the country to develop policies requiring firefighter gear to be frequently washed and stored in a fashion that reduces or prevents these chronic exposures that have led to these increased cancer rate. On scene decontamination after fighting a fire, clean cab efforts (keeping contaminated fire gear out of the passenger space in a fire truck), and detailed medical surveillance programs have all been put in place to reduce the prevalence of cancers. (See Chapter 3, Associated Toxic Conditions, Closed Space Fires).

Sleep

There have been several studies indicating the effects of sleep on toxic events. The lack of sleep has been identified as a factor to produce a greater negative effect during a toxic event. What many of the studies have found is that, as a person becomes tired and fatigued, the enzymes needed to detoxify the chemical may not be produced at a high enough rate. Other studies found that tired individuals do not exercise safety precautions resulting in an exposure. In either case, lack of sleep can affect toxic exposures.

There is still much to learn about these seven factors. Humans are diverse and chemical exposures affect different people in different ways. It is difficult to understand why a chemical exposure can have devastating effects on one person and not have any effects on another. These situations remain a mystery and still require much more study.

For this reason, the concept of medical surveillance to monitor the health of emergency responders and those who are in constant contact with chemicals was established. This one concept is important in terms of prevention and treatment. It may be years and thousands of exposures before there is a true understanding of the effects of chemicals within the human organism. In the meantime, engineering and administrative controls must be employed to reduce or remove exposures.

Biochemistry

The liver, kidneys, and lungs are involved with the detoxification of chemicals. There are two basic scenarios that can occur that can lead to injury or toxic exposure. The first is a non-toxic substance that gains access into the body and, during the detoxification process, is converted into a toxic substance. The second situation involves a toxic substance gaining access and during the detoxification process, the chemical is converted into a different toxic substance.

Liver metabolism is a complex and detailed process. In general, the liver, bile, gastrointestinal tract, kidneys, and blood all participate in the detoxification and elimination of toxic materials and metabolites from the body.

Chemical compounds that are ingested are absorbed into the blood stream from the small intestine and are transported directly to the liver. Once in the liver, chemicals that are unwanted by the body are metabolized or chemically converted. This metabolic process may result in several outcomes when metabolizing a toxic substance. It may metabolize the substance into a nontoxic substance, a more toxic substance, a less toxic substance or, it may not engage in a reaction at all. In all cases, the metabolite being produced either enters the general circulation for filtration or placed in the bile for elimination.

There are two types of toxic substances; one that enters the body as a toxin and one that is converted into a toxin during the metabolic process (protoxic). Chemicals also have the ability to enhance, cancel, support, and/or give an additive response to toxicity. This occurs in the liver during Phase I and Phase II reactions.

Detoxification

Phase I and Phase II Reactions

In discussing a Phase I reaction, an assortment of outcomes can be created out of a single compound because of the other available compounds that will combine with it. These possibilities are multiplied once a chemical enters the body, not only because of the chemical conversions that take place but also because of the presence of other chemicals.

Metabolism of a chemicals takes place throughout the body. Detoxification, which is a form of metabolism takes place most often in the liver but, to some degree, takes place in other body compartments even before it reaches systemic circulation. This pre-liver detoxification is referred to as first-pass detoxification. In fact, in many ways, this process if very efficient. For example, in drug administration, it is the first pass metabolism that necessitates the need to administer a drug intravenously to bypass the other compartments and delay entry into the liver.

When referring to detoxification it is actually a biochemical process that chemicals go through as they move through the liver. Specifically, Cytochrome P450 system or what is referred to as CYP 450*. Figure 5.17 shows how the naming process takes place for these enzymes.

CYP enzymes are found in all animals. Humans have about 60 different CYPs that are specific for different types of toxins. This class of enzymes is called the oxidases, which are abundant within the liver. These enzymes introduce polar groups to increase solubility.

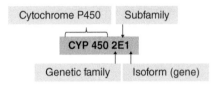

Figure 5.17 Naming sequence for cytochrome P450.

As an example, CYP 450 2E1 is the microsomal cytochrome responsible for the metabolism of alcohol. When the concentration of alcohol in your liver is low, CYP 450 2E1 oxidizes about 10% of the ethanol, but when the concentrations of alcohol increase, so do the concentration of the cytochrome. The other 90% is metabolized by a non-CYP 450 enzyme called Alcohol dehydrogenase, which breaks down alcohols within the system and forms acetaldehyde which is further broken down into acetate.

In Phase I detoxification a chemical enters the liver and is polarized through oxidation, reduction, and hydrolysis reaction. The chemical is essentially attached to a polar group creating a polar chemical that can be removed though the blood circulatory system where it is filtered out by the kidneys and eliminated in the urine. Some chemicals only go through Phase I and do not need to progress into Phase II detoxification.

So, what happens in Phase II detoxification? If the liver is unsuccessful in making the chemical water soluble in Phase I, then the chemical, which is lipid soluble, is sent to Phase II detoxification. In Phase II, the liver uses transferases that place small polar molecules onto the chemical in order to make it more water-soluble. These make the conjugation reactions from internally derived polar compounds. In order to do this, the liver uses a transferase enzyme called *N*-acetyltransferase (as an example), which attaches an acetyl group making it more polar or hydrophilic.

Tylenol is an excellent example. Tylenol is moved from Phase I with very little biochemical movement but metabolized within Phase II metabolism that forms a conjugate to be eliminated by the kidneys. The compound then becomes very polar because there are now more polar attachments. The result is a hydrophilic compound that is placed in the blood stream to be filtered out by the kidneys or place in the bile for elimination through the feces.

*Note: *The CYP 450 is a cytochrome that is heme-containing, which will bond with oxygen. More importantly, the heme enzyme bonds with carbon monoxide that, when measured with light, absorbs at 450 nm, and hence the cytochrome that absorbs at 450 is CYP 450.*

For the most part, the Phase I reaction takes the highly active polar chemicals and converts them into lipophilic toxic chemicals (in some reactions). Once this chemical becomes modified, in Phase I it becomes a water-soluble compound. However, this process is relative to the solubility of the original chemical compound. In other words, if the compound is slightly soluble, then the product will, more than likely, become more water-soluble. On the other hand, if the compound is not at all soluble in water, although the compound may have water-soluble qualities, it may not be as soluble as is necessary for the excretory process to handle (the conjugate does not produce a sufficient degree of polarity). If the substance becomes not as soluble as it should be for the kidneys to take over, then an "active site" is placed on the compound so that the substance may be eliminated through the kidneys. This attachment of an active site is termed the Phase II reaction. Figure 5.18 shows the effects of Phase I and Phase II reactions.

Remember that these reactions are the body's attempt to detoxify an intruding chemical. In some cases, the chemical intermediate is a highly toxic compound, one that can actively destroy cells and tissue. In other cases, the cells are hindered from accepting oxygen and

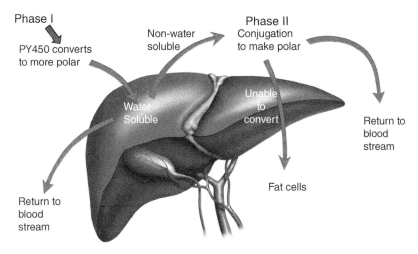

Phase I

PY450 converts to more polar

Non-water soluble

Phase II Conjugation to make polar

Water Soluble

Unable to convert

Return to blood stream

Return to blood stream

Fat cells

Figure 5.18 Liver metabolism is one form of detoxification and represents a complex chemical-physiological process.

nutrients or from releasing carbon dioxide. The intermediates may have a synergistic, additive, potentiate, or antagonistic quality.

Most antidotal treatments are based on the organism's ability to excrete the end product. There are limited amounts of antidotes as compared to the wide variety of chemicals that one may become exposed to. Antidotes that are not discussed in this book but that require mentioning are the chelating agents and chemicals that are used to combine with the toxic complex for biochemical elimination.

Chelating agents are used to form a chemical compound between the insulting chemical and the chelating agent. After infusion of the chelating agent and subsequent bonding, the compound is either eliminated through the process of natural kidney filtration or removed through dialysis.

For an introduced substance to be excreted from the body via the kidney, the substance must be water-soluble/polar. Remember, polarity determines the solubility of a substance. With some chemicals, the body "looks" for a way to convert the nonpolar (thus non-soluble or lipid soluble) substance into a polar molecule. This must occur in order for the excretory process to handle the metabolite. Unfortunately, the body may take a nontoxic, nonpolar substance and convert it to a polar substance for excretion. During this process, a toxic polar intermediate may be formed that can damage tissue.

For the most part, these reactions occur at the enzyme protein level. Actually, during metabolism, the alteration of these chemicals (enzymes) may interrupt the body's ability to recognize particular unwanted chemicals. If this cycle of reactions continues (repeated exposure), the body will be unable to return to the pre-exposure state. Suppose the exposure to the chemical is maintained at a minimum; specific enzymes within the body act on the toxic metabolite. The chemical reactions continue to take place until the unwanted substance is excreted from the body.

It should be remembered that enzymes are chemically specific compounds; that is, they have very precise chemical and physical properties, which give them the ability to carry out the very precise, specific roles for which they were intended. Enzymes are responsible for changing nonpolar and nontoxic substances into the desired substances. If, for example, the body may produce an enzyme to counteract a particular substance. If the intermediate or the original chemical itself hinders the production of this particular enzyme, a toxic state is produced.

Detoxification by the Lungs

The lungs may also play a major role in the detoxification process by removing carbon dioxide along with other products via the respiratory pathway. As we saw in normal metabolism, carbon dioxide is produced from the combustion of oxygen and fuel to create energy, water, carbon dioxide (waste), and heat. It is the lungs' responsibility to eliminate the carbon dioxide while providing oxygen needed by the cells in order for metabolism cycles to take place.

Chemical Toxic Qualities

The chemical qualities of a compound influence the toxic character that a particular compound possesses. It is the shape of a molecule that the body recognizes. This recognition places the molecule into the appropriate chemical reaction. Remember that activation energies that drive reactions are like a hill or a mountain. The lower the required energy needed, the faster and more probable that the reaction will continue. The opposite is also true; the higher the activation energy, the lower the probability of the reaction to finish.

This arrangement in space identifies one structure as polar and another as nonpolar (remember, polarity is another way of saying how soluble or hydrophilic a substance may be). As an example, polar substances are soluble in water (thus excreted out through the kidneys), whereas nonpolar substances are usually soluble in fat. This provides yet another problem when dealing with toxic chemicals. A chemical that is soluble in fat may, through the detoxification process, become stored within the adipose tissue, where it may be released in small quantities over and over again whenever the fat is utilized in a metabolic cycle.

This concept explains why an individual can ingest a poison and not suffer any discernible effect, yet sometime later, during the process of losing weight, may become affected. LSD is the perfect example of this. Here the drug goes through Phase I and Phase II, unable to make it polar, and the drug is sent to the fat cells. The individual revisits his/her "trips" when fat is burned through metabolism, releasing the drug again into the system.

The shape of a molecule may also lead to toxicity within the body. Enzymes utilize their shape to bond at receptor sites. Some chemicals can, through metabolism or by their own chemical shape, "look" like the enzyme and thus occupy the receptor sites meant for certain enzymes. This receptor site bonding can lead to both acute and chronic symptoms.

So far, several problems associated with chemical exposure as it relates to the detoxification process have been discussed. Many processes contribute to the toxic effects of a chemical, but four factors determine the overall response of chemical exposure:

1) The amount and concentration of the chemical.
2) Rate of absorption, which has to do with the shape and polarity of the chemical (absorption and distribution).
3) Rate of detoxification, which is dependent on the organism's metabolism.
4) Rate of excretion, which is conditional to the end result of the metabolic pathway.

Interrupt one or more of these four factors, and toxicity can be lowered or eliminated. For example, if the toxic limits (see earlier in this chapter, Levels of Concern, Chemical Timelines) of a chemical are known, and efforts are made to limit the exposure to below these limits then the amount and concentration of the chemical has been controlled and the toxicity of the chemical has been significantly reduced or eliminated.

This can be accomplished by wearing chemical protection that includes a skin barrier and respiratory barrier that is appropriate for the chemical to limit or remove the level of absorption that may take place (i.e. PPE). We know that detoxification is primarily accomplished in the liver. When administering antidotes following a chemical exposure, this is essentially providing the body an opportunity to metabolize or help metabolize the chemical in question. This helps with excreting the chemical through the renal pathways.

Chemical Excretion

There are two main ways the body can excrete chemicals: (i) The hepatobiliary system and (ii) kidneys. The hepatobiliary system involves the liver excreting the modified chemical from Phase II detoxification into the bile. Bile is then released into the intestinal tract with it is eliminated though the feces. Many drugs are excreted through this pathway.

In the blood, there may be a concentration of the chemical, which in turn the nephron within the kidney filter and excrete through the urine. Remember, the liver, through the processes of Phase I and II detoxification, has made this chemical hydrophilic.

Many believe, and there is mounting evidence this is the reason why we see bladder and kidney cancers with the emergency response services. Here the concentration within the blood is recognized by the kidney, and a certain degree of concentrated chemical is now held within the bladder. Within the excretion process, there is a certain rate at which urine is being produced, allowing the kidney and the bladder to be exposed to the chemical in question or the metabolite of the chemical for an extended period of time. During low urine production especially in a dehydrated state, the longer the chemical stays within the system and allowed to create more damage. During high urine production there is short contact time within the kidney and bladder reducing the chances of tissue damage.

Nanotoxicology

In recent years, there has been a growing concern about the toxicity of nanoparticles, leading to the research and development of a new branch of toxicology, which is the study of nanomaterials on humans and biological systems. Nanoparticles can be defined as materials that have less than 100 nm to their dimension. In the fire service, these extremely small particles are present within carbon soot present in smoke and is pushed out of small openings of a building that is on fire.

The use of nanoparticles in various coatings have increased since their inception in the 1990s. They are seen in crack-resistant paints, coatings for solar cells to increase efficiency, scratch-resistant eyeglasses, and stain-resistant fabrics are but a few of the products that are used on a daily basis. These products improve our modern life but also have unique properties making these chemical particles hazardous.

Due to their size, the volume to size ratio gives these products a large surface area and are found in a variety of shapes (rods, tubes, fibers), all of which have an impact on biological systems. Exposure within the manufacturing process is generally controlled, but as more products are developed for this branch of science, unwanted exposures are bound to occur.

As mentioned earlier in the text, "the dose makes the poison" is a traditional thought within toxicology; however, this needs to be modified because of the shape configurations, size, surface areas, and surface chemistry that these particles have shown.

Take size as one factor; these materials are microscopic and can influence cellular uptake (how these particles move into the cell). They can then influence cellular processes, distribution process, and elimination of the chemicals from the body. Because of this, great care needs to be used as these particles can enter biological components and reach areas sensitive to biological processes.

Examples of this are superparamagnetic iron oxide nanoparticles (SPION's) ,which are commonly used as contrast agents in magnetic resonance imaging, titanate nanotubes (TiONt's) as a versatile platform to deliver optical imaging probes, advanced nuclear imaging, and a combination of therapeutic molecules (antimicrobial as one example).

These particles behave in ways that are not fully understood. It has not been well defined how these particles interact with biological body components. Nanomaterials have demonstrated the ability to institute a significant change to the biological systems. In some cases, the forming of proteins have been influenced by the incorporation of nanoparticles. As an example, these particles can acquire a "corona" of particles that have the tendency to aggregate or agglomerate according to their size, density, shape, and surface chemistry. The cell sees these particles, which create these layers, and responds differently than previously thought. The degradability of these materials for acute and long-term toxicology is also under study.

Traditionally, the concept in toxicology has been given a certain amount (dose) all materials can become averse to a biological system (poison). With nanotechnology, this definition needs to be expanded because of the unexpected and unusual toxicology these particles have. So, the new description may be that the dose that makes the poison should really be the life cycle of a material that makes the poison.

Determining the Level of Medical Surveillance

There is a lot to the toxicology of chemicals and the exposures that can occur at the scene of an event. As you can see, the pathways that a chemical can take are many, and all too often, the effects from a chemical exposure occurs well after the fact. Often, it is difficult to define the exact exposure that resulted in a long-illness or injury.

It is for this reason that an active and on-going medical surveillance program be instituted for existing hazardous materials teams and during the development of a new hazardous materials response team. The agency, along with the medical director, should formulate the level of surveillance that should occur identifying all toxicological considerations. In the medical surveillance chapter, a medical program is identified and special considerations of that program discussed.

Risk Assessment and Detection

This chapter represents the background needed to provide a detailed risk and hazard assessment. These concepts are already found within OSHA documents for scene safety planning and incident action plans but, here the science behind these safety components is described for greater understanding. Evaluating a scene using what is known about the chemical and toxicological principles in mind and from a hazard management perspective is critical to risk assessment.

The method presented in this book is one that looks at the risk assessment as a process utilizing a routine method to evaluate the risk and continuously manage it, through an operational process. It is a method by which the hazard is identified and controlled to conserve resources during mitigation. This includes:

1) Identify the hazards
2) Assess hazards to determine the risks
3) Develop controls to manage the risks
4) Implement controls
5) Supervise and evaluate the process

Identification of Hazards

A hazard can be a condition that surrounds the incident or potentially interfere with the operation. The hazard may be actual or potential and include the entire operational area and create additional danger for the responder. These sources of danger can be identified either through the use of detection equipment and/or through simple observation of the incident. Observation of the incident involves evaluating such clues as occupancy and location, placards and markings, container shapes and sizes, and/or evaluating facility paperwork, bill of lading, or the consists from a train.

Evaluation will also include the location or coordinates (the 3D space) of the incident. The incident located on the side of a hill or mountain presents different dangers that one located on a flat roadway. An incident located in a heavily populated area must be managed differently than one in a rural area. Weather is another observable hazard that should be evaluated. Air temperature, wind, and humidity may all have an impact.

Evaluating all of these factors that can complicate the scene, increase the danger, and increase the overall hazards is important when determining the human and physical resources needed for mitigation.

Assess Hazards to Determine the Risks

As the level of danger and hazard is realized, the risk that is identified should become the focus of hazard reduction and risk management. Developing a ChemLine for the products involved helps to identify the levels of the chemical that are considered safe, unsafe, and dangerous. This allows for immediate action to be planned to reduce the most severe hazards.

If a container is involved, Benner's HazMat Behavior model can be applied to evaluate what event may be proceeding. Where on the spectrum of hazard risk does this chemical or chemical family typically fall given the current weather, the location and placement of the event, and the time of day.

Develop Controls to Manage the Risks

Controls come into three general categories and are dependent on the human resources available and equipment available on scene. These control categories are commonly referred to as defensive, offensive, or non-intervention. Each has its own set of considerations and can present management challenges. However, in general, the controls are to move toward mitigation in a safe and decisive manner.

Implementing Controls

Incident command, along with safety, must ensure that the controls used to manage the hazard are well thought out and employ the best options possible. They must be safe and effectively used in an effort to move forward in controlling the incident.

For example, if bonding and grounding were needed to remove the hazard of the creation of a static shock while transferring a flammable substance, the function been carried out properly to not cause an additional hazard in the performance of this effort. If the identified hazard was toxic or flammable gases inside of a structure and ventilation is the hazard reduction tactic then, all downwind populations must be protected as part of the effort. If ventilating a space was the hazard reduction process, have downwind population been controlled. If additional resources are required beyond the agency's ability, then the hazard reduction effort must be delayed until the resources are available. In an effort to control the hazards and dangers being presented on the scene, there must always be a focus on the safety of the public and the responders operating on the scene.

Supervise and Evaluate the Process

As the incident moves forward, it is important to evaluate the progress. By using the HazMat Spectrum, identified below, along with the ChemLine discussed earlier, an evaluation of the progress can be made. Strong incident management and command and control principles must be in place. The continued evaluation process seems to be one of the hardest steps to consistently follow but is essential to determine if the plan for mitigation is working. Evaluation and re-evaluation are the most important steps to follow to ensure a successful incident mitigation process.

Utilization of the ChemLine model allows the incident commander to identify where in the spectrum of danger (see Figure 5.19 Risk Spectrum) the responders are dealing with. When a high level of danger is involved, it becomes apparent that a higher level of PPE will be involved and more technical equipment will be needed. In addition, additional safety parameters will be put in place to ensure the safety of the entry team. Even with appropriate PPE and safety procedures there always remains the hazard of suit failure or other equipment failure.

When a responder utilizing the ChemLine, they can quickly identify the level of potential exposure that may be encountered. If the numbers are severe to the right, you know that this chemical is an extremely hazardous substance. If toward the left, it may be less hazardous.

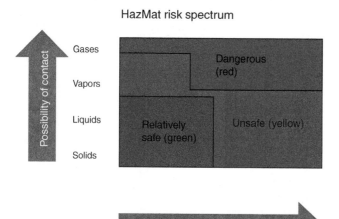

Figure 5.19 Using a hazmat risk spectrum is an efficient means of evaluating hazard-risk assessment. *Source:* Adapted from Dr Kristina Kreutzer PhD, Fire Chef.

- Based on the work by Dr Kristina Kreutzer PhD, Fire Chief

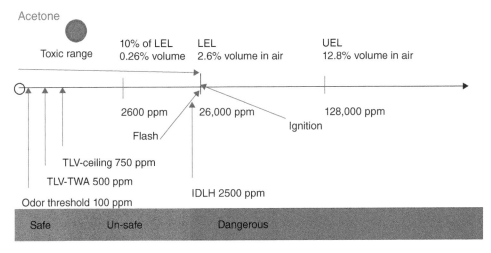

Figure 5.20 Acetone ChemLine for comparison.

However, in both, vapor pressure must be accounted for. As an example, evaluating the ChemLine of Acetone (Figure 5.20) and Hydrogen Cyanide (Figure 5.21) without taking into consideration the vapor pressure, it appears that acetone is more toxic than hydrogen cyanide. But once the vapor pressure is taken into consideration, it become more apparent that hydrogen cyanide is a greater hazard than acetone. Once these two chemicals are evaluated for both vapor pressure and the ChemLine, it can be seen how these chemicals behave.

Here, both evaluated on the HazMat Risk Spectrum which truly identifies the hazards presented by both:

In the risk spectrum, the green region identifies an area that is relatively safe. The yellow region identifies the unsafe area, and the red region signifies the dangerous area with the biggest threat. All this is inherently intuitive once you place these concepts down on paper

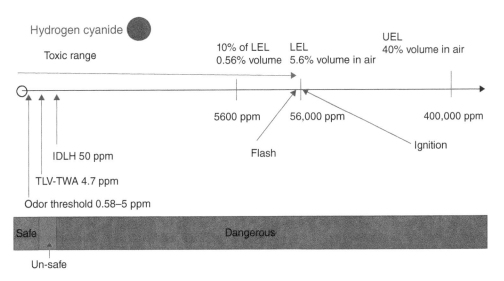

Figure 5.21 Cyanide ChemLine for comparison.

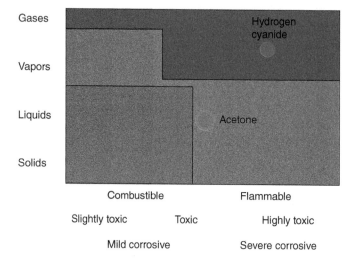

Figure 5.22 By placing both hydrogen cyanide and acetone on the HazMat risk spectrum it is easy to see the danger of both.

and see how they look graphically. Although both models have their pros and cons of application, one can "see" where the chemical in question may be in terms of both flammability and toxicity. (See Figure 5.22.)

Summary

The chemical and physical properties are fairly straightforward when evaluating how these properties affect an injury. However, the toxicology of an exposure is more complex. For example, a chemical with a vapor pressure above 40 mm/Hg will produce more vapors and become a respiratory hazard. This concept is not difficult to understand. But how the chemical affects the body may be more difficult to predict. The physiologic effect from one chemical may be greatly different than another chemical.

The ChemLine allows the responder to place both the chemical properties and the toxicology together on the line to gain an appreciation for how these two concepts work together. Once there is an understanding of how these concepts fit together, a practitioner, can identify the exposure and treat the symptoms of the poisoning in an appropriate and efficient manner.

This book would not be complete without discussing exposure prevention, one of the responsibilities of the HazMat medical responder. A knowledgeable responder can use these skills to identify the possible chemicals on the scene and then apply that knowledge to danger to the responders as well as the symptoms displayed by an exposure victim.

This chapter guides the reader in hazard identification and interpretation. If the hazard is recognized early, the chances of exposures are greatly reduced. But hazard identification also assists the healthcare provider into determining a differential diagnosis, should an exposure occur at a facility. Using these clues, along with the signs and symptoms (toxidrome analysis), the provider can determine a differential diagnosis and begin to identify the appropriate treatment for the patient(s).

6

Team Capabilities

Introduction

The capabilities of any unit or team could be a book in itself. Volumes of literature have been written on everything from PPE, hazard and risk analysis, to response planning, resource management and incident control. The intent here is not to write a work that completes the topic but rather a discussion on these capabilities. This chapter will present team capabilities that can influence the tactical level of a medical operation at the scene of a hazardous materials event. In addition, it represents a highlight of the concepts at the most basic level and compliments the next chapter on HazMat Safety Officer. Together these chapters identify the operational scope and practices used on hazardous materials teams (Figure 6.1).

Case Study – Sodium Nitrate Overdose

At around 10:30 pm a fire department engine and advanced life support rescue were dispatched to a female attempting suicide. The address brought them to a house in an upper-class neighborhood where they were met by the patient's husband. He states that the patient had taken a substance that she purchased off of the internet. The responders were provided a bottle with the label "Sodium Nitrate, 50 gm." He believes that she had taken the substance about an hour before the arrival of the emergency vehicles.

The patient was found in the bedroom. She was unresponsive and frothing from the mouth. The paramedic suctioned the patient and provided an advanced airway and supported ventilations using supplemental oxygen. Following their local protocols, the paramedic gave activated charcoal and transported to a small community hospital emergency room. The vital signs on the scene were:

BP: 78 by palpation,
Pulse: 12 bpm,
EKG: displayed sinus tachycardia with multi focal PVCs
Respirations: were 24 bpm and were deep and clear
Skin: warm and wet
SpO_2: 66%
Capnography: 28 mm/Hg but it was reported to be as low as 6 mm/Hg prior to arriving at the hospital.

Once at the hospital the patient was given 2 prefilled syringed of 8.4% sodium bicarbonate (100 mEq) and an epinephrine IV drip was established to maintain her blood pressure.

Hazardous Materials Medicine: Treating the Chemically Injured Patient, First Edition.
Richard Stilp and Armando Bevelacqua.
© 2023 John Wiley & Sons, Inc. Published 2023 by John Wiley & Sons, Inc.

Figure 6.1 Team capabilities are generally related to funding, staffing, leadership, and training.

By night in ICU the patient became responsive, was signaling for the endotracheal tube to be removed, and the vital signs were reported to be stable.

By morning she had again lost consciousness. Her Glasgow coma score was less than 3, she was suffering from cerebral edema and her pupils were dilated without response to light.

By midday it was determined that she had no brain activity and she was removed from all life support when she subsequently died.

After incident report:

Although in the county where the incident occurred there was an existing "HazMat Alert" policy but, it was never activated. The EMS responders did not interpret this as a hazardous materials incident. Activating a HazMat alert notifies the hospital of a chemical involvement in the incident and also directs the dispatchers to notify the Poison Control Center. Even at the hospital the Poison Control Center was not consulted. The physician and emergency department staff did not understand the physiology of a nitrate poisoning so the methemo-globinemia was never addressed. Interestingly, the neighboring County fire department has a hazardous materials response team and maintains a joint response agreement for hazmat incidents. The county team maintains two hazardous materials medics at all times who are trained to treat nitrite/nitrate poisoning and maintains an inventory of methylene blue used to treat this exposure.

Technician Operational Considerations

Personnel Protective Equipment (PPE)

The use of some level of PPE while at an event will always be part of the mission. Whether the incident requires minimal protection or a level of protection that is required for maximum exposure, some level of protection will be required. The material's chemical and physical properties must be analyzed and a risk assessment completed in order to select the appropriate garment. The determination of the PPE must also be made with the consideration of the physiologic and psychologic impact of the responder wearing the PPE.

Physiological and environmental stressors are the two most overlooked topics of concern when discussing personnel protective equipment. The potential of physiological insults that could arise from hazardous materials incidents is paramount, especially when evaluating the possible amount of time in the protective garment and the environmental conditions, such as heat. For the most part, fire-rescue services do not train for extended times extended periods of time in encapsulating personal protective equipment or heat conditions (see Chapter 2, Environmental Exposures, The Hot Environment for acclimation to heat). This, coupled with the need to maintain hydration, will constitute a most important on-scene concern.

A major relief to psychological stress is familiarization with the use of personnel protective equipment for long periods of time. Acclimation to the equipment is the key to functioning proficiently in equipment with limited communication capability, and decreased dexterity and vision.

Limited visibility, dexterity, and heat stress all add to the frustration and anxiety one may feel when this equipment is donned. Repetitive training becomes a necessity to reduce and manage the stress during the mission.

Heat stress has a large impact on Psychological stress. During normal workloads (not utilizing PPE), one will lose heat through radiation, evaporation, convection, and conduction. However, within a PPE ensemble, this ability to lose the heat generated during work and for long periods of time is severely hindered if not removed completely. In some cases, the symptoms from physiologic stress may mimic those from a chemical exposure making it difficult to distinguish the early effects of a toxic chemical versus the effects of heat stress. These include dizziness, slurred speech, staggered gait, and others. Although heat is always an issue, cold can cause problems as well (see Figure 6.2).

Longer periods of work time within a lower temperature may also have detrimental effects on the responder and may include reduced attention spans and cause an inability to complete

Figure 6.2 Physiological and environmental stressors are the two most overlooked topics of concern when discussing PPE.

detailed assignments. Both hot and cold environments can provide additional challenges to incident commanders and operations chiefs.

In addition to selecting the appropriate level of chemical protection and evaluating concerns wearing the chemical protection over a long term, an Operations Chief or HazMat section must consider the possibility of a flammable atmosphere. The protection that is required at these incident types will vary. Some may require a strong defense against toxic hazards while others need protection from a fire concern. Whatever the threat may be, the responder must use a level of protection to complete the mission.

The world of protective garments has changed over the last ten years. In years past, the material was the basis of protection; however, the interfaces of the gloves, suit, and face shield were not tested, only the base fabric. But, in recent years, much of the testing standards have changed. Now, total ensembles are tested using challenging chemicals and biologicals. However, the old method of PPE nomenclature has remained within the response group lexicon.

When these ensembles were first named, they followed the EPA definitions. Later, most followed the nomenclature found in the OSHA 29 CFR 1910.120, which lists protection that is based upon different levels of engineered safeguards. The nomenclature has been for suits Level A, B, C, D (see Figure 6.3).

Chemical protective suit designation levels

OSHA 29 CFR 1910.120 designation	Chemical protective clothing ensemble	Use conditions	Other
Level A	• Chemical resistant: • Suit • Gloves • Safety shoes • Additional equipment • SCBA/Rebreather • Hard hat • Radio Communications	When the highest level of protection for the skin, eyes, and the respiratory tree is needed, such as: Concentrations of vapor/gas or particulates are considered hazardous to human health.	Flash suit may be placed over this suit for abrasion protection. The suit must be compatible with the materials that are involved
Level B	• Chemical resistant: • Overall or one- or two-piece chemical splash suit or one-piece chemical resistant suit • Gloves • Safety shoes • Additional equipment • SCBA/Rebreather/Supplied air with escape tank • Hard hat • Radio Communications	IDLH environments do not pose a high vapor/gas skin absorption hazard. Include atmospheres that have less than 19.5% oxygen concentrations.	This suit offers the same respiratory protection as level A but less skin protection. This is splash protection against the chemical hazard
Level C	• Chemical resistant: • Overall or one- or two-piece chemical splash suit or one-piece chemical resistant suit • Gloves • Safety shoes • Additional equipment • Air Purifying respirator. • Hard hat • Radio Communications	The level of skin protection is as in Level B. Lower respiratory protection Chemical concentration must not exceed IDLH levels, and the oxygen concentration is at a minimum of 19.5%.	Chemicals that will not insult human health concerns when liquid splashes, or direct contact on any exposure to skin The chemical concentrations/contamination have been identified and measured
Level D	• Coveralls • Gloves • Safety shoes • Safety glasses/goggles/face shield • Hard hat	Minimal skin protection No respiratory protection	This is used when the environment has no known hazard. All functions limit any contact with hazardous levels of chemicals Oxygen concentrations of at least 19.5%

Figure 6.3 OSHA Chemical protective suit level designation.

Although the nomenclature is still used today, the testing and ensemble protection has changed toward a true collective garment, which means that all the components of the suit have good interfaces to maintain levels of engineered protection.

In days past, duct tape® or other adhesive tape has been used to ensure that the interfaces would "match up." This created issues with the protection value of the ensemble. As with any chemical protection system, it is a collection of devices/parts to ensure the responder is protected. By using a component such as duct tape, over the seams of the garment, the responder has now added a non-engineered component into the system. This results in the potential to add a higher degree of hazard to the responder. This may cause cross-contamination. By placing the tape on the suit, there are many unknowns including the rate that a chemical can penetrate the tape as compared to the suit material, or if the tape will maintain the chemical after decontamination furthering the potential of cross-contamination of the responder and decontamination team.

Although this practice is still in use today, it is a very dangerous procedure as no tape has been challenged with the chemical tests, the adhesive can collect the chemical and may transfer during doffing, and the tape adhesive on the chemical protective suit may permit a higher degree of degradation, or permeation.

The terms Level A, Level B, Level C, and Level D are defined throughout hazmat literature as a body garment that is recommended as the level of protection based on the overall hazard. The latest version of NFPA 1990 (2022 edition) has identified these levels using a more descriptive approach to the definition. Here the suits are classified toward performance objectives looking at the entire garment as a complete system, which includes the suit, gloves, footwear, and the respirator.

Although the NFPA description of protective equipment ensembles is a more accurate representation of the level of hazard protection, the old OSHA/EPA descriptions still exist in the hazmat vernacular. So how do these two systems of protection description correlate? Figure 6.4 demonstrates a comparison of NFPA and OSHA definitions.

Rehabilitation

In either hot or cold environments, rehabilitation of the personnel operating at the incident is required in order to maintain the physical and mental conditions of the emergency crews. This is to ensure that any strenuous physical activity associated with environmental exposure does not cause injuries to the emergency workers.

When establishing a rehabilitation area on an emergency scene, it should be in a location away from the incident where rescuers can rest, ingest refreshments and/or hydration drinks that are appropriate to maintain hydration and mental attentiveness. Additionally, members that have been relocated to the rehabilitation area should be medically observed and a post assignment medical evaluation performed. Each rest period after strenuous activity should be no less than 60 minutes between assignments, and if the activity is not strenuous, then the rest period can be a minimum of 30 minutes. The responder's physical condition should also be taken into consideration, and the time in rehab may be extended based on the paramedic's evaluation.

The paramedic's responsibilities should include medical evaluation, monitoring, food support and hydration, and the evaluation of changing environmental conditions of the incident (see Chapter 7 for more details).

Comparison of chemical protective clothing best practices (NFPA 1990) and regulations (US OSHA/EPA)

NFPA 1990 Section	OSHA/EPA Level	Respirator	Vapor Threat	Liquid Threat	Considerations
NFPA 1991	A	NIOSH CBRN SCBA	Ultrahigh (> 10,000 ppm)	High	Designed for use against ultrahigh concentrations of vapors and aerosols indicative of confinement where ventilation is not possible
NFPA 1994 Class 1	A	NIOSH CBRN SCBA	High(> IDLH, tested at 10,000 ppm)	High	Designed for use against high concentrations of vapors and aerosols indicative of areas where ventilation is possible and therefore ultrahigh concentrations are not reached.
NFPA 1994 Class 2/2R	B	NIOSH CBRN SCBA	Moderate (> IDLH, tested at 350 ppm)	High	Designed for use against moderate concentrations of vapors and aerosols or vapor-forming liquids. Ruggedized variants are available for areas where
NFPA 1992	B	NIOSH CBRN SCBA	N/A	High	Designed for use against high concentrations of liquids that are not vapor-forming. NFPA 1992 suits are for use in liquid splash only. Use care when utilizing in situations where neutralization reactions are taking place as the heat of the reaction will increase the vapor pressure of the material thereby producing vapors.
NFPA 1994 Class 3/3R	B	NIOSH CBRN SCBA	Low (< IDLH, tested at 40 ppm)	Moderate	Designed for use against low concentrations of vapors and aerosols and moderate liquid splash risk. Generally used for operations where direct contact with threat material is not intended, such as decontamination. outer perimeter monitoring, evidence collection, and others.
NFPA 1994 Class 4/4R	B	NIOSH CBRN SCBA	N/A	Low	Designed for use against particulate threats followed by decontamination (low liquid threat).
NFPA 1994 Class 5	B	NIOSH CBRN SCBA	N/A	Liquid repellency	Designed for use with gases and vapor forming liquids are primarily flammability hazards and are not dermal hazards at the levels of consideration.
NFPA 1992	C	NIOSH CBRN APR/PAPR	N/A	High	See NFPA 1992 above.
NFPA 1994 Class 3/3R	C	NIOSH CBRN APR/PAPR	Low (< IDLH, tested at 40 ppm)	Moderate	See NFPA 1994 Class 3/3R above.
NFPA 1994 Class 4/4R	C	NIOSH CBRN APR/PAPR	N/A	Low	See NFPA 1994 Class 4/4R above.
NFPA 1994 Class 5	C	NIOSH CBRN APR/PAPR	N/A	Liquid repellency	See Class 5 above.
Work Uniform	D	None	N/A		

Figure 6.4 Chemical suit comparison between NFPA 1991, 1992, and 1994.

Decontamination

As a general concept, decontamination is a process, which is used to protect the victims of the event, the responders, their equipment, the community, and the environment as a whole. OSHA, NFPA, and EPA all have developed rules and regulations pertaining to the subject of decontamination.

Occupational Safety and Health Administration

Occupational Safety and Health Administration (OSHA) identifies decontamination as the removal of hazardous substances from employees and their equipment to the extent necessary to preclude foreseeable health effects. OSHA's focus is predominately the health and safety of the worker, his/her equipment, and how this may affect the environment, i.e. runoff.

National Fire Protection Association

National Fire Protection Association (NFPA) defines decontamination as the chemical and/or physical process that reduces or prevents the spread of contamination from persons and equipment to protect other responders, the public, and equipment.

Environmental Protection Agency

Environmental Protection Agency (EPA) identifies decontamination as the removal of hazardous substances from the environment to the extent necessary to preclude foreseeable effects to biological processes and systems.

For practical purposes, this book defines decontamination as; the physical and/or chemical process of removing a hazardous substance from individuals (responders and civilians) and their equipment to the extent necessary to prevent the spread of contamination and preclude the occurrence of foreseeable adverse health effects to biological systems including the environment.

The main focus for this definition is to prevent the occurrence of cross-contamination between persons, persons and equipment, and the environment. Strict decontamination procedures prevent the contaminants from spreading beyond the initial area and thus causing a secondary contamination and exposure issue.

Decontamination should not be thought of as adherence to federal regulation, law, or rule but rather a system of prevention. The intent is to remove the contamination and to return the people, tools/equipment, and the environment to the level of cleanliness that it was before the incident.

Science Behind Decontamination

The discussion of decontamination cannot progress into tactical goals until the science of decontamination is considered. It is the chemical and physical properties of compounds and materials that come into play when the method and rationale of decontamination are being evaluated concerning a particular substance. Additionally, the biological processes that can occur from exposure to particular chemicals (thus their properties) must be understood and considered. This becomes crucial when emergency medical care is a part of the overall strategic goal. As an example, many documents state that isopropyl alcohol can be used as a decontamination solution. Let's take a moment and analyze this process in regards to the impact medically, isopropyl alcohol could have when used to decontaminate a person.

Isopropyl alcohol mixes well with most chemicals, especially polar and near-polar substances. However, it also has a higher vapor pressure, which means that it will evaporate very quickly and is also absorbed through the skin at high rates. These conditions create several issues: First, due to the high evaporation rate, the use of isopropyl alcohol on the skin creates the potential to significantly lower body temperatures. Second, it is a hydrocarbon that has the potential to affect both the central nervous system and the irritability of the heart. And third, because of its ability to absorb through the skin quickly, isopropyl alcohol can actually serve as a carrying agent to increase the chemical absorption of other chemicals on the skin (see Chapter 3).

Although some might not believe that a discussion on the principles of chemistry and topics of biology is important when discussing decontamination, it is these laws of nature that guide responders through the logical process of decontamination. After all, decontamination is the first step toward patient management and care for those who are contaminated.

Many of the chemicals that are encountered within the emergency response to hazardous materials are organic substances. Unfortunately, these chemicals present additional issues involving decontamination. For example, these specific compounds are, for the most part, non-soluble.

Polarity is the reason for a compound to have the capability to be soluble. If a substance is polar or has a tendency to have polar qualities, then the compound will also have soluble qualities. On the other hand, if a compound is nonpolar and the tendencies of this compound is to maintain nonpolar qualities, then the result is a non-soluble compound.

Note: The ability of a chemical to be polar is a system of forces that exist within a chemical complex. In general, we have four systems that add to the chemicals ability of polarity in pure materials: dispersion (van der Waals), dipole–dipole, hydrogen bonding, and ionic bonding. These collectively are called intermolecular forces.

The dispersion (van der Waals) force is a result of electrons moving about the compound creating a momentary positive and negative end, these interact to combine atoms. These forces occur in all materials are stronger or weaker depending on such things as mass. When we call a compound nonpolar this is the only force that the compound has.

With dipole–dipole force is when the electrons within the molecule produce a negative and positive end but are not temporary are permanent ends that create what is similar to a tiny magnet.

The hydrogen bond is similar to the dipole–dipole is extremely strong as compared with the previous forces and occurs when hydrogen is bonded to nitrogen or oxygen (can occur in ammonia and hydrogen fluoride).

Ionic bonding is the intermolecular force (the force that exists between the molecules) and the intramolecular force (the force that holds atoms together within a molecule) are the same. The intermolecular force holds the atoms together as an actual chemical bond. Intermolecular forces affect the phase of matter, boiling points, and the vapor pressures of chemicals, along with solubility and miscibility.

In an effort to remove substances that are non-water soluble, another substance must be added to water. The use of an emulsifier (such as detergent) is a way that can be used to make non-water-soluble substances, somewhat "miscible," or "soluble." Emulsifiers do not change the substance chemically; rather, they affect the specific gravity of the substance, thus giving the appearance of dissolving in water.

Emulsifiers do this by a process that is similar to adsorption. For example, soap, which is an emulsifier, forms around the non-water-soluble chemical allowing it to be washed away. This occurs through an attachment of hydrophobic hydrocarbon chains directed towards the center of a sphere formed by the emulsifier. In the middle of the sphere is the nonpolar, non-soluble chemical. The surface of the sphere contains polar anions. In polar media such as water, the sphere is removed by water in what appears to be a soluble mixture.

Depending on the chemical properties of a hazardous substance, the compound may be distributed over various components of the environment; water, soil, and air. Where a substance ends up is solely dependent on the substance's vapor pressure (volatility), water solubility, and lipophilic qualities.

This can be related to a hazardous materials event, where the decontamination of individuals will occur. If the chemical product is soluble in water, then water is used in the process. If the product is not water soluble, then an emulsifier such as soap/detergent and water are utilized. However, if the product is not soluble in water and an emulsifier does not produce the desired effects, a non-miscible product may be used, but great care must be taken as to the chemical used.

A good example of this is when a victim is exposed to phenol (carbolic acid). This substance is poorly water-soluble and is not easily removed by soap and water. Instead, the recommendation of using another substance as a mediator is to make decontamination more efficient. In this case, the use of mineral oil or olive oil will help remove the phenol from the skin while mixing with the oil. Then the use of soap and water becomes much more efficient.

Caution always must be taken when using any type of solvent for decontamination. The solvent used on equipment is rather safe, but solvents used on tissue such as skin or eyes can be devastating. Acetone, mineral spirits, alcohol should never be in consideration for the removal of chemicals from the skin. The physiological changes that take place when using these solvents can cause greater damage and greater injury from the chemical.

Types of Decontamination

Gross Decontamination
This is the removal of the chemical alteration of the contaminant. It is assumed that residual contaminant will remain on the host after gross decontamination, and thus the reason for further washing and substance removal. Brushing and vacuuming are examples of gross decontamination or gross contaminate removal. Disrobing an individual is a form of gross decontamination. See Figure 6.5 for an example of gross decontamination.

Figure 6.5 Gross decontamination is the quick removal of the majority of a chemical on the body or clothing.

Figure 6.6 Secondary decontamination includes chemical neutralization, disinfecting, absorption, dilution, adsorption, and emulsification.

Secondary Decontamination

Here the residual product is removed or chemically altered. It is a more thorough decontamination process than the gross effort. Solutions, chemical neutralization, disinfecting, absorption, dilution, adsorption, and emulsification all comprise secondary decontamination. Figure 6.6 demonstrates secondary decontamination.

Tertiary Decontamination

The term tertiary decontamination is a specialized, highly technical level of decontamination that includes the diagnosis and treatment of the poisoning. This type of decontamination is reserved for the hospital environment and is specifically the decontamination or elimination of poison from the gastrointestinal tract. It includes the use of milk, ipecac, gastric lavage, activated charcoal cathartics, demulcents, alkalinization, forced diuresis, acidification, and neutralizers.

Emergency Decontamination

It is a procedure that is in addition to the routine decontamination is to prevent the loss of life or severe injury to emergency response personnel when immediate medical treatment is required. It is a procedure designed to quickly decontaminate and assist the individual's medical needs. The purpose is to remove, as quickly and as efficiently, the contaminants from an exposed or injured person. It is used in the event of an unexpected emergency without concern for providing the capture of runoff if it interferes with the immediate decontamination efforts.

Techniques

The decontamination process is planned and it is a systematic removal of contaminates from equipment and personnel. It utilizes techniques, which are designed to progressively remove the chemical hazards from equipment and personnel. The technique that may be employed depends on the chemical's state of matter, its physical and chemical properties, equipment available, and the procedures that the agency has set forth.

There are two main types of decontamination *physical* (the chemical remains unchanged) and *chemical* (the chemical is removed utilizing a chemical process).

Physical Decontamination

Absorption   This is the process of utilizing equipment that will soak up the contaminate. Wiping or sponging off equipment is an example.

Adsorption   It is when a contaminate adheres to the cleaning material. The chemical does not soak through the cleaning material but rather sticks to its surface. Petroleum specific pads are an example of this.

Dilution   This is the use of water to flush equipment and personnel. It is the most commonly used method, generally in association with soap or some surfactant-based product. This method is most effective due to the abundance of water.

The next is a process in which all of the steps are incorporated with another decontamination procedure. Here the hazardous product is identified and is isolated. If it is on equipment or personnel, it is removed by brushing and or vacuuming. Once the material is contained, it is disposed of. This three-step method may also use procedures from other physical; decontamination methods or chemical decontamination. These are isolation and removal (brushing or vacuuming).

Factors to Consider During Decontamination

In previous sections of this book, solubility was defined as the ability of substance to blend uniformly with another to form a solution. When developing a decontamination plan it is important to determine if the chemical involved is soluble in water. Ideally, the ionic compound, salts, acids, and bases, in general, have the ability to blend with water making water and, in some cases, soap and water as the ideal decontamination solution.

If petroleum products are involved, the decontamination plan should include a solution with an emulsifier such as soap/detergent that will break down non-water soluble and encapsulate them for easier removal. However, some solvents are not easily removed with just water and using soap is not always efficient. Therefore, either repeated washes with soap and water are needed or some type of intermediary must be used.

Chemicals with higher vapor pressures and lower boiling points must be evaluated for causing respiratory injury, especially during the decontamination process. Chemical vapors and gases that are water soluble tend to injure the upper respiratory system where the airways are coated with a water soluble mucous that absorbs these chemicals. Those that have low water solubility tend to injure the alveoli where the inner surface of the alveoli is coated with a lipid soluble chemical; surfactant. Non-water-soluble chemicals are absorbed into the surfactant and either change the chemical make-up of the surfactant, or produce a more toxic chemical capable of causing tissue damage to the alveolar cell wall.

Those chemicals that have low vapor pressures, lower than the atmospheric pressure, are considered to be somewhat safer than the high VP chemicals; however, safeguards are utilized during operations due to the characteristic of the chemical rather than its physical properties. It is important to understand that it is the volatility of a chemical (evaporation) in combination with its toxicity that truly presents the problem.

When evaluating gases, the vapor pressure, volatility, and density all must be considered. This, in association with the uniqueness of the environment, all has a bearing on the toxicity of the atmosphere. These environments are affected by conditions inside the structure vs. outside the structure, closed rooms without ventilation causing positive pressure, or rooms that contain negative pressure.

Volatile liquids may be controlled within a structure using temperature. In some cases, the use of cooling systems may postpone liquids volatility at toxic levels until occupants can be moved. Although this may be a tactic that, in theory, will work, it is difficult to control.

However, positive pressure vs. negative pressure is a consideration when dealing with populations within a structure, and the issue of evacuation vs. protection in place must be considered. In either case, the science of decontamination and the techniques of patient management can be used to identify those strategies that are most beneficial to positive patient care management.

The pH of a chemical is of concern from both the decontamination side and medical management issues. As a consideration of decontamination, water and, in some cases, copious amounts of water can neutralize the chemical enough to prevent further injury. However, personal protective equipment must be able to withstand small quantities of the acid or base during rescue and decontamination efforts. Acidic or basic solutions entering open wounds, abdominal cavities, and neurological endings are detrimental to the patient but may be difficult to control in the field decontamination setting. Additionally, if acid gases are inhaled, aggressive medical management is required to minimize long-term effects.

During the decontamination process there is always a concern about the hazards created in run-off from the decontamination process. Studies have proven that when a person is contaminated with either a dry substance or wet substance a considerable amount of the chemical is removed just by removing the clothing. Some documents suggest that 80–90% of the chemical contamination can be removed simply by removing the clothing.

With that in mind, when a victim goes through a wet decontamination process, the concern is about the hazard of the run-off solution. The question is always: "how much chemical is really in the run-off water?" In theory, very little. But this is not an excuse totally disregard the run-off solution.

As mentioned earlier, emergency decontamination is provided immediately to reduce the medical effects being caused from the exposure. In these cases, OSHA allows decontamination without the capture of run-off because life safety is a higher priority than the environment. But in other cases, it is a requirement to capture the run-off and evaluate it for disposal. In most cases, the run-off does not contain enough chemical to cause additional injury or environmental damage, but it still must be evaluated and disposed of properly if there is a dangerous level of chemical contained in the run-off material.

Equipment Uses

The amount of hazardous material that one drags out of the hot zone on tools and equipment is of more concern than the patients that have ambulated out of the incident. This equipment may have been used in direct contact with the chemical. The direct, purposeful contamination is higher and thus a concern when decontamination of equipment is implemented. Capturing run-off is of a higher priority than what was described above.

Choosing a Decontamination Site Location

Ideally, the area that is utilized should be uphill and upwind of the incident itself. Additionally, the area that runoff is flowing toward should be an area that can be easily isolated, or its flow can be managed. For example, allowing the decontamination solutions to flow into a storm sewer vault may be an option during an emergency decontamination operation. As time and resources permit, this vault can be neutralized or pumped out, lessening the environmental impact and further exposure to the public.

Detection and Monitoring

It is not the intent of this book to identify how and when to use detection and monitoring skills as a part of a medical response to a hazardous material incident. However, this skill has become an important issue to address for the EMS responder. The EMS provider must understand the hazards from which the patient came from in order to provide medical care to an exposure victim. Therefore, there needs to be a review of the concepts of detection at the hazardous materials incident so that the medical care provider can be provided with important information concerning the chemical involved in the exposure.

At the beginning of this book the situational assessment continuum was presented and discussed. This process utilizes three circles that must be assessed to ensure that there is a complete understanding of the chemical exposure. These circles include: patient presentation (signs and symptoms), event conditions (scene evaluation including information about the scene of the event), and finally the scene assessment including what is found by conducting monitoring of the environment where the exposure took place.

It is within the scene assessment that evaluation concepts using detection equipment lies. Within this area, the principles of chemical and physical properties, along with observations and lastly, the direct reading of traditional detection devices, such as a five-gas meter, infrared spectroscopy, or even high-pressure mass spectroscopy, would take place.

This does not mean that the medical branch must go out and purchase this equipment or even learn how to use this technology. What the medical staff needs to understand is: what can the technology provide to the medical officer, and how can the medical officer apply this to patient management?

As stated, the field of detection is complex and diverse; however, it can apply to medical modalities when it comes to patient management. The signs and symptoms of cyanide, hydrogen sulfide, and carbon monoxide are very similar if not exactly the same but the treatment can be significantly different. It is important for the patient care provider to know what chemical caused the symptoms. A responder with a simple five gas monitor can quickly evaluate the location of the exposure and provide a definitive answer so the appropriate treatment can begin immediately. The information gathered from detection and air monitoring can be compared to the symptomology displayed by the patient in order to determine a differential diagnosis and many times to determine the exact chemical involved.

A second example is the use of air monitoring equipment placed on EMS jump bags to warn the responder of a potential threat. Over the past ten years there has been an increase in the use of simple monitors that warn dangerous levels of carbon monoxide. These monitors produce a loud audible warning in the presence of carbon monoxide. Carbon monoxide is colorless, odorless, and tasteless but can kill in high concentrations.

Observations at the scene of an emergency incident are the most powerful clues toward mitigation strategies. Some of the clues are very obvious, and others are more elusive. It is up to the emergency responder to investigate the scene for the presence of these clues.

First arriving emergency response units should take immediate actions on the scene and should focus on the highest priorities of life safety and incident stabilization. Scene control, gaining additional information, and protection of onlookers, workers, and the general public all fall within these priorities. The first arriving unit should report their immediate observations over the radio so other responding units can gain an early understanding of the scene. Observable conditions should be used to develop objectives and request additional resources. Until trained personnel arrive on the scene and make a basic evaluation of conditions, they are relying on information gained by the dispatcher from on scene civilians who are probably untrained and may be unaware of the danger. Those incidents that stimulate multiple calls for help from numerous callers allow the dispatcher together different clues from each caller, adding to the information transferred, via radio, to the first arriving units.

If victims are still in the hazard zone, and in need of rescue, then the incident command must develop rescue parameters. These rescue considerations must take into consideration the chemical properties, physical considerations such as container shape and size, weather conditions, and available resources to determine if a rescue can be safely made.

In some situations, line-of-sight rescues can be accomplished into the hazard zone using firefighting gear and SCBA. This can be done safely if certain conditions can be met and there are signs of life from the victim in danger. These conditions include, a second rescuer dressed in firefighting gear with an SCBA ready just outside of the hazard zone to act as a back-up for the first rescuer. They must be ready in case something happens to the first rescuer. Second, a charged hose line operated by a firefighter, available for immediate decontamination upon exiting the hazard zone. There are, in reality, very few chemicals that would penetrate firefighting gear and an SCBA upon entry into the hazard zone, as long as the time spent in the hazard zone is very brief and only intended to enter and drag out a person in danger.

Typically, law enforcement or independent EMS agencies do not carry sufficient personal protective equipment to make a line-of-sight rescue from a contaminated hazard zone and must wait for firefighters to arrive. There are many reports of law enforcement, EMS, and even off-duty firefighters entering a contaminated hazard zone in an attempt to make a rescue only to suffer from the acute effects of the chemical. These incidents are made more complex and add to the challenges of controlling the scene when the emergency responders arrive.

The Approach

Mitigating a hazardous materials incident becomes much easier once you know the chemical or biological agent(s) that you are dealing with. Many of these incidents, such as fuel spills, become routine and cause a sense of complacency. It is the incidents with complex or unidentifiable chemicals, with extensive hazards, that challenge hazardous materials responders.

Observations alone can provide the clues that are needed to identify the materials involved in the incident. An earlier discussion identified some of the basic clues that emergency responders should be evaluating such as occupancy and location, container shape and size, placards and labels, and facility documentation such as Safety Data Sheets (SDS). These are the basic clues that must be evaluated at every scene, but evaluation can be even more detailed.

Containers for storing and transporting chemicals are designed based on the chemical, physical, and toxicological properties that the chemical possesses. In some cases, the chemical may be placed in another state of matter to make the transport, storage, and use of the chemical more efficient. In all cases, the chemical will be found in either a solid, liquid, or gas. It is these states of matter that may provide the initial clues.

Solids (or solid particulate) are the easiest to manage. They tend to remain in their form unless placed in water for a solution, touched, moved, or generally messed with. Solutions such as bleach, for example, is simply an oxy salt mixed in water. Solids tend not to burn and a very small group may sublime meaning that they change from a solid directly into a vapor state without becoming a liquid in the transition.

Solids are toxic only if they are eaten or make physical contact with body tissue. These are usually seen in dry bulk carriers, super saks, or bulk saks, covered hopper cars, or in drums and boxes. These are inherently inorganic compounds (ones can be found on the periodic chart) or manipulated organic compounds that have very large molecular weight; for the organic compounds, it is this weight that they tend to stay in the solid state. Plastics are an example here.

Liquids are a bit more problematic. These compounds have the tendency to move into the vapor state, release or emit vapors, may not mix with water, and many are toxic. Unlike most solids, liquids go directly into a vapor state. Vaporizing chemicals can quickly create dangerous and toxic environments.

The containers used to store and transport liquids include corrosive liquid cargo tankers, non-pressure cargo tank, low-pressure cargo tank, non-pressure intermodal, portable bins, non-pressure tank cars, corrosive tank car bottles, multicell packages, and carboys.

Gases present the highest hazard. Some gases are placed into a liquid state of matter for efficiency in storage and transport. Gases are changed into a liquid through either pressure or refrigeration. Propane is a good example. Propane exists naturally as a gas but is compressed into a liquid during transport and storage. When a leak from a container takes place, large areas can be affected in a short period of time. Depending on the size of the container, these chemicals tend to displace oxygen and may be flammable. Some are soluble in water, and all are toxic and/or asphyxiating.

The containers used to store or transport gases include the pressure cargo tank, cryogenic liquid tank, high-pressure tube trailer, one-ton cylinders, pressure tank containers (spec 51/51L), cryogenic intermodal, tube modules, pressure tank cars, cryogenic tank cars, cylinders, and dewers.

As demonstrated from this discussion, the container shape and size can provide clues that relate to the state of matter. By evaluating the containers and thus the chemical state of matter, a process is started to identify the chemical and physical properties that one should start to look for once the state of matter is identified.

The System of Detection

Detection and air monitoring are additional methods for the medical provider to isolate the potential chemical or hazard that has caused the chemical injury.

As described earlier, the state of matter can often be determined through simple and informed observation. If the material is a solid, it is most likely an ionic (salt) compound, if

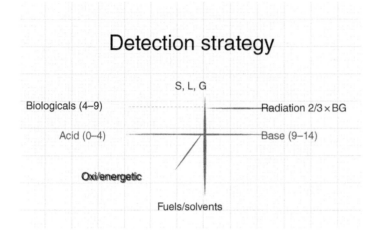

Figure 6.7 When providing air monitoring, a detection strategy must be put in place. Looking at the state of matter and the hierarchy of hazards.

it is a liquid, it is a salt solution or organic compound, and if a gas, it is an organic compound (low molecular weight) or an elemental gas from the periodic chart located above the division line (see Figure 6.7).

Radiation

When dealing with any hazardous materials/WMD incidents, one of the first hazards to evaluate on an emergency scene is radiation. A quick evaluation using a radiation detector (Geiger counter) or personal radiation detector will determine if radiation is presenting a hazard. Radiation does not create an immediate sensation of any kind. There is no smell, no feeling of irritation, and it cannot be seen. There are simple radiation meters that detect the presence of radiation and there are more advanced spectral radiation detectors that can even identify the isotope creating the radiation.

Evaluating the scene with a reading above the background activity indicates that abnormal radiation is involved in the event. Most emergency response agencies use a radiation reading of twice the normal background to determine if a radiation emergency exists; however, there are other response agencies that use a factor of four times background radiation as their benchmark to determine if a radiation emergency exists.

Radiation can be found in a variety of states of matter. Remember that radiation is just one of the hazards that may be associated with that chemical substance. It will also have the inherent properties of the chemical family that the substance belongs to. Additionally, many naturally occurring substances have radioactivity and will alert the survey meter. Granite, as an example, will show higher levels than the background but does not present a hazard and is naturally occurring.

After radiation is evaluated, the next hazard to identify is the presence of an acid or base. pH paper will identify three important points: (i) If the pH paper turns red, the product is an acid, (ii) If the pH paper turns blue, it is a base, (iii) If the pH paper turns white, the product contains chlorine as a potential chemical as chlorine is one of the few products that will bleach out the pH paper. Oxidizers other than chlorine usually do not turn the pH paper white in color. This is primarily done by chlorine or chlorine derivatives, such as bleaching products.

pH

Whenever pH paper is used in air monitoring it should be moistened with water. Water enhances the reaction on the pH paper as it is the water that creates the acid or base from vapors or gases.

In atmospheres that are chlorine-based, it is possible for the pH paper to be white in one area and red in another. Chlorine reacts at one concentration and bleaches the paper, while another concentration creates hydrochloric acid and turns the paper red. It is also possible for two responders using pH paper on the scene of a chlorine release to provide different information: one with red pH paper reading and one with white pH paper reading when in reality; they are both detecting chlorine but at different concentrations.

Note: A quick trick used by responders who often deal with chlorine emergencies is to use a spray bottle containing household ammonia. As the ammonia solution is sprayed around the chlorine a white cloud appears, indicating the presence of chlorine. The white cloud is the formation of chloramines when the chlorine and ammonia chemically combine. Responders use this solution to find even small leaks from chlorine storage cylinders and piping.

Oxygen

Determining the level of oxygen in a space is the next step in the detection process. Oxygen occupies one fifth of the atmosphere that covers the earth. In order to change the normal level of oxygen in an outside environment takes a sizable quantity of another product. However, in an inside environment the level of oxygen can change with a reasonable lower quantity of product.

The average percentage of oxygen in the air is roughly 20.9%. This may vary by very small amounts depending on the environment. Some detection companies use levels from 20.8% to 21% in the testing of their sensors. The level of 20.9% oxygen concentration represents the ideal amount that a healthy human body needs to participate in efficient cellular metabolism.

The first time an oxygen sensor establishes that a change in concentration has occurred is at 20.7%. At this level there is a decrease of 0.2%, which translates to 2000 ppm ($0.2 \times 10,000$ ppm, which is the conversion from % to ppm). Two thousand ppm is multiplied by 5 (remember that oxygen is roughly 1/5 of the atmosphere), which gives us 10,000 ppm. For every 0.1 % deflection, there is 5000 ppm of something else displacing the oxygen in the room! Understanding this concept, it can be seen that the oxygen sensor is not a good indicator of threat as small increments below normal may represent toxic levels of the chemical that is replacing the oxygen. This is especially troublesome when most oxygen detectors alarm at 19.5%, which is the OSHA limit for oxygen deprivation.

Organic Compounds

The next step of investigation are the petroleum products (organic chemicals or volatile chemicals), which are the most difficult to categorize and will take a more investigative mindset in order to place the unknown chemical into a family name. Petroleum products and solvents are all part of the hydrocarbon category and have similar effects on the body.

From a medical standpoint, exposure to hydrocarbons create a group of symptoms specific for these chemicals. They include CNS depression, cardiac irritability, and a decrease in the seizure threshold. These symptoms can appear at level of exposure too low for some detectors to identify. For example, the typical four gas meter does not begin to detect hydrocarbons until they are above 500 ppm. In higher level detection such as infrared detectors, 25 ppm is the lowest level that can be detected.

The last level of detection is that of toxicity of organic compounds. It is important to know that organic chemicals have vapor pressure, thus emit a vapor, which can be acknowledged through the use of detection equipment.

For example, these chemicals are flammable and will activate the lower explosive limit (LEL) sensor on the typical four gas meter. These meters are set to alarm at 10% of the LEL. But because the LELs and UELs are developed from lab studies with a constant temperature of 68° Fahrenheit and the meters are typically calibrated using methane gas, both of these factors can significantly misrepresent the 10% level. Responder should understand that a 1–9% reading on the LEL sensor could represent a flammable environment and above 10% of the LEL a potentially explosive flammable environment.

Another meter that is commonly used to determine the presence of a hazard is the photoionization detector commonly called the PID. The PID utilizes a bulb to determine the ionization potential of a chemical. Typically, these meters contain a bulb rated at 10.60 electron volts (eV). The first three chemical structures within the alkanes and alkenes (organic compounds) have an ionization potential above 10.60 eV. This means that a PID set up with a 10.60 bulb will not see these chemicals. However, these chemicals are flammable and will activate the LEL sensor.

The PID detects volatile organic compounds (VOC) and reads them in ppm. The simple approach to this is that at a reading of 10 ppm there will be the presence of a chemical in the room. This should be evaluated in context to what the occupancy of the building/room is. As an example, in a day care facility you should not have 10 ppm; however, an auto repair should normally have well over 10 ppm of VOCs.

A PID alarms first at 50 ppm; 50 ppm also means that for 50% of the VOCs, the atmosphere at this point is also at 50% of the TWA. Looking back at the chemical timeline that was discussed on page XXX, this level also represents a safe region but is approaching an unsafe area. At 100 ppm we have exceeded all TWAs and at 200 ppm we are at second alarm and are approaching IDLH and or LEL.

The hydrocarbons are nonpolar, thus do not mix with water. It should also be understood that the longer the chain of hydrocarbons the heavier the product becomes. These facts can help in evaluating some of the characteristics of unidentified chemical(s). For example, a small sample of a hydrocarbon is placed in water; if it floats it is a relatively short chain, if it sinks it is a relatively long chain. Adding into the detection strategies the PID and four gas meter, responders may quickly identify the hazards presented

Looking at these numbers from a medical application should be noted that when any of the meters alarm there is already significant amounts of chemical within the room. However, determining the amount of the exposure may become a skill the responder may need at the scene of an event.

Many times the toxicological data is referenced as mg/m^3. In order to place the numbers into a frame of reference that are interpreted by monitors, these numbers must be converted to ppm. The formula to make this conversion is:

$$PPM = mg/m^3 \times 22.414 / \text{molecular weight}$$

So as an example, Ethyl mercaptan has a NIOSH REL of 1.3 mg/m3 and a molecular weight of 62.1. By plugging these values into the formula the ppm value can be reached.

$$PPM = 1.3\,mg/m^3 \times 22.414/62.1$$
$$PPM = 0.469 \text{ or } 0.5 \text{ therefore the value is } 0.5 \text{ PPM.}$$

When there are mixtures and a need to understand the exposure concern for one or more vapor/gases, the following formula can be used to evaluate the environment:

$$\text{Exposure Concentration } (E_c) = (C_n = \text{on scene concentration})/$$
$$(L_n = \text{reported exposure limit value})$$
$$E_c = C_1/L_1 + C_2/L_2 + C_3/L_3 \ldots + C_n/L_n$$

Biologicals

The detection and analysis of biologicals came to the forefront in 2001 when America Media Incorporated, in Boca Raton Florida, received a letter filled with weapons grade anthrax. Since then, HazMat teams have had some more and some less capabilities for detection of biologicals. Recently, with the outbreak of COVID 19, these capabilities have been revisited as a public service. Clearly, the role of emergency responders has changed and now includes response to biological emergencies.

Detection and Monitoring Responses

Detection and air monitoring of an environment has been an art form for hazardous materials teams for many years. Although most do not see this segment of the operation as an art form, it is truly has been and will be into the foreseeable future. It is the technician's job to look, analyze, and interpret the information that the instrumentation is displaying. The reality of current-day operations is the fact that responders must have the ability to logically identify the answers to the questions such as reoccupation of a facility, office space, and/or relocation of a population. In order to provide these answers to a variety of difficult questions, the first responder must understand the advantages and limitations of each instrument. All technology has its advantages and limitations; it will be this understanding, which the technicians must apply toward the event, device information, and their field observations. Issues such as sick building syndrome, accidental chemical release, and of course, the intentional use of hazardous substances to thwart society, all have led to response agencies toward looking at equipment that was once considered as a "laboratory" practice, which has now moved into the emergency response arena.

With the move of high-tech equipment into HazMat response, the tactical application of specific equipment should not lie in the technology but rather in the responder's ability to translate the information into useful action.

As HazMat responders when given a name of a chemical, DOT number, CAS number, or physical property can identify the chemical family or group from which the chemical in question is derived and thus define the hazards. The power that instrumentation gives the HazMat community is the ruling out of one family or group over another. How and why, one would use a specific piece of instrumentation is based upon the incident, the chemicals involved, and the family or group they belong.

As with any tactical deployment, the understanding of the chemistry and the science behind the technology should guide the technician. HazMat technicians are not chemists nor are they industrial hygienists. They are a group of individuals that can look at the scene and analyze divergent conditions of an incident and make some logical deductions. This reasoning is gained through experience, learned facts, and intense training. However, some common sense should have been sprinkled into the responders' mind during their training events.

Current Detection Technologies

Over the past few years, the amount of equipment being pushed out into the response community has been overwhelming. Some responders stay with the old tried and true while others have ventured out into the higher levels of technology. Whatever side of this spectrum the responder is on, the simple fact will remain; what information can be gathered to prove or disprove the level of chemical (or biological) that is potentially present on the scene that you are dealing with? In order to provide this simple yet frustrating answer, one must be able to look at the event and logically process what the instruments are informing you of (see Figure 6.8).

Instruments talk to us; yes, they speak to us. The unfortunate part is that responders are not listening. No, let me rephrase that responders are sometimes not listening in terms of their level of speech. What is meant by these statements is that the limitations and the advantages are the way the instrument "speaks." Combining the advantages and the limitations of multiple technologies can give the responder a "view" of the world that they are trying to seek answers from.

When a responder walks onto a scene with only a combustible gas indicator (catalytic bead), expecting that one instrument to say the atmosphere is safe! But what about the corrosively, or toxicity, or radioactivity? The point here is that detection and monitoring must be conducted in a systematic approach and a logical approach to the question, "Is there something in this environment that can hurt me!"

Radiation Detectors

The term radiation is a broad term that identifies energy transmission. Radiation is the spontaneous decay of radioactive materials that emit particles and/or energy from the atom. Ionizing radiation produces a wave of energy that disrupts that atom and bombards the matter around it. This is the cause of radiation damage to the cells, tissues, and organs.

There are four forms of radiation; alpha, beta, gamma, and neutron radiation. Alpha particles are positively charged and are made up of two neutrons and can travel approximately 4 in from the source. Beta particles can be positive or negative and are approximately the size of an electron; they can travel up to 30 ft from the source. Gamma radiation is an electromechanical wave of energy that can travel great distances and penetrate materials great distances from the source. The neutron is a particle that is spontaneously released when the atom is trying to manage its atomic weight (proton/neutron ratio).

As far as detection goes, this hazard is the easiest to detect, qualify, and quantify. The technologies in this area are many. Some can only identify that you have been in a radioactive field, while others can identify the isotope that you are dealing with.

Dector technology table		
Technology	**Use**	**Medical modality application**
Radiation detector	Detection of radiation sources	To establish cleanliness of the patient
pH paper	Detection of Acids and Bases	Determination of acid/base after decontamination and during ocular irrigation
Oxidizer papers	Detection of explosives	Identify nitrates and nitrites comparing to cardiovascular presentation
LEL	Detection of flammable gases and vapors	Compared to respiratory/cardiovascular presentations (Hydrocarbons)
Oxygen sensor	Detection of oxygen deficiency and enhancement	Compared to respiratory/cardiovascular presentations (oxygen deprivation)
Carbon monoxide	Detection of carbon monoxide gases (cross sensitive to hydrogen and hydrocarbons)	Compare to respiratory/cardiovascular presentations (pulse oximetry/ capnography, hydrocarbons)
Hydrogen sulfide	Detection of Hydrogen Sulfide gases (cross sensitive to hydrogen cyanide)	Compare to respiratory/cardiovascular presentations (pulse oximetry/ capnography)
Volatile organic compounds (PID)	Detection of organic compound gases and vapors (limited to Ultraviolet light bulb, usually 10.6 eV)	Compare to respiratory/cardiovascular presentations (pulse oximetry/ capnography, hydrocarbons)
InfraRed. Raman	Detection of organic compounds liquids and solids - Can give you the chemical family or substance (limited to simple or non-mixtures)	Compare to respiratory/cardiovascular presentations (pulse oximetry/ capnography, hydrocarbons)
High-pressure mass spectroscopy	Detection of organic compounds gases and vapors above 44 and below 350 molecular weight units. Can give you the chemical family or substance.	Compare to respiratory/cardiovascular presentations (pulse oximetry/ capnography, hydrocarbons)

Figure 6.8 Quick reference of available technology for hazardous materials response.

Geiger–Muller tubes, traditionally a handheld device, have been made small enough to work as a dosimeter, and this technology is also within handheld survey meters. The meters contain two electrodes with a gas-sensitive barrier between the electrodes. This gas-sensitive space provides a medium for ionization to take place when radiation is introduced. This "ion chamber" has three general regions and has to do with the differences in voltage and electrons that are collected within the chamber. In general, alpha particles are detected in all three regions, beta particles are detected at higher voltages, and gamma will saturate the chamber. This type of detector can read very high levels of radiation.

Advantages:

- Rule out radiation as your first step in the unknown analysis.
- Most detectors are very intuitive, and attention must be made to the scale of detection.

Limitations:

- Ranges of the instrument must be identified for the initial survey.
- Some technology CsI as an example will not give good information in higher radiation fields.
- Neutrons, due to their tremendously fast speed, require a special instrument.

pH Paper and Impregnated Papers

This technology is pretty simple. Papers are impregnated with indicator solutions and/or salts that react in the presence of an acid or alkaline condition. This color change occurs over a range of hydrogen ion concentrations and is expressed as a pH range.

Advantages:

- Excellent gateway test in order to rule out acids or bases from the unknown solution or environment.
- Quick application.

Limitations:

- Buffered acids, in many cases, will not change the indicator and give a false negative.
- Should wet the paper with water to increase the potential reaction.
- May not indicate oxidizers.

KI Paper or Oxidizer Paper

Here the indicator is potassium iodide which in the presence of hydrochloric acid and an oxidizer will turn color, indicating the oxidizer.

Advantages

- A quick test to look for potential oxidizers
- Easy way to rule out potential oxidizers

Limitations:

- To provide an accurate reading, it requires a specific concentration of hydrochloric acid on the test paper.
- Only looking for oxidizers.

Wet Chemistry

There are several types of wet chemistry procedures. This technology uses the principles of physical chemistry in an algorithm to "decide" the chemical group or chemical family the unknown may fall into. These use a system of reactions and observations to determine the potential outcomes. This style of identification can determine low concentrations and even mixtures of chemicals. It is extremely useful when trying to identify certain metal combinations in solution, and chemical families.

Advantages:

- Wet chemistry provides an excellent gateway testing procedure before electronic equipment is deployed.
- Can provide long-term testing of the environment.

Limitations

- The algorithm can become very time consuming, with consistent training events.
- Reagents must be maintained, along with frequent training cycles.

Electrochemical Sensors

The basic principle of operation is a tub of an electrolyte solution that has electrodes (the solution can be alkaline or acidic and is specially designed to react with the target gas). The gas passes into this tub of solution and creates a chemical reaction (oxidation or reduction reaction). The electrode "counts" these reactions (actually current is created) and gives the responder a reading.

It is a simple tub of solution with a positive (anode) and a negative (cathode) electrode. The target gas passes through a semi-permeable membrane (sometimes a sensing electrode is placed here) and into the solution. A chemical reaction starts between the electrolyte solution and the target gas. This sets up a greater or lesser degree of positive or negative charges on the cathode or the anode, which produces an electrical signal, which the instrument through the use of a resistor gives a proportional relationship to the target gases concentration as parts per million or percent.

Advantages:

- Very selective toward the target gas that the specific sensor was designed for – Oxygen being the most stable sensor.

Limitations:

- Low concentrations can cause cross-interference
- Ozone, chlorine, fluorine, carbon dioxide, bromine, and certain concentrations of hydrogen sulfide can neutralize the alkaline type of sensor.

Catalytic Bead

Flammable and combustible gases will not burn until the ignition temperature is reached. However, if we place a catalyst into the system, we can start the burning process at lower temperatures. This technology has exploited this phenomenon. By placing a coated metal oxide on a coiled platinum wire and then coating this system with platinum, palladium, or thorium dioxide, we have created a catalytic bead. Placing these beads in pairs, one which will eventually be exposed to the testing environment and the other bead as a reference, one can now "test" the environment for potential flammable or combustible atmospheres.

As the instrument is exposed to the environment that we would like to evaluate, the atmosphere is drawn into over the active bead and burns, while the reference bead is also burning, it is the relationship between the actual "active" bead and the reference bead that the instruments "reads" the environment.

Advantages:

- Gas/vapor burns can be adjusted to evaluate the flammable/combustible environments through bump testing and calibration.
- The catalytic bead detector is easy to use and understand technology, very accurate if calibration and maintenance are consistent with use.

Limitations

- The bead can become "poisoned" in certain environments, such as oils, chlorine, heavy metal fumes, sulfur compounds, and silicon, to name a few.
- Halons will cause the beads not to work for upward of 48 hours after exposure.
- Oxidizing gases can heat up the sensor array faster, giving a false high reading.
- Correction factors are needed for precise readings; however, the gas of concern should be calibrated if precise for a specific gas is operationally required.
- Higher humidity or cold weather can mask the sensor, giving a slower response or a false reading – warm-up time is essential.

Colorimetric Tubes

Detector tubes are a hermetically sealed glass tube that contains specific ionic salts or reagents to "indicate" a specific chemical or chemical family. The ends of the glass tube are broken off, and a known volume of suspected air is pulled into the tube and over the chemical reagents. The contaminated air reacts with the indicating chemicals, and a change in color occurs. The reaction of the indicator chemicals has sensitivities to other chemical families and can be used to "investigate" other potential hazards in the unknown environment. This can be useful when trying to distinguish between similar unknowns.

Advantages:

- Able to measure a wide diversity of chemicals
- Very easy to use in high levels of protection
- In some cases, this technology is the only technology that will detect certain chemicals

Limitations:

- Tubes can be used only once and should be considered expendable equipment.
- Lack of a color change does not mean that a chemical is present or not. It may mean that a chemical was not present in the quantity that was within the measurement range.
- These tubes can be affected by temperature, humidity, pressure, storage conditions, shelf life, and interference gases, and all play a role in its potential error rate of 50%.

Photoionization Detection (PID)

PIDs utilize ultraviolet light to bombard chemicals in order to ionize them. The matter is made up of atoms that have electrons rotating around in a cloud about the central nucleus. These electrons within this outer cloud are susceptible to electron bombardment. Electrons being negatively charged, when ultraviolet light stimulates the electrons, this light "pulls" the outmost electrons away from their original orbit. The energy that is released when the electron returns to its original path is termed ionization potential (IP) and is measured in electron volts (eV). Materials have IP, some more than others, which can be measured by the PID. This is all dependent on the level of energy the lamp within the PID is capable of.

The lamps that are available for PIDs are 9.5, 10.2, 10.6, 11.7, and 11.8 eV. Each chemical that has an ionization potential will ionize if the lamp energy is high enough. For example, cyclohexane has an IP of 8.95 eV; any of the above lamps produce enough energy to pull the outer electron into a higher energy path. As opposed to cyclohexanol, which has an IP of 10.00 eV, would need the lamps of 10.2 eV and above. The 10.6 eV bulb is the most practical lamp to have within the PID. Above this energy level, the lamps have a tendency to expire very rapidly. Relative response ratios must be known for a specific chemical for interpretation within a given environment.

Advantages:

- Able to detect a substance in an environment very quickly.
- Most detectors in a general sensing mode are intuitive to "see" if a contaminate is within the environment.

Limitations:

- Humidity, interference gases, and particulate matter can limit the detection capabilities of the lamp, giving false or inaccurate readings
- Overall concentrations of PID's are 50–500 ppm
- Ionization is dependent on the lamp eV 10.6 is the most practical, leaving IP's above 10.6 to about 13 eV "unseen" by the PID

Flame Ionization Detection (FID)

This technology is very similar to the PID; however, it uses a hydrogen flame to produce ionization. The FID is sometimes referred to as an organic vapor analyzer or OVA and has the capability of higher eV and very low detection ranges (up to 15 eV and range of 0.2–50,000 ppm). Here as well as with the PID relative response ratios are required and matched to the specific chemical in order to qualify the parts per million (with this instrument, parts per billion are attainable).

Advantages:

- Extremely low detection capabilities
- Full-spectrum of eV's (up to 15 eV) are possible
- Able to detect a substance in an environment very quickly.

Limitations:

- FIDs use hydrogen flame; oxygen must be present within the atmosphere.
- Its sensitivities to aromatics are extremely low.
- It cannot be used for inorganic compounds like sulfides or for inorganic gases.
- Functional groups attached to the contaminate may reduce the sensitivity of the instrument.

Ion Mobility Spectroscopy (IMS)

In IMS, a heater and a radiation source are used to vaporize and ionize the chemical, respectively (some use an intense light discharge called a corona light). As a chemical moves into the instrument, a heater coil increases the temperature of the unknown in order to increase its vapor pressure. Once the vapor pressure of the chemical has been raised, a radiation source ionizes the chemical as it enters a chamber. These ions move down the chamber in a path, which is a signature of a specific chemical. The ion moves down this chamber based upon the charge of the ion (negative ions move toward a positive area, and the positive ions move toward a negative area). The time distance and speed of the ion's movements are measured as a mathematical algorithm or pattern (spectra) and are compared to known patterns within its library. When the instrument "recognizes" the specific pattern, the alarm sounds and/or an identification light/readout is given.

Advantages:

- Very specific toward certain chemicals.
- The chamber can be modified to "alarm" towards certain chemicals of concern.

Limitations:

- The sensor pathway can become saturated or overexposed.
- Cross sensitivities are very high, giving false positives.
- Recovery rates when saturated or between survey activities vary.

Infrared Spectroscopy (FT-IR)

Infrared has a variety of applications; the specific IR we use in the field is FT-IR (Fourier transform infrared spectroscopy). The instrument produces light at a very specific range of light frequencies. The specific range of frequencies called wave numbers are generated by a light source and shined toward the chemical in question.

All matter consists of molecules, some are simple, and some are complex. Each has its own characteristics and level of inherent vibration. When a light source of specific frequencies passes through a chemical, some of the light is absorbed, increasing or modifying the natural vibration of that molecule. Some may be absorbed; some is reflected while some of the light is refracted.

Each section of a molecule absorbs this energy at different but specific wave numbers at the points of enhanced vibration. Some absorb at low energy levels (water), while others need more energy to create this vibration (aromatics), creating a spectrum. The frequency of vibrations (due to the light energy absorption) is compared to the reference spectra and compared to the reference library.

Advantages:

- Very specific spectra can be shared between response agencies and resource scientists.
- Spectra is independent of temperature, pressure, humidity variations.
- Spectra is a property of that specific molecule or family of molecules.
- Excellent at identifying functional groups and hydrocarbon chains.

Limitations:

- The molecule must have a covalent bond.
- Cannot detect elemental substances, ionic compounds, and diluted compounds.
- Mixtures are very problematic.

Raman Spectroscopy

Raman spectroscopy is similar to infrared in that a beam of light is pointed at a molecule. Raman uses a specific frequency of light and "looks" for the scattering of electrons that is created when that specific frequency of light is shined at the molecule. It is from this scatter of electrons at specific wave numbers that a spectrum is generated and can be read similar to the infrared spectra.

The wave numbers at which scatter occur are very precise and, again, are specific to the molecule's construction. Each chemical compound has a unique vibration, and thus a unique scatter pattern or spectra.

Advantages:

- The bond must be a polarizable covalent bond (organic compounds, hydrocarbons, organic solvents)
- Some inorganic compounds (metal oxides, oxy salts)
- Some pure elements (sulfur, carbon, phosphorous)
- Ramon active chemical in a water solution (Ramon does not "see" water)
- Acid / base solutions

Limitations:

- Ramon cannot "see" pure elements, ionic salts, highly polar compounds, fluorescent compounds from coloring or contamination, individual components of any mixture,
- Concentrations less than 5%

Positive Protein

Positive protein tests use an indicator solution that identifies the amino acid peptide bond within a protein. Bacterial membranes and viral coatings have protein. The protein is made up of amino acids that are connected with peptide bonds. This test (as a wet chemistry test for biologicals) is specific for these bonds. However, many proteins exist in the environment, as well as a variety of bacterial materials that are not harmful to humans. However, it is a good gateway test after the pH of a product or solution has been identified. The bacterial material that we are concerned with as a responder is Anthrax, Plague, Ricin, smallpox, etc. The pH range that these "bugs" can exist in is a range from 4 on the acid side to 9 on the alkaline side.

Advantage:

- A quick, simple test that, when applied with the pH, can identify or rule out biological material.
- Simple and easy to use one-time use.

Limitations:

- Are not a definitive tool for bacteria or viruses of concern
- Must be used with other technologies
- Cannot differentiate between different bacterial or viral materials

Handheld Immunoassay (HHA)

A handheld assay is a plastic-encased card with several windows. The first window is where the sample is placed. The sample has to be prepared with a buffered solution, which acts as a transport fluid down the card, through the sample window, and toward the test area. Within the test, the area has specific antibodies that have specific antigen characteristics. Once the sample (unknown) in solution passes over the labeled (with the antibodies) area, a reaction takes place. The presence of the specific antigen causes a color line to appear, identifying the presence of a specific bacteria or virus.

Although very simple and straightforward several problems can arise during the use of these assays. If too much of the material is placed within the buffered solution, it can clog the travel path. Additionally, too much solution can cause the material in question to "run" past the labeled area. Dilution of the initial solution may be required, and this procedure should exist in the standard operating guidelines for this test.

Advantage:

- The use of this test in combination with pH and positive protein can categorize a potential biological threat.
- The technique is important when using this procedure; however, it is easily mastered.
- Tactical decisions can be made within about 30 minutes.

Limitations:

- The temperature of the sample, cross-contamination and interfering microorganisms can cause inaccurate readings.
- Are not a definitive test to determine pathogen or toxin.
- Should not be used to make patient care or prophylaxis decisions.
- Sample must be sent to an accredited lab for further analysis.

Polymerase Chain Reaction (PCR)
All bacteria and viruses have a form of deoxyribonucleic acid (DNA) and ribonucleic acid (RNA). DNA and RNA are molecules that contains genetic instructions used in the development of living organisms. DNA contains the instructions for the construction of cell components, protein, and RNA. Chemically the structure is a long strand of material made up of smaller units called nucleotides that are held together by sugars and phosphate groups. Segments within the structure carry the genetic material. Each sugar has one of four types of base molecules (adenine, cytosine, guanine, and thymine). It is this specific sequence of these bases that encodes the information of genetics. This sequence specifies the amino acids within proteins as an example. This code of genetics is read by copying segments of the DNA.

Because DNA is the template of genetics and organism outcomes, we can strip the DNA and make copies of this strand. It is from this macromolecule that this instrument is able to detect bacteria by making copies of the unknown DNA. The instrument looks for a strain of DNA and strips or unzips the DNA molecule. Because the base pairs of the DNA have very specific comparative parts, we can reconstruct the original DNA through this unzipping and through the use of a primer. This primer is placed into a solution, which assists in making copies of the base-pair pattern. It basically takes the DNA and makes repetitive copies of the original DNA strain. It then analyzes the DNA for specificity and organism typing.

Advantage:

- Can give a very precise evaluation of the target biological material.
- Within 45 minutes, a confirmation can be made.
- The technique is important when using this procedure; however, it is easily mastered.

Limitations:

- Cannot differentiate between living or dead biological material
- Not widely used in emergency response but gaining momentum

Mass Causality Incidents

Triage Considerations (Non-START Triage)

Over the past 55 years of EMS, how to manage patients at the scene of a mass causality has ebbed and flowed. From concepts that foster the idea of moving the first patient that you come to or do you sort the injuries. In both cases, the concept and the intent are to provide the best and most efficient method of treatment for those that need medical assistance. When historically evaluating hazardous materials incidents, the frequency of having a large number of patients is extremely rare. Even when large releases have occurred, the number of casualties from these releases are low. However, there have been instances where the potential patient load on the EMS systems has been overwhelming and, in some cases, beyond capability. This is further complicated by the fact that intentional releases have the potential to occur anywhere.

Over the years, EMS responders have used a variety of tools to score and sort multiple causalities in an effort to provide the most immediate care to those in the most critical condition. All of this is done with the intent to properly utilize limited medical resources more efficiently. These scoring and sorting mechanisms have always been based on traumatic situations where shock and impending shock is the determining factor to for a victim to receive emergent care. These systems that are widely used throughout the United States but are not efficient or effective after a chemical exposure and injury has occurred.

The answer is not found with the traditional triage methods but rather a method that is born out of the basic tenants of toxicology and pathophysiology. Here the scoring mechanism looks at three physiological responses with an immediate treatment protocol in combination and used over a time frame; one can "measure" the severity of chemical exposure.

Stilp and Bevelacqua Exposure Score

As with anything in the response, world observations can lead to a recognizable conclusion. This method came from years of recording and analysis of hazardous materials incident reviews. Over time it started to make sense that the chemical insult was body system-based, i.e. cardiovascular, respiratory, and neurological. It presented as a patient would when exposed to a chemical.

The intent of this exposure scoring system is to evaluate mass casualty patients who have been exposed to a toxic chemical, with the purpose of providing an assessment of multiple patients by developing a triage score that will guide which patients need immediate treatment and transportation and which ones can be delayed.

To make this system easy to remember, the CBRN acronym has been incorporated. These four areas of evaluation are to determine the degree of severity of injury to the exposure. These areas are cardiovascular, breathing (replacing respiratory), Rx (quick initial treatment), and neurological, the CBRN acronym. In each of these categories, a list of easily assessable signs have been identified, while the Rx step is identified as quick treatment options that may affect a positive outcome (see Figure 6.9).

The application is that you take a score upon patient contact as soon after the chemical insult as possible and again at five minutes. In each category, there are assigned numbers that will be assess and reassessed after 15 minutes to determine how aggressive your treatment should be based on how fast the chemical is interacting with the patient's body systems.

C Cardiovascular status		Max of 4
Capillary Refill +2	Greater than 2 = +1 Less than 2 = 0	
Weak Pulse	Weak/Tach/Brady (or difficult to find) = +1 WNL = 0	
Skill Temperature Cool/Hot	Cool/clammy to touch or is hot/dry = +1 WNL = 0	
Skill Red/Flushed	Yes = +1 WNL = 0	
B Breathing status		Max of 4
Chief Complaint	SOB or displays difficulty breathing = +1 WNL = 0	
Respiratory Rate	<10 or >24 = +1 WNL = 0	
Respiratory quality	Deep accessory muscle usage, retraction of the neck muscles, intercostal muscles and/or diaphragm breathing = +1 WNL = 0	
Skin or Mucosa Color	Cyanosis, pale, ashen color of the lips and/or mucous membranes = +1 WNL = 0	
R Rx - Immediate Basic Treatment	Immediate treatment is focused on maintaining breathing by placing the airway into a position that will assist respiration. Maintain blood pressure by positioning and removing patient from the chemical which may include a quick decontamination.	
N Neurological status		Max of 4
Alert/Oriented	Determine A/O if < 3 = +1 A/O x 3 = 0	
Anxiety	Fidgeting or any signs or patient states a level of anxiety = +1 No = 0	
Seizure	Involentary muscle spasms, active seizure, post ictal, or severe fasciculations = +1 No = 0	
Unconscious	Loss of consciousness not related to trauma = +1 No = 0	

All persons exposed to a chemical need to be evaluated, all normal values receive a 0. All elderly and young patients if vitals are normal must be evaluated continuely, and should be placed in the next level of criticality.	Moderate	0–3	Yellow
	Serious	4–7	Red
	Critical	8–12	Dark Red

Figure 6.9 The SB triage system is based on chemical influences on the body unlike other triage systems that evaluate trauma-related issues such as hemodynamic activity.

Note: Remember, in the traditional triage systems, all victims who could walk off of the scene were considered to be category "green" and treatment could be delayed. In a chemical environment many of those victims who could walk off of the scene could be dead in the next 10-15 minutes. There should never be a determination of category "green" in a chemical environment.

Each category has an assigned number of concerns. The minimum is 0, and the maximum is 12. If a patient states that they have been exposed to a chemical, using the scoring mechanism will provide a degree of concern when dealing with exposure. This information can then be relayed to the hospital staff for further evaluation and assessment.

For triage purposes, the numbering system is outlined as follows – the degree of chemical exposure score:

- Moderate score 1 to 3
- Serious 4 to 7
- Critical 8 to 12.

Cardiovascular
- Capillary refill: Greater than 2 + 1
                        Less than 2 _ 0

- Weak: Weak/Tach/Brady (or difficult to find) = 1 Within normal limits = 0
- Skin Temp Cool/Hot: Cool/clammy to touch or is hot/dry = 1 Within normal limits = 0
- Skin Red/Flushed Yes = 1 Within normal limits = 0

Breathing (Respiratory)
- Chief complaint – Shortness of breath or displays difficulty breathing = 1, Within Normal Limits = 0
- Respiratory rate – Less than 10 or greater than 24 = 1, Within Normal Limits = 0
- Respiratory quality – Deep accessory muscle usage, retraction of the neck muscles, intercostal muscles and/or diaphragm breathing = 1, Within Normal Limits = 0
- Skin or mucosa color – Cyanosis, pale, ashen color of the lips and/or mucous membranes = 1, Within Normal Limits = 0

Rx – Immediate Basic Treatment
Note: Immediate treatment is focused on maintaining breathing by placing the airway into a position that will assist respiration. Maintain blood pressure by positioning and removing the patient from the chemical, which may include quick decontamination.

Neurological
- Unconscious – loss of consciousness, receives a full category score of 4
- If not unconscious:
  - Alert and oriented - alert and oriented to person, place, time, situation. If less than 4 = 1 If A/O×4 = 0
  - Anxiety – Pt. fidgeting or states a level of anxiety receives a score of 1. No signs of anxiety = 0
  - Seizure – Involuntary muscle spasms. Active seizure activity, postictal or fasciculations receive a score of 1, none of the above = 0.

Mass Decontamination

Initial Operations

As with any emergency scene, the initial operations are vast and many; size-up can provide clues toward a plan and the implementation of that plan. It is this analysis of the problem that gives rise to the plan of attack, and in many cases, if done properly, a positive outcome.

Over the years of hazardous materials, response agencies have been preaching the need and effects of run-off, the need to capture this runoff, and the use of specialized equipment for the decontamination procedure. This, in effect, has crippled the modern fire service and its strategic goals when presented with a mass causality decontamination event. In essence, what has occurred is a paralyzing effect on the system as a whole rather than the smooth proactive motion that accompanies a well-oiled machine. OSHA has stated that when presented with an event that decontamination of people is provided for, runoff is not an immediate concern. However, at some point, this will have to be addressed during the recovery phase of the incident.

The idea of the first-in fire engine company as the primary decontamination source is one that should dovetail with the normal operating procedures of the fire service. It also should give a level of practicality, customary response, and relative function toward the operation. This idea is one that is novel yet functional, quick, yet efficient; however, you consider the problem, the idea is practical.

In any decontamination process, available human and physical resources are your restrictive commodities. As with any growing incident, these resources are the commodities that you seem to run out of fairly rapidly. It is the management of these resources that become a necessity. The management of resources, or rather a resource allocation, should be a primary focus of the incident commander.

The decontamination corridor and the support that is required of this corridor only grow exponentially as the number of victims increases. This is inclusive of supportive functions for the decontamination section, decontamination evaluation section, medical evaluation section, causality collection points, areas of safe refuge, field hospitals, and medical facilities.

Additionally, there are the issues of scene security and the maintenance thereof. Law enforcement will be needed to secure supply corridors, causality collection points (CCP), field hospitals, and hospitals. The law enforcement component will also be needed for crime scene investigation that may need to occur at the same time mitigation and rescue efforts are underway.

What about the runoff? What do we do with the patients? The next few sections will discuss these issues, along with considerations for patient care, scene security, and scene managment. In addition these sections will define and discuss the following areas that are typically assembled during large multiple casuality incidents.

a) The CCP (Field treatment site) is a predesignated site for initial medical treatment, triage, and preparation of the inured for evacuation to the mobile field hospital or definitive care facility.

b) Areas of safe refuge is a designated or preplanned area specifically intended to hold people/victims as safe as possible until movement into a defined safe area can be established or the people/victims can be moved.

c) Mobile field hospital (Alternative Care Facilities) is a medical center that provides services under critical condition in order to release patient load pressure of the established definitive care facility. It is movable and can provide a variety of services dependent on the human and medical resources that are available.

d) Communication Considerations. An effective command and control center will need to be established for any and all the above. Communications management is the key to logistical support, appropriate patient movement, and scene safety.

HazMat Alert

There are stroke alerts, trauma alerts, and even alerts for mass casualty incidents. Why wouldn't you have a HazMat Alert? Its sole purpose is to provide a framework by which the hospital under a variety of conditions can be alerted to the arrival of persons that have been exposed to a chemical.

First developed by a student during a Hazmat Medic class, it was first implemented within the Orange County Fire Rescue Department, in Florida, and rapidly grew to a policy embraced by all the departments and systems within the Central Florida region. It has now been placed within the state SOP, and several areas around the country have since implemented it as well (see Figure 6.10).

The idea is to create a standard method of patient care for an incident that has involved patients exposed or is suspected of having been exposed to a hazardous substance or material. By declaring a HazMat Alert, a pre-planned series of events will take place to protect the patient, response personnel, and the receiving facility's personnel. In essence, it provides:

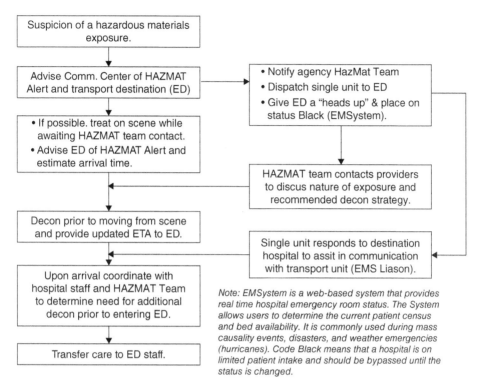

Figure 6.10   As with STEMI, stroke, and trauma alerts, the HazMat alert gives prewarning to hospitals to allow reaction time to prepare for an inbound patient requiring specialized care.

- The HazMat Alert provides early notification to the receiving hospital(s) of an incoming HazMat Patient.
- Allows early involvement of the closest HazMat Team to assist in the decision-making process.
- It also involves the Regional Poison Control early in the event.

The HazMat Alert is initiated at the time of dispatch when a caller reports a medical emergency involving a chemical smell/hazmat exposure. Or when the first arriving units suspect a hazardous materials exposure due to odor, history, or other clues presented at the scene, or by the hospital emergency department in the event, a hazardous materials exposure is suspected from a walk-in event such that additional specialized resources can be utilized.

Hospital Interface

Both the fire department and emergency medical services are staffed and equipped based on the predicted and calculated needs of the community they serve and the funding available. Some areas of the country have well-funded emergency services, while others struggle to fund even minimal services. In an attempt to augment a community's inability to fund more equipment and staff, they enter into mutual aid agreements that allow them to tap into the surrounding community's resources or to combine resources to improve both communities. The minimal funding of emergency services may work on a daily basis, but when and if disaster strikes, these minimally funded services fail, and the community suffers from an

Figure 6.11 In most communities, hospitals have the ability to set up mass decontamination units to provide an additional tier of decontamination.

inability to respond and care for the citizens under their responsibility. Unfortunately, many live in an environment of emergency resource limitations.

Hospitals are no different. They build, staff, and equip a hospital based on the average daily need. In addition, they plan and equip for a medical surge. A medical surge, in the eyes of the hospital, includes any days that are uncommonly busy and any days with a significant emergency intake (mass casualty). But just like emergency services, they have adopted the idea that partnering with neighboring hospitals is a way to improve the services to the community. Figure 6.11 demonstrates the hospital's ability to provide emergency decontamination.

The federal government realized that hospitals that planned and prepared individually and not as a group was not in the best interest of the community they served. So, as the government is known, they decided to tie emergency preparedness funding to a coalition of hospitals instead of individual hospitals. By requiring a coalition to be formed in order to receive government funding, a panel of representatives from all of the hospitals now make decisions on government-funded equipment, staff funding, exercises, and training based on the needs of the community and not on the needs of an individual hospital.

But this alone does not solve the issue of overwhelming a hospital when a large mass causality incident occurs, especially in smaller communities where an individual hospital may have to take the brunt of the victims.

In addition to the institution of a coalition to manage government funding and thus emergency preparedness in a holistic manner, most emergency services have assisted hospitals with common communication systems where they track hospital census data in real time so that decisions concerning mass casualty patient transports can be made without overwhelming any single hospital. These systems make information concerning hospital census, bed availability, and critical services (such as ICU, Neurologic, Orthopedic, etc.) immediately available to the field units so transportation destination decisions can be made in the best interest of the patient's needs and condition.

These systems and processes can certainly assist in better coordination of patients from the scene to the hospitals but do not help control the issue of self-evacuations that can make up the majority of patients treated at the hospital. History has proven that most self-evacuees are those with relatively minor injuries. The Sarin attack in the Tokyo subway generated more than 5000 patients. Many went directly to the hospital but had only minor symptoms, such as "eye pain." These were non-life-threatening and relatively minor issues but served to clog up the medical system delaying treatment to those victims suffering much greater effects from the exposure.

Most hospitals have developed plans for dealing with a large population of minor injuries, which include alternate care facilities within walking distance of the emergency department or converting cafeterias and auditoriums to mass causality treatment areas for minor injuries while keeping the high-tech areas of the hospital (trauma rooms, operating rooms, and X-ray/CT scan areas, etc.) available for the more severe injuries. In most hospitals, triage is conducted well away from the emergency department doors with the intention of keeping the critical areas clear for arriving critical patients.

These efforts do not mean that hospitals cannot be overwhelmed. This country has witnessed large events producing hundreds of patients. Incidents such as the Oklahoma City Murrah Building Bombing, the World Trade Center Aircraft Terrorism Event, and New Orleans after Hurricane Katrina are examples of this, but there are so many more examples where the hospitals have put emergency procedures in place and were able to step up to deal with a large emergency intake of injured patients. This was very evident when the bombing of the Boston Marathon took place and large numbers of both severe and walking wounded required medical care. Hospitals have internal plans to cancel elective surgeries, discharge patients who do not need hospital care, use unconventional areas as patient care locations, and call lists to get employees who are off duty, to respond to the hospital to shore up personnel needs.

The idea of regional preparedness is to manage the entire incident as a system-wide emergency. This concept includes immediate emergency care at the scene and a continuation of that care through transport and transition into definitive care at the hospital. By developing a system following a planned emergency response that continues all of the way to definitive care, weak links will become evident and remedied to ensure that the victims will receive consistent quality care.

It is known that a certain percentage of victims will self-evacuate the scene and arrive at a medical facility for care. This percentage is greatly dependent on the type of disaster that has occurred. During the Sarin attack in the subway in Tokyo, as many as 80% of the minor injuries arrived at the hospital using their own means of transportation. But, in most cases, the victims who self-evacuate are a much smaller percentage and are easily triaged out of the main flow of more critical patients who take priority.

Most hospitals identify a staging location for minor illnesses and injuries during a mass casualty event in nontraditional treatment areas such as auditoriums, classrooms, and the cafeteria. These pre-established areas keep the noncritical patients out of the higher acuity areas of the hospital like the emergency department, X-ray and CT scan area, and surgical areas.

Those patients that are transported from the scene by prehospital care providers arrive in a much more coordinated manner. The scene is quickly divided into critical areas. First, the incident area is identified and quickly secured. Then teams of triage personnel enter to provide efficient and timely triage. Those that are determined to be urgent or critical are immediately brought to a treatment and transport section within the casualty collection point. Those that are less critical are moved to a safe area where any minor injuries can be treated, and they can wait until the more critical patients are transported. These non-critical patients will eventually be transported to a definitive care facility after the urgent and critical patients have left the scene.

The triage process also identifies those patients who are without vital signs or are too critical to attempt to save. Using valuable resources on futile efforts must be avoided. It is always best to concentrate on the greater good by treating those who can be saved. This has remained a difficult process for emergency responders whose criteria for determining who gets extensive care is greatly different when dealing with a single patient.

Those transported from the treatment and transportation area are transported based on the seriousness of their condition. Hospitals know they will receive the most critical patient first from the scene, which is much different from the walk-in patients where the least critical generally arrive first.

Casualty Collection Points (Field Treatment Site)

The Casualty Collection Point is used for the assembly, triage, field medical treatment, and transportation of casualties. This area may also be used for the check-in and receipt of medical resources (physicians, nurses, and medical supplies). Ideally, the collection point should be located outside of the danger zone of the incident and near an open field or parking lot that can be used for a medivac helicopter landing pad. The new language used by both State and Federal Emergency Management Authorities to describe the Casualty Collection Point is the "Field Treatment Site."

Temporary Medical Care Units (Alternate Care Facilities)

Many hospitals have developed what they call "Temporary Medical Units or Alternate Care Facilities" for large surges of mass casualty patients. These units are designed to be set up within a few minutes and operate to care for a less critical patient. These are not to be confused with Field Hospitals that are set up before a scheduled disaster such as a hurricane, flood, or other weather emergencies (i.e. severe heatwave). Figure 6.12 shows the institution of a Temporary Medical Unit set up on the campus of a hospital system expecting a large intake of patients. These units accommodate the overflow of more minor patients that would tend to slow or hinder emergency care to more critical patients.

The temporary medical units are generally staffed with personnel from the Medical Reserve Corp (MRC). The MRC is comprised of a group of physicians, nurses, and other professional healthcare providers who are retired or currently not working full time. Those serving in the MRC have agreed to immediately break away to serve during a disaster.

The Medical Reserve Corp

The MRC is a network of over 200,000 volunteers located in around 800 community-based units in the United States. These volunteers have agreed to donate their time to respond to emergencies and to support ongoing preparedness initiatives. These volunteers include medical and public health professionals as well as other members of the community who can provide service during a disaster/emergency. They participate in emergency exercises, planning, and actual response. With very short notice, many of these groups have been trained to respond to the hospital, obtain needed equipment, portable shelters, and set up these temporary medical units to take the burden of the less critical patients off of the hospital resources. These temporary medical units:

1) Provide for the walking wounded
2) Moderately affected patients functionally treat within a controlled environment
3) Become a resource cache for the system in a time of need.

Figure 6.12 Alternate/Temporary Medical Care Facilities can be set up anywhere and can assist local hospitals by providing care and releasing minor injuries.

The benefit of the field casualty collection point, hospital-based temporary medical units, and the medical reserve corps are to keep the hospitals functional during a mass casualty incident and allow these medical facilities to do what they do best, treat serious and critical patients. Although hospitals across the United States have written policies, purchased surge equipment, trained, and exercised their medical surge plans, it is critical that the pre-hospital personnel do what they can to relieve the burden of the less seriously injured by providing care before they reach the hospital. By achieving this goal, the hospitals can efficiently treat those victims who are more critical.

Hospital Decontamination Considerations

In many states, the regulatory state agency over hospitals now requires that all new hospital emergency departments design and build decontamination capabilities. These capabilities are for the intention of taking care of a limited number of chemically contaminated patients and are not intended for mass casualty situations. These built-in decontamination areas generally contain a privacy area, protected with shower curtains or doorways, drainage into a holding tank, and additional ventilation to evacuate the air out of the contaminated area.

For larger mass casualty incidents, most hospitals have portable (often inflatable) decontamination showers. These portable showers have a disrobing area followed by the showers and then a redress area. Hospitals have planned for a coordinated arrival of patients directed into the hospital property than to the decontamination area. The decontamination area is capable of handling both ambulatory and non-ambulatory patients. Following decontamination, hospital triage is performed, and acuity levels are set prior to the patients entering the facility.

Even on single patients, the hospital's staff will decontaminate a chemically injured patient before entering the hospital. This includes patients who were already decontaminated on the scene. The hospital staff's intent is to protect the hospital from secondary contamination. Since they cannot confirm the quality of field decontamination, most provide it to all chemically injured patients rather than take a chance on allowing a contaminated patient to enter a critical care area of the hospital.

The Joint Commission, a national accreditation agency for healthcare organizations, sets standards and expectations for disaster preparedness for hospitals. These standards mandate that patient decontamination be available in emergency department settings. Much greater detail is outlined for radiation decontamination. Although a certification ("Gold Standard") is a voluntary standard for hospitals, the benefits of being certified to a "Gold Standard" are great. The Joint Commission has always had a big focus on patient safety and emergency preparedness, and it is this focus that has pushed hospitals to develop detailed emergency preparedness plans, purchase equipment, and train staff members to respond in times of emergency.

The decontamination teams are usually made up of non-clinical staff. They are generally environmental, maintenance, and other support service personnel are purposely avoiding using clinical personnel who will be needed for emergency medical care. Many hospitals pay bonuses or other incentives to become part of the disaster team.

The biggest challenge in healthcare is the continuous turnover of the healthcare staff. Hospital workers, in particular, are constantly drawn to hospitals paying better wages or providing greater benefits. Therefore, to keep the program efficient and viable, training must be offered on a regular basis, and the recruiting of new members should be done on a continuous basis.

Call lists to employees who are off duty are critical. On any given day, the hospital intentionally staffs the hospital based on the patient census. When a mass casualty incident occurs, hospitals must have a way to supplement the current staff. This is especially true on weekends, nights, and holidays when even the daytime administrative staff are off and cannot support patient care activities.

PPE in the Hospital Environment

Hospitals have approached the PPE needs of hospital employees who are working with contaminated patients in a general manner. Their focus is to provide the level of protection to keep their employees safe while allowing them to complete their duties. They have accomplished this by purchasing Powered Air Purifying Respirators (PAPR) that function utilizing a hood and positive pressure. By using this type of respiratory protection, a fit-test does not need to be performed on each hospital responder. In addition, the filter units are powered by a lithium battery pack that is rechargeable and lasts for years of service. This respiratory device, fitted over a non-encapsulating chemical protective suit with gloves and boots provides a high level of protection commonly referred to as Level B protection.

This PPE is easy to wear and requires minimal training to don and doff safely. It is very efficient for decontamination, triage, and even initial patient care. The filter cartridges used for these PAPRs are generally the 3M™-FR57 which protect the user from most toxic industrial chemicals and all biological agents. Through federal grants, state grants, and internal resources, most hospitals maintain a large inventory of this PPE for a mass casualty incident and even for high hazard patients such as Ebola.

Hospital Isolation Rooms

In most states, all emergency departments/rooms have a requirement for at least two isolation rooms. Throughout the hospital there are additional isolation rooms. These rooms are used for patients with infectious diseases that may be spread via cross-contamination to other people within the hospital facility. Isolation rooms are built under very strict standards with the intention to separate the air environment from other areas of the hospital. An isolation room may or may not have an anteroom. An anteroom is a separate room between the isolation room and an adjacent room or hallway. The anteroom is used by healthcare providers as a location for donning personal protective equipment upon entry, doffing personal protective equipment when exiting, and storing both reusable and disposable medical supplies.

Isolation rooms are often called negative pressure rooms. These rooms maintain pressure less than the surrounding rooms. This intentional difference in pressure is intended to keep any infectious agent in the room and out through dedicated exhaust systems instead of it possibly infusing into neighboring rooms or hallways.

When an anteroom is required, it must have a self-closing door and sufficient room to allow for the donning or removing of PPE. Because of the negative pressure in the isolation room, the airflow must be in the direction from the hall into the anteroom, then from the anteroom into the isolation room. This is achieved through the use of a separate ventilation system specific to that room. The ventilation system must maintain a minimum of 12 air exchanges per hour of supplied air. In addition, the ventilation system must maintain a minimum of 0.01 – inch of water positive-pressure differential.

Although these rooms are intended to prevent the spread of biological material, they are also an excellent room to be used for preventing the spread of chemical gases or vapors. Patients entering the hospital must be decontaminated prior to entering, but, in some cases, a patient can off-gas, vomit, urinate, or defecate large amounts of chemicals capable of causing injury to unprotected people nearby. An isolation room is an ideal area for these patients to receive treatment in a safe manner without the danger of causing injury to others.

Notification and Preparation

The notification of an incident is critical for hospital preparation. With little or no notification, emergency departments and the hospital's infrastructure cannot prepare for the sudden intake of patients. This often happens in urban settings where hospitals are located in high population areas. Most often, mass causality incidents take place in these areas, and the first sign of a disaster is the arrival of self-evacuees at the emergency department door.

This sudden intake of patients without prior notification automatically places the hospital facility and staff at a disadvantage. Although a hospital may have excellent disaster plans, these plans need some time to be initiated in order for them to work. Depending on the disaster, the influx of patients may put the hospital well behind in preparing for treating multiple patients arriving suddenly at the hospital.

The same situation will occur if the responding agency does not immediately notify the hospital of the occurrence of a disaster. Unfortunately, emergency responders are dispatched to an incident that generates multiple patients and fail to notify the hospital immediately. If the incident generates 10, 20, 50, or 100 or more patients, the hospital needs time to put emergency plans into place to ensure that equipment, personnel, and hospital rooms and equipment are available for the intake.

Earlier in this chapter, the HazMat Alert System was discussed. This procedure must be in place to ensure that the hospital receiving facilities are notified and can prepare for the sudden increase in the inpatient population. Then, if decontamination is needed, the process can be put in place prior to the arrival of patients.

Responding to mass casualty incidents, especially hazardous materials mass casualty incidents, is always a team effort. The efforts to prepare, train, exercise, and fine-tune a mass casualty response can be restricted and sometimes overwhelmed by not communicating the incident to the hospital early in the response

Hospital Scenario Possibilities

For a hospital, there are four possible scenarios:

1) HazMat incident with trained and prepared responders
2) An incident with untrained responders
3) An incident with expedient responders providing transport
4) Injured contaminated patient providing self-transport

The type of incident classification is irrelevant; rather, the potential hazards that each present will be discussed. Here the HazMat incident with trained, prepared responders is the most common; however, it may present to the hospital as an incident with untrained responders due to the communication failures that may occur between the field units and the medical system, as an example. The injured contaminated patient providing self-transport or who is aided by a good samaritan is the most dangerous and the least prepared for within the medical facility.

Type 1   The events that occur with the HazMat incident with trained responders is one that is the most common and thus the most often planned and practiced. Here the incident occurs, and 911 is activated; the response is in relation to the potential harm toward the community vs. the hazard at hand. The field units respond, control, and contain the incident. On-scene decontamination and field medical treatment is provided with transportation to the medical facility. These incidents are rarely beyond the capability of the local resources so that attention to hazard potential is available and the focus is directed toward a minimal level of patient load. Hospitals are notified in advance of patient delivery, and the systems respond to the incident very much like the normal alarm.

Type 2   An incident with untrained responders or a situation in which the patient load is beyond the response unit(s) capabilities produces a dangerous scenario. Once the initial dispatched units arrive on scene, it is determined that the situation is beyond the unit's capability, or the understanding of the situation is limited, resulting in a reactive control option and thus inappropriate management. Within this scene type, the patient(s) are inadequately decontaminated and thus delivered to the emergency department as a contaminated victim.

Additionally, the resources on the scene cannot manage the patient load, causing a contingency of individuals that will self-evacuate to the hospital. These are examined within the next two incident types.

Type 3   The event has occurred, and the patient(s) do not feel that activation of the emergency services will provide the level of care they wish to receive. Or they feel that their exposure is best managed by a hospital or that they are not affected and go home. Whether the patients feel that EMS/Fire cannot give the care they are seeking, or the patient has no symptomology and will either wait to see what happens or, would like to be checked out by the local hospital, transportation is done by an untrained individual who may, in fact, become part of the problem, and now both may be contaminated. In either case, no decontamination has occurred with entrance into the hospital, local private physician, or local medical clinic, exposing a limitless number of additional victims. Much of which is an integral part of the medical system.

Type 4   The unannounced arrival of the contaminated patient from a type 3 incident or self-evacuation directly into the emergency department gives the highest potential for cross (secondary) contamination. These are the most dangerous and the most likely incident types within the WMD-MCI incident.

Hospital Decontamination Corridor
Decontamination for the hospital must be systematic, easy, and mobile. This is to allow for system deployment flexibility and ease of storage. The decontamination system itself must be easy to set up, expand, and use, with the minimal level of training.

Emergency Departments have the highest level of employee turnover. Because of this one simple fact, education for the incident that may occur is difficult for hospitals to sustain. This is in light of medical system regulations and the needs of the hospitals. This diminishes the functional training hospitals have within this area of concern.

With this in mind, the general staff should have a cursory understanding of the decontamination corridor, with supervisors and department managers having comprehensive knowledge of the system's capability and decontamination deployment. Because of the high turnover in the clinical areas, many hospitals have reached outside of these areas for their trained decontamination staff. Environmental workers, engineers, and even administration has been tapped into to provide this resource.

Because the hospital is charged with the responsibility of finding out about the incident prior or simultaneously with the field units, decontamination and treatment deployment must be quick and easy with limited steps and functional daily operational application. These patients must receive decontamination prior to the application of medical assessment and life support procedures. Remember that gross decontamination can provide a high level of rescuer protection against secondary contamination.

The procedure is the same as previously discussed. Begin with gross decontamination (removal of the clothes) and a soap and water wash. Eyes, open wounds, mucus membranes should receive saline irrigation, with the concern of the run-off as a secondary contamination issue. Non-soluble chemicals may require additional decontamination solutions such as the addition of a detergent solution (soap and water).

Hospital Decontamination Sequence Model

Primary or Gross Decontamination   This is the removal of the clothing, which can remove upward of 80% of the contamination. This is dependent on the type of clothing and the amount of body surface area that is covered. Water is the universal decontamination solution, which must be readily available directly outside the medical facility.

Runoff is a consideration as a potential source for secondary contamination to the hospital staff and rescuer. However, capture should not risk patient care but rather isolate from staff, limiting further contamination.

Secondary Decontamination   This is the washing of the individual. This concept is not foreign to hospitals; however, the application is. Ensure that all area of the body is washed in a head-to-toe fashion. Paying particular attention to the head and face (airway), under the armpits, and groin. These areas can absorb chemicals efficiently. Move the patient to a clean surface using an additional backboard, stretcher, wheelchair, or other patient conveyance.

Decontamination Confirmation   This is the one area that hospitals will have the challenge to perform. In the field, in association with the local hazardous materials team, confirmation of decontamination is accomplished with the use of monitoring and detection devices. Hospitals do not have these devices; furthermore, it is not advantageous for the institution to purchase, maintain, and train their personnel.

In this light of the need to confirm decontamination and the ability, hospitals may employ a second level of decontamination to ensure contamination reduction, with the local hazardous materials team confirming once able to break free from the incident.

At this point, medical intervention is applied. During a medical procedure, vomiting or lavage patient may interject contaminated fluid into the clean area. Procedures must be established for the removal of this material immediately.

Evidence   Clothes, body fluids, and runoff may be required as evidence. These issues must be preplanned and procedures established with the local law enforcement agency. The patient's clothing and valuables must be bagged and sealed, with a chain of custody established. Runoff, although not a system priority, should be contained and maintained for two reasons. One is the movement of hazardous material into the common environment and also as an evidentiary solution. Hospital policies should identify the use of disposable equipment whenever possible. Any reusable equipment must be tagged and bagged appropriately until decontamination can be performed.

Security Issues   The most hospital has security measures in place for common issues such as unauthorized removal of infants, undesirable individuals, and so on. These issues are common for modern medical facilities. Additionally, most have corridors for employee entrance and exit that are semi-secure. Within the context of the WMD-MCI, security is challenged with two major roles. First is the movement of personnel in and out of the facility and personnel accountability. Second is the lockdown of the institution including controlling all access points into and out of the hospital facility. Accountability can be managed by the utilization of one or two pre-designated points, with security monitoring these entrances and exits. This action in itself provides access control.

Hospital Incident Command   One of the big differences in the hospital incident command structure is that many hospitals include a separate Section Chief for Medical Staff Operations (for physicians). This is driven by the politics of the hospital. Instead of the Medical Staff being part of operations, they are elevated to a section chief position to be included as part of the senior decision-maker which include command staff and section chiefs. In some hospitals, the medical staff is included as part of the command staff along with the Safety and Liaison. But, for the most part, hospital incident command follows the same structure as the National Incident Management System (NIMS) – ICS.

In addition, most hospitals have identified an Emergency Operations Center (EOC) within the hospital. The hospital EOC conducts business much like a county or city EOC and usually communicates and coordinates with the community's EOCs. In hospital systems where a number of hospitals fall under one management structure, they will have an EOC for each hospital and a Corporate EOC who will coordinate resources between the multiple hospitals.

Although different in some ways, the command function from hospital to the first responder fits together well and allows both organizations to work together without gaps when disaster strikes. The EOC set up in the hospital is very similar to the community EOC. This works well when they must interface and coordinate resources. In a disaster situation, both the emergency response side and the hospital emergency receivers must be able to flex both staff and equipment to respond to the needs of the community.

In both hospitals and community, preparedness will ultimately fail or respond poorly if the following are not followed:

- Develop policies for mass casualty and mass intake that will be put in place when disasters occur. Without preplanning, the response will be slow, inefficient, and ultimately fail to meet the community's needs.
- Identify additional staff that can be made available during a mass casualty incident.
- Ensure that there is equipment on hand to support mass casualty response and mass casualty intake.
- Train all responders on mass casualty response. Provide annual training and exercises.
- Conduct exercises and drills with after-action plans to identify shortcomings and failures.

Experience: Hospital CO   *The Police Department receives a call from the local hospital that a man on the 5th floor is spreading chemicals and saying everyone is going to die. Law Enforcement arrived and secured the scene and called for Fire Department and HAZMAT. Upon arrival Law Enforcement wanted air packs to make entry to "subdue" the person. After several minutes of discussion, it was decided to have two fire personnel and two Law Enforcement officers make entry with air packs and ballistic vests.*

It was a cool night at about 64 degrees outside slight wind from the south at 4 miles per hour. Upon entering the stairwell, with a five-gas meter, they started to get readings of 100–150 ppm of CO. Making their way to the 5th floor the readings got higher. Opening the door to the 5th floor, the meter maxed out at 499 ppm of CO. The "suspected terrorist" was in fact a physician who was staying on the 5th floor with patients. They were able to move the doctor and patients to another area of the hospital, however, after checking the other areas. All floors in the hospital had elevated readings: the 4th floor was over 400 ppm, 3rd floor was over 300 ppm, 2nd floor was over 200 ppm, and the ER and first floor were between 100 and 150 ppm. Doors were opened on the roof and stairwells were opened up to get ventilation into the building. An electric fan was utilized to assist in ventilation.

Patients on oxygen were not affected; however, several staff were feeling symptoms and had signs of CO poisoning. Due to the mass of people in the hospital, some rooms were found to have low levels (less than 20 ppm) and people and patients were moved if they could be to these pockets of refuge. The heating system was shut down and maintenance was called. It took several hours to reduce the levels of CO especially when windows above the 1st floor do not typically open. It was a systematic slow removal of the CO.

It was determined the hospital had an old boiler system with an exhaust removal system that did not "talk" to the boiler. The gas removal systems motor burned out but the boiler kept running. Filling the hospital with CO.

Summary

Team capabilities covers a whole set of issues that may not be considered during a response but are often identified after the fact. The role of the HazMat Medical Officer must be defined and their responsibilities identified. As highlighted in this chapter, the capabilities of the team will allow them the ability to overcome the changing environment of a disastrous incident.

Regardless if the responding agency is a fire department, third service EMS provider, emergency department, or a transport agency within a hospital system, the training, equipment, and personnel must match with the mission, goals, and objectives of the agency, no matter what. It should also be realized that some events, regardless of how prepared or equipped the response agency is, are beyond their control. In these cases, a less than perfect outcome will be expected.

At times, a less than desirable outcome of an emergency response can be related to personnel provided with a lack of clear direction, inappropriate or not enough equipment, or lack of training to achieve the mission.

One such event is the activation of a "MayDay" during an emergency incident. Maydays occur when an emergency responder is trapped or injured and cannot escape the hazardous environment without help. Although Maydays do not happen often, they do occur and have been occurring more frequently over the past few years.

During a review of several Mayday cases that recently took place, found that, although a pre-entry medical evaluation was performed and heat exhaustion was suspected, there was a failure to act upon the findings. Other times, there was a lack of communication to the proper officer within the command structure or the information was ignored. Whatever occurred, a Mayday response took place and an injury resulted.

On another reviewed incident, proper safety procedures were not adhered to resulting in a potential career ending injury. In both of these cases there was a lack of training, equipment, or personnel resulting in injury to the responder.

It is also interesting to note that in the last version of NFPA 470*, a rescue team is identified. This concept is new within the standard, but not new for practical application. Within the definition of the rescue plan is the statement that the rescue team will hand over the injured responder to on-scene EMS agency. This highlights the fact that the receiving agency should have their staff appropriately trained in the handling of a chemically contaminated patient and training surrounding other hazardous response-related injuries such as heat-related stress and heat stroke.

The clear and defined roles and responsibilities during hazardous materials medical emergencies must be outlined, personnel trained, equipment purchased, and additional resources identified in order to have a team to have the capability that is expected.

---

* Information gathered from Project MayDay; Don Abbott

*NFPA 470 A.3.3.71

7

HazMat Safety Officer

Introduction

The HazMat Safety Officer's role is vast and can include many different components. This chapter reviews one of the most important roles a safety officer can have. That of looking after the health and well-being of the hazardous materials response team. It may also include overseeing all department members who may be exposed to a chemical on the scene of any kind of emergency incident.

Medical Surveillance is not just reviewing exposure reports. It is a complete package of caring for emergency responders before, during, and after the emergency response involving chemical, biological, and radiological materials. A true medical surveillance program starts on day one and continues until the end of employment. Then, the medical records must be maintained many years after employment.

This chapter follows in great detail the ideal medical surveillance program. A program this detailed is expensive to maintain and often goes beyond the requirements of OSHA. But, this type of program may actually save budgetary dollars by identifying medical issues early allowing intervention before the injury/illness becomes more critical.

Medical surveillance of all levels is important. OSHA identifies minimum requirements but a department wishing to ensure the best safety and well-being for its members will look at improvements beyond what is law.

Case Study – Lieutenant Dan

Summer of 1996 Ocala, Florida

During the years 1996–1998, the authors had the distinct opportunity to teach first responders across the county. Most of the classes were first responder operations along with Chemical Site operations, radiation for the first responder and EMS's role in hazardous materials response. During one of these training events in Ocala, Florida, a strong discussion about medical surveillance ensued. The department had not started a program, and now the responders were becoming educated in the type of program they should investigate and the merits of such a program.

Because of the work that the authors had done for the HazMat Medical program they had developed, they were well versed in type, management, and frequency of such a program. During one of the discussion topics, was that of annual physicals. A Lieutenant was

Hazardous Materials Medicine: Treating the Chemically Injured Patient, First Edition.
Richard Stilp and Armando Bevelacqua.
© 2023 John Wiley & Sons, Inc. Published 2023 by John Wiley & Sons, Inc.

questioning how and where can he find more information on development, and implementation. The authors provided the information from OSHA law and NFPA consensus standards. Not long after the coursework was completed, a medical surveillance program was put in place in the department.

A couple of months later, the authors were contacted and advised that the lieutenant, who had worked so diligently to get the medical surveillance program in place, had received his first medical physical as part of the hazardous materials team. When his chest X-ray was read by a radiologist, he was advised that the top part of his kidneys that were caught as part of the chest X-ray found irregularities. During follow-up and after a biopsy was preformed, it was found to be cancerous. He soon underwent a radical nephrectomy of the right kidney.

Six years later still at the fire department and still within the medical surveillance program, he started presenting with dizziness and nausea. He was then diagnosed as having acoustic neuroma, a rare brain tumor located near the ear. This new finding was determined to be the result of the previous surgery. After removing the tumor, he was left with hearing loss in the right ear.

Twelve years after the diagnosis of his first kidney cancer and, while still serving on the fire department, he was diagnosed with a second primary kidney cancer. At this time, he underwent a partial nephrectomy of the left kidney. Leaving him with only 80% of the left kidney. After retiring from the fire department and seven years from his last cancer diagnosis, he was diagnosed with prostate cancer for which he was treated.

Currently, he is enjoying his life with his wife and grandchildren. He credits the initial medical evaluation as part of a medical surveillance program as saving his life and allowing him to enjoy his wife, kids, and grandkids.

Medical Assessment

A health assessment should be well-rounded in order to establish a good medical "view" of the employee. A medical surveillance physical is a requirement in the OSHA HazWoper standard (29 CFR 1910.120). The standard is very specific and is required under these specific conditions:

- Are exposed to hazardous substances or health hazards above the permissible exposure levels (PELs) set by OSHA,
- Wear a respirator for 30 days or more a year,
- Become sick from possible overexposure, or
- Work on a HAZMAT team.

Although not specifically stated by OSHA, the annual physical should consist of:

- Vision screening
- Spirometry (checks lung function)
- Audiometry
- Liver function tests
- Complete blood count
- Urinalysis
- EKG
- Chest X-ray

However, after reviewing the suggested standards, one may want to incorporate all or components of the following medical assessment, either annually, biannually, or as necessary depending on potential exposures. Some components may be performed to provide a health baseline at initial employment, then again only after retirement, termination, or if a significant exposure occurs.

The content of the physical is totally the responsibility of the medical control officer (physician), and should be reviewed on an annual basis. It is suggested that the individual overseeing the program be a physician or occupational hygienist with an understanding of chemical exposures or an occupational health background. The team safety officer should also have input into the development and maintenance of such a program. In instances in which a qualified medical health provider is not available, a consultant with occupational medicine background should be considered as a part of the advisory board to the medical surveillance committee.

The analysis of these tests should be performed by a licensed physician with medical oversight of the hazardous materials team. This can be an additional requirement of the EMS medical director who provides the mandatory medical oversight for the EMS service. And, because of the more recent expansion of hazardous materials team's responsibility involving infectious diseases, an infection control officer will have to be part of this process.

The individual should receive the conventional age, sex, height, weight, temperature, blood pressure, pulse, respirations, head, nose, and throat evaluation, neurological responses (inclusive of reflexes), and evaluation of the musculoskeletal system. Also included should be an exam of the genitourinary system, abdomen, rectum, vagina, and testes. Integumentary, peripheral neurological, and vascular system evaluation are also suggested with a complete fit for duty (ability to wear required PPE under the job description) exam as identified in 29 CFR 1910.120 (f).

Note: Guidance for medical surveillance and acceptable medical conditions can be accessed through NFPA 1582: Standard on Comprehensive Occupational Medical Program for Fire Departments and NFPA 1500: Standard on Fire Department Occupational Safety, Health, and Wellness.

A *vision test* corrected and/or uncorrected should test for color perception, depth perception, and refraction. It has also been suggested that a test for the degree of night vision be conducted. As suggested by some physicians, a test for dyslexia may also be conducted.

The *auditory test* should reflect the hearing capacity of the individual. Levels that should be tested are 500, 1000, 2000, 3000, 4000, and 6000 Hz. In association with the auditory test, adequate hearing protection is required by the employer in order to maintain the level of acuity that is required of the employee. The employee at all times must protect his or her hearing against high-frequency noise and noisy environments.

A *chest X-ray* should be performed in order to establish any pre-existing abnormalities. Exposure to a chemical can cause a minor injury, and the employee may not remember an insignificant exposure. The damage of this "insignificant exposure" may have been substantial enough to result in a slight injury but may not have a high enough level of discomfort to present a concern to the person (or this individual did not recognize the cause and effect of the incident). This can be identified with a chest X-ray.

In association with the X-ray, a *pulmonary function test (Spirometry)* should be done. This test establishes the lung capacity of the individual and serves as the baseline of total pulmonary function. Each pulmonary test should encompass, but not be limited to, a forced expiratory volume (FEV), forced vital capacity (FVC), and the FEV to FVC ratio. This should be compared to normal values with respect to age, sex, weight, and height. In addition, this establishes a wellrounded baseline of the individual. Total lung capacity, residual volume, forced expiratory flow, maximal expiratory flow, and functional residual capacity should all be calculated and recorded for annual comparison.

Some agencies have even gone as far as to require pulmonary diffusion tests in which a short-lived radioactive substance (xenon gas) is introduced into the lungs. The individual has a series of scans while the substance is being inhaled. Then, a radioactive substance is injected into the vein, and a second scan is done. The individual's lung capacity is monitored, and diffusion through the lung tissue is viewed. By scanning both the gas side of the lung (alveoli) and the vessel side (capillary bed), the lung can be evaluated for scarred lesions. Poor diffusion areas can then be established.

Pre-exposure diffusion tests are usually not a part of the medical surveillance program. Most physicians in industrial medicine agree that this test should be limited to post-exposure after the pre/post chest X-rays have been done, evaluated, and compared. The diffusion test is an additional tool for the physician to evaluate the extent of injuries after exposure to certain chemicals. The primary concern is the health of the respiratory system.

The cardiovascular system can be assessed by comparing the resting EKG 12 lead to a stress EKG. This graded assessment can give the cardiovascular baseline that is needed for employment. This comparative EKG is then used to collate the post-exposure injury with the pre-exposure findings. Any suspicions about the electrical conduction system then can be further analyzed.

Urine is used to testing for multiple system functions to include color and appearance, pH, specific gravity, and glucose levels. Protein, bile, sediment, and glucose tolerance are all additionally suggested. Metabolites from a previous exposure may process through the liver and are excreted in the urine. It is these metabolites that are searched for during a urinalysis test.

Complete workup including blood, urine, EKG, and neurological testing should be performed after all exposures of known problem chemicals such as pesticides, heavy metals, and aromatics. In these cases, a comparative analysis may provide information on exposure review.

Medical Surveillance

Medical surveillance and medical screening have often been used interchangeably, but, in reality, the strategies are significantly different.

Medical screening is part of medical surveillance but most certainly not all-inclusive of a medical surveillance program. Medical screening has a medical focus and includes routine physicals and periodic medical exams. The focus of medical screening is to discover, diagnose, and treat a disease process that is already underway.

Medical surveillance is much more complex and involves strategies to reduce exposures, and has a prevention focus. The purpose of having a surveillance program is to eliminate negative trends, identify underlying causes of exposures and hazards.

Both are important components of a successful health and safety program. In this section, the focus will be on the components of a quality medical surveillance program, including routine physicals, work histories, physical assessments, and biological testing.

Initial Baseline Physical and Annual Physical

OSHA Standard 29 CFR 1910.120 establishes mandatory medical testing for all hazardous materials team members. But the standard goes further and actually identifies a process for a more comprehensive medical surveillance program. This program includes a four-part process including:

1) Baseline physical conducted when a person becomes part of a hazardous materials response team member.
2) Annual physicals are conducted each year that the person is a member of the hazardous materials response team.
3) An exit physical is conducted when a member of the team leaves the team.
4) Post-exposure-specific physical/follow-up exam only when deemed necessary.

In addition to these mandatory steps, medical monitoring is also required on the scene of a hazardous materials incident.

The baseline physical should include a comprehensive medical health history including any recent or chronic diseases, prior chemical exposures followed by a physical examination that includes hearing, sight, detailed blood laboratory testing, and other system specific diagnostic testing (i.e. Chest X-ray, Spirometry).

Lab work should include a complete blood count, kidney and liver function, blood sugar and urea nitrogen, creatinine, sodium, potassium, chloride, magnesium, calcium, inorganic phosphorous, total protein, albumin, globulin, total bilirubin, alkaline phosphate, lactate dehydrogenase, gamma-glutamyl transpeptidase, aspartate aminotransferase, alanine aminotransferase, uric acid, and urine tests. In addition, heavy metal testing should include mercury, arsenic, and cadmium. The physical testing should also include an EKG, pulmonary function test (spirometry), and chest X-ray. The results of these tests provide a baseline for future reference should a chemical exposure take place.

It should also be noted that detailed records of all medical evaluations are to be kept on file and must be easily accessed should they be needed to compare with findings after an exposure takes place. All records should be maintained for 30–40 years after termination or retirement. This time is dictated by federal and state laws or local ordinance which can require the retention of records for up to 70 years (see Figure 7.1).

A comprehensive metabolic panel (CMP-14) includes 14 different blood tests used to determine kidney function, liver function, fluid and electrolyte balances, and diabetes. The CMP-14 was formally known as the SMAC (sequential multiple analyzer chemistry). The CMP-14 includes:

1) Glucose – to evaluate for diabetes.
2) Creatinine and
3) Blood Urea Nitrogen (BUN) to evaluate kidney function and damage.
4) Albumin is produced by the liver. Low levels indicate poor liver function.
5) Total Bilirubin is an assessment of liver function. High bilirubin could indicate liver damage.
6) Total Protein is the combined measurement of both albumin and globulin. This test helps evaluate both kidney and liver function.
7) Alanine Aminotransferase (ALT) is produced in the liver, kidneys, heart, and pancreas. High levels indicate damage to the liver.

MEDICAL HISTORY FORM

Details about your medical history is confidential and handled under the laws governing HIPPAA regulations.

Last Name: _____ First Name: _____ Middle Initial: _____

Date of Birth: _____ Gender: M / F   Emergency Contact: _____

Home Address: _____ City: _____ Zip: _____

Home Phone: _____ Work Phone: _____ Email: _____

Occupation: _____ Insurance: _____

Past or present medical history have you ever had any of the following:

☐ Asthma	☐ Irritable Bowel Syndrome	☐ Fractures: _____
☐ Heart Disease	☐ Ulcers	☐ Eczema
☐ High Blood Pressure	☐ Celiac Disease	☐ Psoriasis
☐ High Cholesterol	☐ Crohns Disease	☐ Hives
☐ Diabetes	☐ Liver Disease	☐ Do you drink alcohol: Y / N How much: _____
☐ Thyroid disorder	☐ Convusions/Seizures	☐ Do you drink alcohol: Y / N How much: _____
☐ Thyroid disorder	☐ Convusions/Seizures	☐ Do excercise: Y / N
☐ Kidney Stones	☐ Migraines	☐ Do you take recreational drugs: Y / N
☐ Chronic Kidney Disease	☐ Stroke/TIA	☐ Endometriosis
☐ Chronic Sinus Infections	☐ Anemia	☐ Weight loss surgery
☐ Eye Disorders	☐ Blood clots	☐ Tonsillectomy
☐ Hearing loss	☐ Cancer	☐ Gall bladder removed
☐ Allergies/Hayfever	☐ Arthritis	☐ Ovarian cyst removal
☐ Allergies Medication: _____		

Does your immediate family have any of the following: Adopted ☐

	Mother	Father	Siblings	Grandparents
Alcoholism				
Blood Clots/Clotting Disorders				
Breast Cancer				
Colon Cancer				
Melanoma				
Other Cancers (list)				
Diabetes				
Drug Dependency				
Epilepsy				
Heart Disease				
High Blood Pressure				
High Cholesterol				
Hepatitis or other liver diseases				
Mental Illness				
Stroke				
Cardiac Arrest under age 50				
Other (explain)				
Deceased				

Your Signature: _____ Date: _____

Figure 7.1 Utilization of a medical history form will ensure that all pertinent information concerning an exposure is captured.

8) Alkaline Phosphatase (ALP) is an enzyme produced in the organs and bones. Low levels indicate poor liver function or bone-related disorders.

9) Aspartate Aminotransferase (AST) is an enzyme found in the red blood cells, liver, heart, pancreas, kidneys, and muscle tissue. High levels of AST are produced when there is damage to one of these areas.

10) Sodium is a mineral that helps to keep a proper balance of water and electrolytes in the body and is important to muscle and nerve function.

11) Potassium serves much the same function as sodium. Both sodium and potassium compensate for each other.

12) Carbon dioxide (CO_2) is a waste product of metabolism. Kidneys and the liver maintain the levels of CO_2.

13) Calcium is the most prevalent mineral in the body and is used in the functioning of the muscles, nerves, and brain.

14) Chloride is an electrolyte important for blood pressure, volume, and blood pH levels.

Testing for heavy metals usually contains tests for lead, mercury, arsenic, and cadmium. It may also include copper, zinc, aluminum, and thallium. These are found in the environment, foods, water, and medicines. High heavy metal concentrations in the blood cause organ damage, CNS damage (behavioral changes), and cognitive difficulty. Removal of high blood levels of these metals is difficult and often requires the use of a chelation agent and dialysis.

The annual physical can be the repeat of the initial baseline physical or a modification of that physical. The information from the annual physical is used to determine the member's fitness for duty.

When the member's tour on the hazmat team ends, an exit physical must be completed. This physical should mirror the initial baseline physical completed when they started on the hazmat team. This physical is the endpoint of physical monitoring for chemical exposure. The exit physical helps to determine if a workplace exposure has occurred or if a chronic illness has changed the overall health of the team member.

The exposure-specific physical may be performed after an exposure has been experienced/documented, whether symptoms are present or not. This specific physical includes components from the routine physical examination and laboratory tests that are customized toward the chemical involved in the exposure. This examination may take place on a given schedule, such as every two weeks or once a month, and continue until the physician determines that a complete recovery has occurred.

Pre-Entry Physical

Pre-Entry Physicals have continued to be an area of contention. Other documents and books list blood pressures, pulse rates, temperatures, and other criteria used to exclude someone from making either initial entry or re-entry into the scene. Many believe that there can be black and white exclusion criterion established. This can never be true. Although guidelines can be established purely for reference purposes, the true criteria should be the good judgment of the paramedic conducting the physical.

Hazmat scenes are stressful situations. Placing any emergency responder into a level A, fully encapsulating suit is also stressful. Combine this with the thought that they are entering an atmosphere that could lead to injury or death without the suit, and then expect them to not have high blood pressure or a rapid pulse is unrealistic. Expecting pulse rates below 100 is ridiculous. So, why has anyone established exclusion criteria at all? Somehow having hard guidelines gives paramedics and incident commanders a sense of comfort. Paramedics simply do not have to rely on their assessment skills but, instead, rely on nothing but taking a pulse and respiration and saying yes based on hard numbers. This is not the belief of the authors of this book. Both are long-time patient care providers, and find it impossible to accept the thought of making a decision based on black and white numbers. Instead, it is expected that a paramedic evaluates the diagnostic findings along with paramedic intuition to determine if a team member is ready for entry.

Pre-entry and post-entry physicals conducted at hazardous materials incidents should provide a somewhat detailed observation into the physical and mental capability of an entry team member. Some departments take this task lightly and only conduct a quick pulse and blood pressure, as a cursory step in monitoring their entry team members. But the most comprehensive and safest pre-entry monitoring is done as a 10-step process. This provides complete medical monitoring information to help a paramedic determine the fitness of an entry team member to make an initial or second entry.

The pre-entry and post-entry physicals should include:

1) Blood pressure
2) Pulse
3) Respirations
4) Pulse oximetry/Capnography
5) Lung field auscultation
6) 12 lead EKG
7) Temperature
8) Weight
9) Reinforced hydration
10) General sensorium

Some departments conduct the pre-entry physical upon assignment in the morning of the shift, after the shift briefing. The best time to determine the real value of blood pressure, pulse, and respiration is before significant work is performed. Even checking out the equipment on the truck, washing the truck, and other morning duties can elevate these vital signs. The best practice is to conduct these tests right after the morning briefing. Then, record these findings on a confidential sheet that is carried on hazardous materials calls and can be used as comparison data with the post-entry physical.

The one data point that cannot be measured during the AM meeting is the pre-entry weigh-in. It is critical that this be done prior to donning PPE and post doffing at a hazmat scene.

Note: The determination to allow the team member to make entry or re-entry into the hot zone should be made by a HazMat Medic who has appropriate knowledge of hazardous materials response and toxicology. Ideally, the HazMat Medic should be informed of the chemical involved and the signs and symptoms of exposure. This will allow the medic to pay particular attention to those pre-entry data points that could change if an exposure takes place. Knowledge of the physiology of both hydration and heat-related illnesses is also imperative for an accurate assessment of the team member.

It cannot be stressed enough that entry criteria must remain flexible based on each entry team member and are meant to provide guidance (not absolute exclusion) to the HazMat Medic. A HazMat Medic can make a determination of the current physical status of the entry team member based on these findings, patient history, pre-existing knowledge of that team member, and overall paramedic intuition.

Considerations of the Entrance Physical

1) ***Blood pressure.*** Blood pressure is a measurement of working pressure (systolic) and resting pressure (diastolic). This relationship is shown by the formula CO =SV × HR; B/P = CO × PR, where CO is the cardiac output. The CO is dependent on the heart rate (HR) times the stroke volume (SR). In other words, for every beat of the heart, a finite volume of blood is ejected from the heart. Plugging this into our second formula, the cardiac output times the resistance (peripheral resistance, PR) gives us a blood pressure (B/P). The systolic is the working pressure of the heart, whereas the diastolic is the resting pressure found in the blood vessels between heart beats. This pressure allows us to recognize the individual's heart function and make a decision about the entry. The assessing paramedic must be concerned with these pressures and how they compare with age/sex norms and values, and the historic values of the individual.

   Be alert for both hypotensive and hypertensive members. The blood pressure should be compared to normal values and/or documented norms for that individual. The paramedic should be concerned with diastolic values at or above 100 mm Hg. If there is a diastolic pressure above 100 mm Hg, the heart during its contraction phase must overcome this pressure in an effort to circulate blood. This high of diastolic pressure puts extreme stress on the heart muscle, increases the oxygen need of the heart muscle, and can contribute to injury of the heart muscle.

   Low blood pressure with a systolic pressure in the low 100 second or below, especially with an increase in pulse should be of concern. When this occurs, an orthostatic blood pressure test should be conducted. If there is a slow recovery during the orthostatic test, the paramedic must consider excluding this responder from entering the hot zone and wearing PPE.

   Be sure to obtain a blood pressure prior to the SCBA harness being placed on the individual. Constriction of the thoracic cavity by tight straps can cause an abnormally high BP. Although not supported by any medical research, it has been suggested that the weight of the SCBA, while the wearer is sitting, applies pressure to the subclavian artery, which in turn reduces the blood flow, causes backpressure, and elevates the BP. In order to obtain accurate vital signs is to provide the medical evaluation prior to donning the SCBA and after it has been removed.

2) ***Pulse.*** Simply counting the pulse rate is not enough. When assessing pulse, the quality and rate must be considered together. The quality and rate of heart contracture may provide clues concerning hidden anxieties. An accelerated heart rate with a bounding pulse may be a sign of high anxiety or a fear that the team member may have been hiding. Compare pulse rates with normal values. A pulse rate above 100 bpm at pre-entry, while in a relaxed state, may be a cause for concern.

3) ***Respirations.*** Respiratory rate and auscultation of the lungs should be conducted on all entry team members. The rate and depth of each breath are a component of the respiratory assessment. Any responder with a resting respiratory rate above 24 a minute, or any abnormal finding on auscultation such as rales, wheezing, or rhonchi prior to entry should represent a concern to the assessing paramedic who may choose to exclude the entry team member from entering the hot zone.

4) ***Pulse oximetry/Capnography.*** Pulse oximetry provides the status of oxygen combined with hemoglobin in the blood. Using the rate and volume of respirations, lung auscultation, capillary refill, and mucosa color give the assessing paramedic information

concerning the ventilation status of the entry team member. Pulse oximetry adds to these data points. SaO_2 below 95% should be a cause for concern for the assessing paramedic. This level, especially when combined with other assessment criteria such as abnormal lung sounds, may be reason to exclude this team member from entry into the hot zone.

Capnography provides information concerning the healthy metabolism of the cells. Normal readings indicate the level of CO_2 on exhalation at the end of the expiratory cycle, called end tidal CO_2 or $ETCO_2$. The normal readings are from 35 mm/Hg to 45 mm/Hg. Although pulse oximetry provides an indication of the amount of oxygen available for cellular metabolism, capnography provides an indication of how well the cells are using that oxygen and producing carbon dioxide. Capnography also provides a capnogram that indicates the flow of carbon dioxide through the lungs during respiration.

A paramedic seeing a shark fin pattern on the capnogram indicating constriction of the bronchioles or an end tidal CO_2 level below 35 mm/Hg are both cause for concern. Both would indicate that there is an issue in the area of respirations and oxygen use and, therefore, should provide concern to the paramedic about the health of the team member being assessed. In addition, an $ETCO_2$ of greater than 45 mm/Hg may indicate the onset of heat stress or heat exhaustion.

5) ***Lung field auscultation***. Auscultation is an important aspect of the cursory medical exam. If at all possible, have the same individual who listened on the entrance physical listen on the exit physical. Continuity is important, especially if an exposure has occurred. While auscultating the lungs, listen posteriorly and anteriorly in the bases and in the apex of the right and left lung field. Be alert to sounds such as rhonchi, wheezing, or rales, especially when assessing the team member after exit or if the exposure occurred.

6) ***12 lead EKG***. 12 lead EKGs have become a normal assessment tool in the pre-hospital arena. When assessing entry team personnel including those working in the decontamination corridor, the EKG should be performed and interpreted prior to entering the hot/warm zones. Elevated U waves can be an indicator of dehydration or associated with other abnormalities that may be present without the benefit of symptoms. Any dysrhythmia that has been established as "normal" for that individual *must* be cleared by the medical director. If not cleared, all abnormal dysrhythmias should be considered exclusion criteria by the assessing paramedic.

7) ***Temperature***. 98.6 °F is the established norm for an oral temperature. It is proven that an oral temperature is more accurate than other more passive means of gaining body temperature. Rectal temperature is actually the most accurate representation of core temperature but it is not likely that entry team members would tolerate this type of temperature assessment. The next most accurate is the tympanic thermometer. At a hazardous materials incident, responders are placed in partial or fully encapsulating suits that restrict the body's normal cooling potential. This creates elevated body temperatures by removing these cooling mechanisms.

The entrance temperature is used to establish a baseline temperature for the entry team member for that particular day and time. When this temperature is compared with the temperature taken during the exit physical, a comparison can be made to determine the severity of heat stress the emergency worker has endured. If there are *any* signs of heat-related illness, then efforts should be taken to rapidly cool the team member. On the contrary, in atmospheres that present with low ambient temperatures, be prepared to warm the emergency responder. The tympanic temperature at entry should be between 97.0° and 99.5 °F. Any temperature above or below this range must be treated appropriately.

8) **Weight.** The difference of body weight from the entrance physical as compared to the exit physical can give valuable information on the hydration status of an entry team member.

Comparison of pre-entry and post-entry weights establishes hydration levels. A 5% loss can represent a responder who is in danger of a heat or dehydration injury. By dividing the exit weight by the entrance weight, a quotient is given. Subtract that from 1, times 100, will give the percent weight loss. Be sure to weigh the entry team member wearing the same clothing at entry and exit. With a weight loss of greater than 5%, aggressively treat utilizing an IV may be warranted.

9) **Reinforced hydration.** Hydration is a concern for anyone wearing chemical protective PPE. During a hot day or an extended event, sweating can cause dehydration. Providing the right amount of water prior to donning PPE can be tricky. Providing too much will cause an entry team member to be very uncomfortable as their bladder fills. Not providing enough may cause premature dehydration while in the hot zone.

Caffeinated drinks are strongly discouraged. Beyond the fact that they slow absorption, cause diaphoresis, and increase metabolism which increases heat production, they also have other psychological/physical contraindications. Research by the International Association of Firefighters and those involved with CISD has indicated that caffeinated drinks add to the anxiety of the emergency worker. If a negative situation is occurring, the stress level will increase and caffeine will add to these levels.

10) **General sensorium.** General sensorium is an evaluation that begins at the entrance physical and does not stop until the individual is in rehabilitation. Evaluation of the conditions of alert and oriented to person, place, time, and situation, along with the appropriateness of conversation, gait, and verbal response. Assessing symptoms of reported headaches, blurred vision, numbness, or ringing in the ears along with following simple commands such as performing grip strength assessment, touching the tip of the nose with arms extended, head back with eyes closed, are simple assessment steps that should be taken. A cognitive evaluation must also become part of the orientation process. Spelling a word backward and counting by 3's, 7's, or 9's (serial assessment) can be used. Based on the findings of these assessments, the paramedic can determine the fitness of an entry team member.

Post-Entry Physicals

This physical is used to establish rehabilitation criteria or guide us through our treatment modalities. It is an additional tool to establish rehabilitation and compare with the baseline. By acquiring extensive information at the time of the incident, this provides the medical community information that can be used later. In addition to this, by medically evaluating on-scene personnel, normal stressors, such as heat stroke can be responded to appropriately.

The post-entry physical is a part of the termination of the incident, but it can also guide the command staff toward future personnel needs. Generally, when an incident arises and the entry team is asked to perform its duties, the exhaustion level that is reached in a short period of time is quite high. In some situations, it would be advisable not to send the original entry team back in for an hour or until readiness can be ensured. However, most hazardous materials teams only have limited resources and manning. The post-entry physical can aid command staff in making such decisions.

The post-entry physical is composed of all the components of the entrance physical. This physical should be taken at one minute after decontamination and again at the five-minute mark. At the five-minute mark, all vitals should be within 10% of the entrance vitals.

The post-entry physical should include:

- Blood pressure
- Pulse
- Respirations
- Pulse oximetry
- Lung field auscultation
- 12 lead EKG
- Temperature
- Weight
- Reinforced hydration
- General sensorium

In all cases where entry into a hazard zone is the mode of mitigation, all team members entering the warm or hot zone shall have the entry physical performed.

Handling the scene of a hazardous materials incident takes planning, with the combination of a strong strategy, followed by concise tactics. In order to perform all the rudimentary functions without forgetting a component, data sheets should be generated. Just like the foreground tactic board, the medical status sheet gives EMS responders a systematic approach to the pre-entry and post-entry physicals.

Personnel should be familiar with all data sheets and their appropriate use. Additional copies should always remain in a file, somewhere on the hazardous materials response unit. A complete kit containing all the needed paperwork that would allow personnel to start medical evaluations should be placed in a conspicuous area of the response unit.

As with all scenes, documentation must be generated. This status sheet enables the response team to start the paperwork within the medical sector. If an exposure should occur, preliminary paperwork has already been started, so that post-exposure assessments can be compared with the pre-exposure assessments.

**Note: Figure 7.1 is an example of such a form. Each form should be simple enough so that anyone possessing medical training can "plug" in the facts.

Use of Findings

The values found on pre-entry evaluation are used simply to make a judgment call. It is not recommended that a hard line be drawn concerning exclusion from entering the hot zone. It is best to have a paramedic evaluate the responder and review the findings of the assessment, then make a determination based on the overall information and not on one value. With that said, here are some guidelines that can be used for reference.

- An irregular pulse not previously diagnosed and any atrial or ventricular ectopy that is multifocal in nature. Heart rate exceeding what would be expected based on the circumstances. One rule of thumb used by some departments is a resting heart rate above 70% of the maximum heart rate as determined by age ($220 - age \times 0.7$).
- Obtaining a tympanic temperature of greater than 100°F may indicate an infectious process or higher metabolic rate.

- Abnormal lung sounds or a capnography less than 35 mm/Hg, more than 45 mm/Hg or SOB.
- Recent onset of medical problems such as diarrhea, dehydration, and vomiting.
- Alcohol consumption in the last 24 hours.
- Blood pressure (undiagnosed) with a systolic greater than 150 or diastolic greater than 100, especially if other signs/symptoms are present (headache, dizziness, etc.)
- Pulse oximetry of less than 95%.

These are provided as guidelines and should never be used individually as specific exclusion criteria. All gathered data should be carefully analyzed before determining the capability of the entry team member. Common sense and paramedic knowledge can and should be used to determine the physical and mental ability of an entry team member to don chemical PPE and make entry into the hot zone.

Monitoring of the entry team should not stop after the pre-entry physical is completed. While in the hot zone, an entry team can still be monitored but on a limited level. Any member of the entry team who states a complaint such as a headache, shortness of breath, or has difficulty answering questions, or displays slurred speech should be immediately removed from the hazard zone. Signs of fatigue such as a staggered gait or propping on handrails or walls should be addressed immediately.

There is some recent technology that allows real-time medical monitoring while inside of chemical protective PPE. The new technology uses a monitoring device that is strapped to the chest of an entry member. A monitoring station located in the command post or other safe location can then monitor the heart rate, activity level, and temperature of the entry team member.

The post-entry physical is essentially the same as the entry physical. Vital signs including pulse, BP, respiratory rate, oximetry, capnography, temperature, weight, and EKG's are taken and compared to pre-entry data. The properties of the chemical involved should be assessed, and the signs and symptoms of an exposure checked against the entry team members. Signs of dehydration and heat-related injuries should be noted. These include a high temperature, low blood pressure, rapid breathing, and heart rate.

The values should be reassessed every 10 minutes until they are within 10% of the pre-entry physical. Taking an unusual amount of time to recover or an inability to recover must be brought to the attention of the paramedic and the HazMat command officer.

Post-entry weight must be taken and recorded. Post-entry weight should be subtracted from pre-entry weight and the change should be calculated into the amount of water lost during entry. Remember that a body can sweat up to 2000 cc an hour. A significant loss of body weight must be considered as possible exclusion from re-entry especially when tied to a low BP and fast pulse weight. In these situations, an IV should be considered to replenish the vascular fluid loss. As discussed earlier in the text, the body can only absorb about 800 ccs an hour through ingestion. Rehydration using oral intake only will work if the loss of weight is not significant, but if the re-establishment of pre-entry weight will take hours through ingestion, then and IV for rehydration should be considered. An IV will hydrate a body quickly and allow a previously compromised firefighter to go back to work in a much shorter period of time. Replacing electrolytes is important after hard work and extensive sweating; most adults maintain electrolytes in reserve. Any symptoms associated with electrolyte loss occur, such as muscular cramping, usually they can be easily corrected through oral or IV hydration alone and electrolyte replenishment can take place later.

Preventive Health Screening

A service that has come up in the discussion of medical surveillance is that of preventative screening, sometimes called wellness screening.

Preventative screening is typically a measure of weight vs. height parameters and where you may fit into the spectrum of the population. Along with weight, blood pressure, vision, blood count, hearing test, eye exams, along blood chemistry are the usual parameters taken and then discussed with the individual. Very similar to the typical entrance medical exam a responder may receive prior to working within the emergency services. These take a snapshot of the health of the individual at that specific point in time. At the beginning of employment, it may be considered as a baseline. As of late, this term has been applied to electronic screening and full-body MRI screens (LifeScan®).

Wellness screening or sometimes called the employee health screening, and it is predominately a screening similar to the preventative type of screening, however, has a dietary component in addition to the above medical tests (LifeScan is sometimes added into this style of surveillance).

Post-Exposure Physicals

If exposure occurs, a set of follow-up physicals are indicated. Again, like most hazardous materials mitigation strategies, preplanning must occur. Standard operating guidelines must address the likelihood of accidental exposure, and the policies that are required are in place. How to deal with it and who is responsible for post-incident physicals must be addressed. Potential exposures are dependent on the products that are in your jurisdiction. While it would be nearly impossible to identify all substances, certain types and/or certain harmful chemicals can be identified as potentially hazardous. These substances, along with any exposure, should activate a mandatory post-incident physical schedule.

The following is a list of hazardous compounds that have a potential to cause severe health effects. Therefore, anytime these chemical categories are encountered, a post-incident physical may be indicated, provided that the quantities and concentration of the material were significant.

1) Aromatic hydrocarbons
2) Halogenated aliphatic hydrocarbons
3) Dioxin
4) Organophosphate and carbamate insecticides
5) Polychlorinated biphenyls
6) Organochlorine insecticides
7) Cyanide salts and their related compounds
8) Concentrated acids and alkalis
9) Nitrogen-containing compounds (organic nitrogen)
10) Any compound that was in fume, dust, or airborne configuration
11) Asbestos

One must be realistic in the approach. For example, the laws that surround the removal of asbestos if one must replace old siding on the house are strict, and the removal process is costly, lengthy, and quite detailed. However, the amount of airborne asbestos that one may be subjected to is nominal when an appropriate level of PPE is used along with administrative controls. Considering that the same monitors that are used in the asbestos removal industry may indicate high levels if placed within a few blocks of any major interstate (asbestos is used within the brake lining, which is released during the application of braking). In general, the surveillance of these chemicals is such that if the quantities and concentrations are high, medical surveillance may be indicated. This level should be addressed during the planning stages of the program and evaluated as needed. The post-incident physical should take place immediately after the incident and every three months thereafter or under the advisement of the attending physician. This is continued until the next annual physical or until deemed unnecessary by the medical director/control.

If an unknown was encountered after the incident, the chemical(s) should be identified (see Chapter 6, Detection and Monitoring). Once this occurs, specific tests to identify target organs can be assessed appropriately. Comparison of sequential medical examination can identify the medical course of action. In any case, a review of the incident should occur. This is done to identify all possible engineering controls and their use and how they worked under real-time conditions (administrative and engineering controls should be reevaluated at this time).

Note: Administrative controls are work place practices; they are clearly written policies and procedures that outline how the agency will do business. It is as detailed as providing guidance on frequency, duration, and severity of the exposure that is allowed under a variety of conditions. Standard Operating Procedures, policies, and procedures are examples.

Engineering controls are procedures or practices that reduce or remove the hazard from the worker. Ventilation systems for confined space, PPE donning and doffing are examples. It is when we place a barrier of some sort between the responder and the hazard.

By using both administrative and engineering controls a level of protection is provided. These are the issues that a Health and Safety Officer (HSO) will look for during a hazmat event.

Biological Monitoring

Beyond the annual physical, a few additional components may be necessary to incorporate within the agency's standard operating procedures for exposures or potential exposures. In cases where exposure has occurred, it may be necessary for the individual to go through extensive exposure monitoring over an extended period of time. This monitoring process measures the biological fluids for the suspected chemical such that damage to a target organ can be identified early on.

Chemicals, metabolites, and tissue sampling are but a few components of this type of long-term medical monitoring. In cases where the hazardous materials responder has the potential of becoming exposed, a different approach is used with components of biological monitoring and the annual physical in mind. For example, the hazardous materials technician should in addition to the routine physical have a heavy metal, cholinesterase, and aromatic hydrocarbon (included in this are PCBs and asbestos) workup. This is done at the time of team selection.

After the individual has spent two years on the team, another analysis is performed (provided that the individual has responded to calls involving these materials or the suspicion of these materials is high). High-level medical surveillance should be done every two years until the emergency worker retires from the team, NOT the agency. All medical information is maintained and analyzed for possible exposure effects for the length of his/her employment, whether on the team or not. Once on the team, surveillance should be maintained throughout his/her career. The same procedures should also be performed for dive rescue and confined space team members. The level of chemical exposure can be far less than for the members of the hazmat team, but the chance of contaminated water entry is quite high, or contaminated confined space is always present.

Team Exit and Retirement Physicals

Exit physicals from the team are just as important as entrance physicals into the team. From the time a responder enters the emergency services, they enter a work environment that has the potential for exposure to a variety of biological and chemical substances. The magnitude of these exposures is what needs to be evaluated long term. In order to quantify these exposures, one must have an entrance physical before team activities begin, annually, and upon retirement from the team. From this information, cause and effect can be analyzed. This would include retirement from the agency as well.

Program Review

A periodic and systematic review of cases should be developed and used in conjunction with total system analysis. Comparison of personnel, equipment, and procedures is essential to determine trends. These trends may mark early signs of system failure, thereby establishing appropriate corrective measures.

Several items must be addressed in order to start this review process:

1) Review of incidents when exposure occurs.
2) Analysis of exposure versus incidents.
3) Protective measures established and reviewed such as procedures, PPE used, and tactics employed.

ADA, Civil Rights, and Health Insurance Portability and Accountability Act (HIPAA)

HIPAA laws play into consideration any time a response team is gathering medical information. However, in order to perform medical evaluations effectively, consent from all personnel is required. Annual physical data, along with additional physical information, can be placed into a relational database. From this, a medical trend can be analyzed. Everything from chemical contact, type and concentration of the chemical, the temperature in the suit, and ambient air temperatures should also be placed in our database.

Analysis of exposure versus incidents is a capability of computerized recordkeeping. Once a database has been established, questions about the engineering or administrative controls can be asked. Under the guidance of industrial hygienists, toxicologists, and medical personnel, trends can be established, and the problems surrounding the trend can be changed or modified.

Protective measures can be established once we know what controls are failing. Procedures can be analyzed for review, and in some cases, models of historical scenarios can test these new procedures. In every step of the way, it will be important for the medical director to become involved with the surveillance program and the intricacies of such a program.

It is important that the attending physician (and medical director) is knowledgeable of hazardous materials operations, the training, and the tactics employed. The exact frequency and contents of the exam are truly the responsibility of the physician; however, a current copy of the OSHA standards and NFPA documents will enable the physician to organize the appropriate surveillance program.

Critical Incident Stress Debriefing

During the early 1980s and into the present, stress management has been a focus of emergency response services thanks to the work of Jeffrey Mitchell PhD and Grady Bray PhD, who are noted for their research in determining the effects of stress on the emergency worker. In recent years, emergency services have placed special emphasis on the management of stress and the related outcomes. In the early days, the term "burnout" was applied. Typically, this term referred to those individuals who, during the course of their career, had been subjected to an event or groups of events that would leave lasting scars on their psychological well-being.

For some, friendships and peer support helped them recover from these psychological injuries including depression, while others sought out other methods to cope with the feelings and emotions, such as alcohol and drugs. Stress in life is something from which we as human beings are not going to escape, especially those involved in emergency services.

The stress that any emergency worker can encounter is wide-ranging. On one side of this double-edged sword are the anxieties that make this line of work so interesting, while on the other side is the reality of pain, emotions, and human suffering.

Each one of us differs emotionally. Some can handle stress very well, and it is only with high levels of emotion that these individuals seem to crumble. Others have a more difficult time and seem to keep the emotions in or talk about them freely with close friends.

It has become generally accepted that law enforcement, emergency medical services, firefighting, emergency room nursing, dispatchers, and aeromedical response all can provide stress that may interfere with the emotional status of the worker. Hazardous materials events, at any level, are no different..

At the hazardous materials incident, chemical properties, environmental conditions, and sheer quantities severely impact the incident-all factors that we cannot control. Emergency workers all have one thing in common; their physiological makeup is more or less the same. The function of emergency responders is to control the scene or incident. They are taught from day one that control of the incident is a means to the end. They practice and encourage incident command systems which in turn organize and control the incident. However, at the hazmat scene, control is only temporary and can be lost at any moment. Chemical properties, environmental impact, and quantities of the materials that are being handling cannot be changed or sometimes even controlled.

All responders seem to have a higher or lower degree of obsessive-compulsive behavior and a need to control, uncontrollable incidents. In training, responders are taught to be perfect and beat themselves up emotionally when something occurs that is uncontrollable.

How do these simple traits affect emotions at the hazardous materials incident? Why do they occur? We know by virtue of the dangers that hazardous chemicals present, that the incident will be difficult if not impossible to immediately control. Unfortunately, something can go wrong. For these reasons and countless others, the hazardous materials team must have, within its team structure, a program that can assist in the emotional well-being of all members.

In order to do so, a policy and procedure for hazardous materials medical surveillance and safety management must include the components of critical incident stress management (CISM). This type of need will not often occur, which makes the necessity of such a program a vital component of the medical surveillance policy.

This policy should provide several levels of emotional support for the affected responders and their families. The first level of assistance should come from peer support personnel who have been educated in hazardous materials mitigation and the principles of CISM. They should have experience with both the CISM format (defusing and debriefing) and be involved with a peer support group.

The second level of the program should be the CISM team. Here, defusing and debriefing along with mental health professionals can establish criteria for when the defusing and/or debriefing will take place. The mental health professional must become involved with the hazardous materials response team, understanding all the components of defensive and offensive roles. Evolutions within encapsulating suits will give the mental health worker a firm understanding of what it is like to be a hazmat team member.

The overall structure of the CISM component to hazardous materials is the same as for all emergency scenes. First, the defusing can take place at the scene or shortly thereafter. These defusing should be done by the individuals that have been trained in hazmat mitigation and CISM principles. They should have the highest regard for confidentiality with respect to the emotions and feelings of others. The defusing should consist of but not be limited to:

- An atmosphere of positive reinforcement.
- Defusing process not longer than 60 minutes.
- The session was as informal and informative as possible.
- No operational critiquing of the incident.

If the incident warrants a complete debriefing, it should occur within 48 hours of the incident. This should transpire away from the incident site, station, hospital, or work-related area. A neutral, non-reminding locale should be sought. The support group should be made up of a broad spectrum of personnel, possibly organizing it by educational disciplines (i.e. EMS, police, fire, and hospital), with the one common thread being the hazardous materials technician. Each group will need one or two mental health care workers and one to three hazardous materials peer supporters.

The debriefing process is within a positive reinforcing atmosphere, without the pressure from administrative authority (all ranks and positions are equal). This session is formal and should consist of the following components:

Introductory phase
Fact-finding phase
Thought phase
Reaction phase
Signs and symptoms
Teaching phase
Reentry phase

Do not, under any circumstance, allow this to become an operational critique. This process should be followed whenever there is:

- Injury and/or death of a child or coworker
- Exposure to emergency workers
- Injury, hospitalization, and/or death of an emergency worker
- Any of the above during night operations and/or large-scale incidents
- Incidents affecting family members
- Any incident which has excessive media
- Or as deemed necessary by personnel
- The operational critique should be done 5–10 days after the debriefing.

Developing a Medical Surveillance Program

As one might suspect, a medical surveillance program can become costly. Here, as with medical control, the medical director can play a significant role in providing some of, if not all, the services required. The medical director can oversee the EMS functions of the department as well as the medical surveillance program.

The first goal to accomplish, which should be done by a committee, is to write down all needed components of the program, including resources that are already available to the agency. In unionized agencies, one member should be the union president and/or a representative. The various ranks and disciplines, along with medical staff, all should have direct input into the program. Only allow five to seven individuals in the committee, each having an area of responsibility and expertise. For example, the hazardous materials coordinator, medical director and EMS coordinator, middle management, and high-level management, along with a hazmat team member, would give a well-rounded committee.

The committee must identify the goals and objectives of the program while identifying direct needs from a medical aspect that could be employed with the least amount of training and cost. Identify strengths and weaknesses of the system and look for avenues of correction. For example, the committee can contact other agencies that have medical surveillance programs, acquiring their SOGs. The committee then can identify those areas that would work within their system and incorporate them within local guidelines.

Once the program is established, program review and system problem identification is a future goal consideration. At this point, a new committee may be indicated, or a subcommittee appointed to the goal of the review. However structured goals with realistic time frames are a must.

Utilize existing services within your system. Have a group of agencies working toward a common goal. Pooling resources and money can provide strength to all agencies at the scene and during training activities. All personnel that would possibly respond to such an incident should be knowledgeable of the course of action that will take place prior to, during, and after the incident.

One area of consideration is that of a relational database. A dedicated computer must be purchased with the manpower to input the medical and incident information. At predetermined time frames, a correlated question to the database is asked, such as liver, kidney, blood abnormalities detected within the emergency response force? How many incidents of exposure to a particular chemical were encountered last quarter? Last year? For a program to exist under emergency situations, preplanning and a rehearsal must be a part of the training activities.

The exposure records and cursory medical evaluations must have a mechanism for the proper routing of this paperwork. All forms must be maintained and held in confidence.

This, along with the computer analysis of the information, will present a challenge to any emergency response service.

The issues that surround medical surveillance are far-reaching. The ideas and controls presented in this chapter viewed the issues strictly from a long-term medical approach. However, these ideas and controls are not without their respective shortcomings. For example, the hiring physicals and the ADA (presented briefly within this chapter) can present the management of any service with difficult problems when keeping medical monitoring in mind.

The whole premise of medical surveillance is that of health maintenance within a field that can have adverse environmental conditions. The issue of medical records and the maintenance of such records are issues that each individual service will have to find viable solutions to.

It was not feasible nor the intention of the authors to discuss all the issues that surround this complex and ever-expanding discipline. This information was taken from health surveillance programs within the industry, suggestions by the United States Coast Guard, EPA, OSHA, NIOSH, and nationally recognized safety programs the information was intended to place in order the progression of what is realistic and of normal practice within the medical/industrial hygiene industry.

The ideal reason for medical surveillance is not the political issues that surround us but rather the medical justifications for biological indicators of toxic agents. Based on our knowledge of analysis, the detection of physiologic change, and the findings of sensitivity in individuals, we can make our jobs easier and safer for future generations of emergency responders.

Computerization of records, maintenance of the cursory physical, and periodic interval monitoring are all issues that border on the invasion of privacy, patient confidentiality, and the breach of civil rights. However, the major issues here are the concerns we have for emergency responders. Only through intensive research, monitoring, and tracking of exposures, will we find the solutions to employee safety. Only through a program that identifies the issues and ethically incorporates, these concerns toward administrative and engineering controls will we as a group have the ability to guide future health issues within the emergency services.

To this end, it must realistically manage all of these issues with the health of the emergency worker in mind. These issues might include hiring physicals, interval and annual physicals, interval histories, maintenance of medical records, confidentiality, and accessibility of medical records, abnormal findings and the effect on medical pensions, retirement, and worker's compensation, informing the worker of the abnormalities, and the certification for employment once exposure has occurred to name a few. All have an overall deliberate impact on employee health. The importance is that the emergency worker is provided with the proper education, resource, and management if the detection of a disease process occurs.

Summary

The topic of medical surveillance sometimes creates a sense of ambivalence, when in fact it is an important program to understand and embrace. A medical surveillance program is a complex and detailed program that must be developed into a policy and procedure within every department and must be reviewed and updated on an annual basis. Some believe that a medical surveillance program is only needed if the department maintains a hazardous materials team. This can never be further from the truth. Chemical exposures can take place on fire scenes, medical calls, law enforcement activities, and trauma-related incidents. A medical surveillance program is important to keep all emergency responders safe and to ensure that appropriate care takes place should an exposure occur.

Every responder should invest in their own health and safety by understanding the policies and procedures of the agency and the state and federal laws that govern such a program.

Every authority from the federal government to the local agency have addressed these issues differently. However, your agency or state has chosen to address the medical surveillance issues, the emergency responder must understand the program/policies, so if the unthinkable occurs and an exposure takes place, then the proper measures are taken.

Ensuring that medical records of all medical surveillance activities are kept and available are not only the responsibility of the agency but also, the ultimate responsibility of the employee. Unfortunately, there has been too many occurrences of lost medical records that happen during employment that prevent employees from collecting the benefits that they are due after exposure. The employee should gather all the documents policies, procedures, state, and federal mandates into a personal notebook. The responder should review this information and understand the proper procedures in place every year. Furthermore, if an agency or state has a program that is not meeting the needs of the responder, then the programs must be updated with the most current level of medical practice.

The development and maintenance of appropriate policies and procedures will help ensure that there will be rapid assessment, treatment, and follow-up should an exposure occur. Departments must consider a medical surveillance program to both ensure the health and safety of their responders and to reduce their own liability should an exposure take place.

8

Terrorism

Introduction

This book could not be considered complete without taking some time and energy to talk about the terrorism aspect of a hazardous materials response. The use of explosives, chemical agents, and biological weapons remains a threat in our world.

Awareness of a developing threat of terrorism can assist public safety in preparing for and mitigating the activity. Public safety responders should understand that many of the planned terrorism attacks takes place in stages and the ability to recognize these stages may serve to reduce the effects from an attack or even prevent it from occurring.

Stages include:

1) **Decision.** The decision is made to carry out an attack. This decision is usually based on a grievance or an ideation.
2) **Planning.** This stage involves the terrorist conducting research, surveillance, acquisition of materials, supplies, and/or equipment, and rehearsal. These indicators may or may not be observed.
3) **Execution.** This includes the actual attack. If stages 1 and 2 are completed, the attack will certainly take place but, if stages 1 or 2 can be identified and disrupted, then the attack can be mitigated before it occurs.
4) **Escape and Exploitation**. Once escape has been accomplished, then the group will state their responsibility for the event giving them a stage to state their grievance and ideation.

According to the FBI, the definition of a terrorist event is "the unlawful use of force against persons or property to intimidate or coerce a government, the civilian population, or any segment thereof, in the furtherance of political or social objectives." (PDD-39; FBI). The choice of how terrorists use force is solely dependent on the affordability of the weapons, their ability to move and disperse the weapon, their level of technology (usually low), and their ability to deny the results if the intended objective was not achieved.

Although many consider the use of chemical and biological weapons as being inhumane and barbaric, there is historical evidence of using these weapons as far back as the Middle Ages. During WWI, chemical irritants were used on the battlegrounds, and since that time, chemical weapons have been improved to ensure even better effectiveness. Because terrorists learn much from the military, responders should understand agents used by military forces in order to gain insight into these unconventional weapons.

Hazardous Materials Medicine: Treating the Chemically Injured Patient, First Edition.
Richard Stilp and Armando Bevelacqua.
© 2023 John Wiley & Sons, Inc. Published 2023 by John Wiley & Sons, Inc.

The chemical antipersonnel weapons are classified in military terms describing their effect on the enemy. It is clear that the intention of the agents is twofold: to incapacitate and to kill. For the most part, chemical agents are broken down into neurotoxins (military nerve agents), chemical asphyxiants (military blood agents), respiratory irritants (military choking agents), skin and eye irritants (military blister agents), and antipersonnel agents (riot control agents). Because terrorists may use many other chemicals to complete their tasks, responders should understand that other chemicals may be used as weapons. Among these are anhydrous ammonia, chlorine, hydrogen fluoride, and sulfur dioxide. All responders must be aware of the typical military chemical agents and the more common toxic gases and poisons used in industry.

Case Study – Salmonella Salad Bar

There are many cases of terrorism found in the history of the United States. Some are well known while others did not make the front page of the news. This case was an act of domestic terrorism that unfolded slowly and remained largely out of the news spotlight.

In 1984, there was a large outbreak of *Salmonella enterica* Typhimurium in Oregon. The initial investigation by the CDC indicated that the outbreak was related to food handlers. The outbreak eventually caused sickness to 751 people with 45 of those requiring hospitalization. At the time, this was the largest foodborne illness outbreak in the United States.

Four months later, Representative Jim Weaver would rise on the floor of the U.S. House of Representatives and charge the Rajneesh with intentionally contaminating the salad bars in eight different restaurants causing the outbreak. Representative Weaver had followed closely the daily reports from the CDC that blamed unsanitary food handlers for the outbreak. He believed that it was impossible for the food handlers to be responsible. In one restaurant, the same food handlers worked in the restaurant and in the private banquet hall. Both had salad bars. Weaver stated that there were no Salmonella infections from the banquet hall but numerous ones from the open restaurant.

Three years before this incident a self-described mystic named Bhagwan Shree Rajneesh purchased a large ranch in Wasco County. He successfully led a religious cult that quickly grew to over 4000 members, all living on the ranch or in the surrounding area. The followers became known as the Rajneesh. The cult eventually took over the village of Antelope. As they grew, the leader's goal was to take over Wasco County by electing members of the cult into both the Sherriff and County Commission political positions. When it became obvious that they did not have enough votes to complete their goal, a plan was put in place to make a large number of non-Rajneesh voters sick on election day so the majority of votes would be cast by cult members.

The Rajneesh purchased Salmonella culture from the American Type Culture Collection in Seattle and began to grow it. Although over a dozen members were involved, the two leaders of the plot were Ma Anand Sheela (Sheela Silverman) who was a trained nurse practitioner and Ma Anand Puja (Diane Ivonne Onang) who was the on-site nurse at the Ranch's medical clinic.

The group entered eight restaurants with salad bars and sprinkled Salmonella over the open food. Within a day, the reports of Salmonella infection were being reported from across the county. Although over 700 became ill, it was not enough to change the outcome of the election. Later, it was discovered that the contamination of the salad bars was intended to be a trial run for a larger attack involving placing Salmonella into the water supply, but this phase was never carried out.

In September of 1985, Bagwan Shree Rajneesh came forward and admitted involvement in the Salmonella attack. He stated that Ma Anand Sheela and 19 others were responsible for the attack. All had fled to Europe. He then invited federal investigators to the ranch. A task force was assembled under the Oregon Attorney General. They entered the ranch in October of 1985 and, in a lab, discovered glass vials containing Salmonella. The CDC found it to be an exact match to the bacteria that sickened people who ate from the salad bars. The investigators also found evidence that the Rajneesh were experimenting with other poisons, chemicals, and bacteria.

Bagwan Shree Rajneesh received a 10-year suspended sentence, fined $400,000, and was deported. He died in India in 1990 at the age of 58. Sheela and Puja were arrested in Germany and extradited back to the United States. They received multiple sentences ranging from 3 to 10 years but were released after 29 months for good behavior. Sheela was deported to Switzerland.

A high suspicion of terrorism utilizing chemicals or biologicals should always be considered when there are multiple patients with similar symptoms. This is even true for seasonal infections (like the flu) that occurring out of season. Without a high degree of suspicion, a terrorist event could be missed until there are numerous deaths or severe injuries.

Terrorism Using Chemical Warfare Agents

Nerve Agents (Cholinergic Toxidrome)

Nerve agents are probably the most common agents selected for use in wartime activities. These agents are very effective because they can enter through virtually any route and cause severe incapacitation and death. They have been formulated to be extremely toxic to the intended target but to break down rapidly so that invading troops can inhabit the area within days after an attack. Similar compounds used in civilian society are organophosphate pesticides; some of these possess extremely toxic qualities. Terrorists may choose to use commonly found pesticides rather than military agents to injure or kill intended targets. Parathion and tetraethyl pyrophosphate (TEPP), both commercial-grade insecticides, are readily available and could be used as military-style nerve agents. Whether military or civilian chemicals are the weapons of choice, the physiology of poisoning is the same.

Military Nerve Agents

Sarin (GB), soman (GD), tabun (GA), and VX (V-agent) are the most widely known military neurotoxins. These agents are organophosphate compounds similar to the pesticides parathion and TEPP.

G (German) agents (GA, GB, and GD) are volatile and evaporate slightly faster than water. This makes them very dangerous as inhalation hazards. Terrorists can use a heat source with these chemicals to increase their volatility, thus increasing the hazard. Both tabun (GA) and sarin (GB) are fairly synthesized, making them the terrorist choice for military-type nerve agents. Soman is not easily formulated and is the deadliest of the G agents. The strong bond formed between acetylcholinesterase and soman is a strong covalent bond (as opposed to an ionic bond). This bond is considered "aged" (strongly formed) and irreversible within two minutes of exposure.

VX ("V" for venom) is not as volatile as G agents, evaporating only as rapidly as motor oil. This makes VX primarily a skin absorption hazard. For VX to become a respiratory hazard, it must be mechanically aerosolized or heated to increase volatility. Because of the viscosity of this agent, its effects last longer, causing injury or death days later.

All of the military nerve agents are odorless. But because terrorists will not care how pure these substances are, a nonmilitary chemical usually has an odor. Clandestine G agents have a fruity odor, while VX has a sulfur odor. The military agents are also pH balanced to allow easier storage and to assist in their breakdown once dispersed. Nonmilitary nerve agents, typically no pH balanced, are very acidic. Depending on the pH value of the acidic nerve agents, air-purifying respirators (APRs, specialized filter masks) will tend to break down rapidly when exposed. Also, military personnel protective equipment (PPE), including APRs designed for war agents, may not work as well during terrorist attacks.

Neurotoxin (nerve agent) overview	
	TWA(ppm)
Sarin (GB), isopropyl methylphosphonofluoridate, $C_4H_{10}FO_2P$	0.000 02
Odor: Fruity	
Route of entry: Primarily inhalation, skin absorption, all possible	
Boiling point: 297 °F	
Vapor density: 4.86	
Soman (GC) pinacolyl methylphosphonofluoridate $(CH_3)C_6H_{13}O)POF$	0.000 004
Odor: Camphor	
Route of entry: Primarily inhalation, skin absorption, all possible	
Boiling point: 225 °F	
Vapor density: 6.33	
Tabun (GA), *O*-ethyl *N,N*-diamethylphosphoramidocyanidate, $G_5H_{11}N_2O2P$	0.000 01
Odor: Fruity	
Route of entry: Primarily inhalation, skin absorption, all possible	
Boiling point: 475 °F	
Vapor density: 5.6	
VX, *O*-ethyl *S*-(2-diisopropylamino) ethyl methylphosphonothiolate, $C_{11}H_{26}NO_2PS$	
Odor: Sulfur	
Route of entry: V	
Boiling point: 568 °F	
Vapor density: 9.2	

Physical Properties and Routes of Entry

For the most part, these nerve agents are clear, colorless, and odorless. They are all found in liquid form with low vapor pressure and viscosity ranging from water-like to motor oil-thick. The most volatile of the group is sarin, which can evaporate at about the rate of water. These agents can enter through all routes, but inhalation causes the most rapid effects. Because an enemy cannot guarantee that inhalation will occur, some nerve weapons have thickeners, ensuring that the agents will remain on objects for long periods of time and, thus, be transmitted to the victim. Once on the victim, the agent is absorbed through the skin or even ingested through contaminated food sources.

Exposure to vapors usually generates symptoms ranging from mild to severe within seconds or minutes. Mild symptoms include those defined in the acronym DUMBELS. Severe symptoms include all of the mild symptoms, with the addition of loss of consciousness, seizures, cardiac arrest, and apnea. When there is a liquid exposure, the onset of symptoms is slower, ranging from 5 minutes to 18 hours.

Decontamination

The military has performed many studies on the decontamination process for nerve agents. Water, both fresh and sea, has the ability to decontaminate nerve agents through a process of hydrolysis. The process works well, but when thickening substances are added to the nerve agents, they are not easily soluble, and hydrolysis occurs slowly. The addition of an alkaline soap has increased the efficiency of the decontamination process. Ideally, a sodium hypochlorite solution (household bleach) works best, causing oxidation of the nerve agent, which deactivates it.

Treatment

The treatment for nerve agent poisoning is the same as treating organophosphate insecticide poisoning. Refer to the Cholonergic Toxidrome section located in Chapter 3 for treatment for nerve agent poisoning.

Refer to Chapter 3, Toxidromes, Cholinergic Toxidrome, Organophosphate Insecticide Poisoning.

Blood Agents – Asphyxiants Toxidrome

Chemical asphyxiants, those used commercially and those produced as by-products of industry, can be easily obtained by terrorists to inflict harm. The most commonly used chemical asphyxiant is cyanide. Cyanide is used for heat treating and plating, fumigation, and chemical synthesis in the production of plastics. It is found as a gas (hydrogen cyanide, AC), as a solid (cyanide salt), or as a liquid and is a common component of many compounds containing carbon and nitrogen. Other common chemical asphyxiants in this group include hydrogen sulfide, carbon monoxide, and organic nitrogen compounds (nitrates and nitrites).

Military Blood Agents

Military chemical asphyxiants consist of two chemicals, hydrogen cyanide (AC) and cyanogen chloride (CK). These agents are identical to their civilian counterparts used in industry. For this reason, terrorists may choose these agents to inflict harm. Cyanide has a long history of use for both executions and homicides. For example, Jim Jones chose it to kill 900 people at Jonestown, and during an incident when Tylenol tablets were tainted with cyanide, others lost their lives. Even the chemist credited with the discovery of cyanide, Karl Scheele, died in a laboratory accident involving this chemical.

Chemical asphyxiants	
	TWA (ppm)
Hydrogen cyanide (AC), HCN	4.7
Odor: Metallic, a person poisoned with HCN will smell like almonds	
Route of entry: Primarily inhalation, skin absorption if liquid, all routes	
Boiling point: 79 °F	
Vapor density: 0.93	
Cyanogen Chloride (CK), CNCl	0.3
Odor: Metallic, musty	
Route of entry: Primarily inhalation, if in liquid form (<55 °F) skin absorbable	
Boiling point: 55 °F	
Vapor density: 2.1	

AC is a liquid at less than 79 °F but vaporizes quickly. Because its vapor density is less than one, the gas rises and dissipates rapidly. For this reason, AC does not remain in surrounding areas. On the other hand, CK is a liquid at less than 55 °F, so in higher temperatures, it is a gas. Its vapor density is 2.1, making it twice as heavy as air. Because of its weight compared to air, CK has a tendency to linger in low-lying areas, possibly inflicting harm for long periods of time.

Physiology

Just like the cyanides described earlier in the book, military blood agents have the same physiologic effects. They cause asphyxiation of the cells by inhibiting the ability of the cells to use oxygen. Following exposure, the cyanide ion enters the cells, binging with the enzyme cytochrome oxidase. This enzyme, in its original form, is necessary for cellular respiration (the use of oxygen to convert glucose to energy). By binding to cytochrome oxidase, these ions cause paralysis of the cells' ability to carry out aerobic metabolism. Without cytochrome oxidase, a cell cannot use oxygen from the bloodstream. The process eventually causes the cells to attempt to function under an anaerobic metabolism, and this inadequate anaerobic state ultimately causes decreased cellular energy production, metabolic acidosis, cellular suffocation, and death.

The half-life of cyanide in the body is only about an hour, but during a true exposure, death takes place well before the body starts to detoxify or excrete the chemical.

Physical Properties and Routes of Entry of Cyanide Agents

Cyanide is one of the most rapidly acting poisons. It gains access into the body most often through inhalation but can also be ingested and absorbed through the skin and mucous membranes. It causes death within minutes to hours, depending on its concentration, the amount of time the victim is exposed, and the route of entry. The speed at which cyanide gas works is evidenced by how rapidly (usually within a minute) a death row prisoner dies during a gas chamber execution.

The victim will present with a wide variety of symptoms because cyanide poisoning affects virtually all of the cells in the body. The most sensitive target organ is the central nervous system, where the urgent need for oxygen is first sensed. Early effects can include headache, restlessness, dizziness, vertigo, agitation, and confusion. Later signs are seizures and coma.

Because both AC and CK rapidly vaporize, they readily mix with air and are, for the most part, nonpersistent. Since respiratory system exposure is the most common route of entry, a cartridge mask or self-contained breathing apparatus (SCBA) is needed if a rescue is attempted in a contaminated atmosphere. CK can break down charcoal filters if exposed for a long duration or repeatedly. Frequent changes of filters are recommended.

Decontamination

If decontamination is needed (exposed to liquid or solid), a mild bleach solution is recommended. Exposure to the gas will warrant the removal of clothing prior to treatment.

Treatment

The treatment for these cyanide-based chemicals is exactly the same as described earlier in the book. The Cyano Kit is recommended and should be rapidly provided.

Choking Agents – Irritant Gas Toxidrome

Strong respiratory irritants have a long history of use by military forces. Many World War I documents are filled with incidents of chemical warfare involving chlorine (Cl) and phosgene (CG) gases. Today, these gases remain in military arsenals around the world. When discussing terrorism, we cannot discount these agents.

Many communities use Cl gas in its pure form for chlorinating drinking water. It is also used as an anti-mold and fungicide agent. Compounds containing chlorine are even used for chlorinating home swimming pools and cleaning toilets and showers. There is no doubt that this chemical is easily available.

CG, although not as common as chlorine, is found in industry and used for organic synthesis during the production of polyurethane, insecticides, and dyes. It is also a by-product of burning freon, which has led to many injuries among firefighters.

Although not thought of as a military agent, anhydrous ammonia also fits into this category because of its ability to cause severe respiratory irritation and injury. Anhydrous ammonia is a common industrial refrigerant and is also used as a fertilizer. Again, it is easily available to anyone.

Respiratory irritant overview	
	TWA (ppm)
Phosgene (CG), carbonyl chloride, $COCl_2$ Odor: Mown hay, sweet Route of entry: Inhalation Boiling point: 45 °F Vapor density: 3.4	0.1
Chlorine (Cl), Cl_2 Odor: Bleach Route of entry: Inhalation, moist skin irritation, eye irritation/burn Boiling point: −29 °F Vapor density: 2.49	0.5
Anhydrous ammonia, NH_3 Odor: Sharp, intensely irritating Route of entry: Inhalation, moist skin irritation, eye irritation/burn Boiling point: −27 °F Vapor density: 0.58	25

Military Choking Agents

CG and Cl are typically stored as liquids but rapidly become gases once released into the atmosphere. Their expansion ratios allow them to be transported in a smaller container, and once released, they fill a large area.

These agents were used on the battlefield to incapacitate an enemy force so that it could be overrun by advancing troops. This strategy worked well because both of these gases dispersed rapidly into the environment, leaving no contaminated objects behind to cause injury.

Once exposed, victims are overcome with severe, uncontrollable coughing, gaging, and tightness in the chest. Bronchiole swelling and laryngeal spasms are common, causing apnea and unconsciousness. Other injuries include tissue sloughing, localized edema, and pulmonary edema, all contributing to obstruction of the airways. CG, in particular, is not easily soluble in water and advanced directly into the alveoli, where tissue destruction is severe. The breakdown of these cells allows fluid from the bloodstream to advance into the airways, a condition called noncardiogenic pulmonary edema, difficult to treat. The victim of a severe exposure literally drowns in his own body fluids. When the injury is not fatal, the victim is left with destroyed lung tissue and a lifetime of respiratory disease ranging from chronic obstructed pulmonary disease to chronic pneumonia.

Physiology of Respiratory Irritant Injury

Injuries to the upper respiratory areas are usually a result of exposure to water-soluble chemicals that readily dissolve into the moisture-coated airways. In the case of chlorine, this results in the production of hydrochloric acid and chemical burns at the site. Ammonia, in reaction with water, forms an alkali, which causes a severe, long-lasting injury, Laryngeal edema, and laryngeal spasms should be expected and treated aggressively.

Injuries to the lower regions of the respiratory tract usually result when the chemical inhaled is not water-soluble, is in a high concentration, or is inhaled over an extended period of time. This deeper injury causes swelling of the finer bronchioles, sloughing of damaged tissue, and damage to the alveoli. Cilia that may be damaged are unable to ride the fine bronchioles of sloughed cells and increased mucous production caused by the damaged airways. The lower airway obstruction caused by this exposure adds to the complexity of the injury and increases the chance of death.

Noncardiogenic pulmonary edema is the result of a chemical irritant reaching the alveoli and causing damage to the cellular walls, and disrupting the surfactant. The damage interferes with the alveoli's natural ability to keep fluids out of the alveolar space, which results in fluids filling the injured alveoli and fine bronchioles, eventually advancing into the upper airways. If the exposure causes severe alveolar damage, the end result will be death.

Physical Properties and Routes of Entry

Military chemical irritants are heavier than air and are able to seek out low-lying areas and persist for long periods of time. This was the case during World War I when troops would dive for cover from gunfire in trenches only to choke in their own foxholes. In the civilian world, hazardous materials responders have been dealing with these chemicals for years. The typical responder would point hose lines toward the cloud of gas and disperse it using a combination of air currents and water absorption.

All of the agents are referred to as respiratory irritants, but the respiratory system is not the only body surface affected. If the irritant is water-soluble (ammonia and chlorine), it will cause moist skin and eyes to become inflamed and burn. Both of these effects are unpleasant and could put troops or civilians out of action. Even at low concentrations, the release of these chemicals into a crowd could send multitudes of patients to the hospital, triggering a panic scenario in line with the needs of a terrorist.

Decontamination

Decontamination with water may be necessary. All of these chemicals are gases and will disperse rapidly into the environment. Clothing removal is important if the patient is to be placed in a closed ambulance for transport.

Treatment

The treatment for respiratory irritant exposure is twofold. The first part is to open the airways to allow the free movement of air into the alveoli. The second is to ensure that the injured alveoli remain open, free of fluid intrusion from the bloodstream, and ready for gas exchange. Getting oxygen to the alveoli is extremely important. The treatment of injured airways and alveoli was discussed earlier in the book when respiratory irritants were covered. The treatment mentioned in this area of the book is exactly the same for respiratory irritants used intentionally against a larger group of victims.

Vesicants – Corrosive Toxidrome (Military Blister Agents)

These chemicals were originally developed by the military because enemy troops could protect themselves from respiratory irritants with masks. Instead, skin and eye irritants could affect the respiratory system and additionally unprotected skin. These irritants are harmful in liquid form but vaporize to become airborne contaminates as well. Most date back to World War I; some have been further refined through the years to become even more efficient.

Many chemicals in the industry are capable of causing skin irritation, but none to the degree that military blister agents can. It is for this reason that this chapter will only focus on military agents and not on civilian industrial chemicals.

Military Blister Agents (Vesicants)

These types of blister agents are only used by the military. These agents include mustard (H) and related variations of mustard (HD, HN, and HT), phosgene oxime (CX), and lewisite (L). For the most part, the agents are liquids that vaporize slowly, causing an inhalation hazard. Skin and eye exposure is the most common effect that results from direct contact with the liquid. CX is normally a solid that will vaporize into a respiratory hazard.

H was developed during World War I and has continued to be a major chemical warfare agent since that time. This agent was reportedly used in 1960 by Egypt against Yemen

Skin irritant overview	
	TWA (ppm)

Mustard (H, HD), bis (2-chlorethyl) sulfide ($ClCH_2CH_2)_2S$ or $C_4H_6CL_2S$
  Odor: Garlic
  Route of entry: Primarily skin absorption; eye and respiratory
  Boiling point: 442 °F
  Vapor density: 5.4
	0.0003–0.0004

Phosgene oxime (CX), dichloroformoxime, $CHCl_2NO$
  Odor: Penetrating irritating odor
  Route of entry: Primarily skin absorption, eye and respiratory
  Boiling point: 264 °F
  Vapor density: 3.9
	0.086

Lewisite (L), dichloro-(2-chlorovinyl) arsine, $ClCHCHAsCL_2$
  Odor: Geraniums
  Route of entry: Primary skin absorption, eye and respiratory
  Boiling point: 374 °F
  Vapor density: 7.2
	0.003

and again during the 1980s as a weapon between Iran and Iraq. The United States has destroyed or is in the process of destroying all stockpiles of blister agents left in this country.

Physiology of Blister Agent Exposure

These chemicals are capable of causing extreme pain and large blisters on contact. If the vapors are inhaled, the lung tissue will form large obstructing blisters. Once the blisters break, a large open would result that allows the establishment of overwhelming infections, a condition that will eventually cause death.

After H gains access to the body, it cycles in the extracellular water forming extremely reactive alkali binding to both intracellular and extracellular enzymes and proteins. Once exposed, a victim may manifest a latent period when no symptoms are present. Later, between 2 and 24 hours, the reaction appears with the formation of blisters on the skin, necrosis of the mucosa of the airways with later involvement of the airway musculature, and severe irritation to the eyes, including swelling in the cornea and related tissues that lead to permanent scarring.

L acts similar to H but has additional effects that are systemic. Symptoms beyond the blistering effect may include pulmonary edema, diarrhea, vomiting, weakness, and low blood pressure. The irritation liquid L causes to the eyes is devastating. If not decontaminated within one minute, the damage will probably be permanent.

Physical Properties and Routes of Entry

H has a freezing point of 57 °F, so it solidifies at temperatures less than this. For this reason, pure mustard may not be a good choice in colder climates. H also vaporizes slowly, making it primarily a skin contact hazard. At temperatures greater than 100 °F, it will vaporize rapidly enough to be a respiratory hazard. Mixing H with some other agent like L lowers its freezing temperature and allows the use of this chemical in colder climates.

Both L and CX vaporize more readily than H, making them more of a respiratory hazard. All of the blister agents have vapor densities much greater than air, allowing them to stay near the ground and not dissipate quickly. L has a vapor density of 7.2, which allows it to persist in low-lying areas for long periods of time.

Decontamination

Decontamination for all of the blister agents must be immediate. Each one harms tissues on contact. H differs from the other agents, as it does not produce symptoms for several hours, leaving the victim without a clue that exposure has taken place. L and CX both cause irritation almost immediately, which alerts the victim to exposure and allows earlier decontamination. If a victim notices liquid on his/her skin, he/she should blot off the agent, taking care not to do this with contaminated material. If the agent is wiped instead of blotted, it will be spread the length of the wiped area, extending the injury.

The decontamination solution of choice is chloramine solution. CX is adequately decontaminated with water, but more efficient decontamination includes soap and water followed by a chloramine or bleach solution, followed again with soap and water.

Treatment

There is no field treatment for blister agents beyond good decontamination. It is important to know that the blisters formed by these agents do not pose a significant secondary contamination danger. The fluid in the blisters is not contaminated with the agent, but typical blood and body fluid precautions should be exercised.

Lacrimators (Riot Control Agents)

Riot control agents are used to incapacitate enemies and make them unable to fight. They are not intended to cause mortal injury and have only rarely caused a severe,

---

Synonyms for typical riot agents

Chloracetephenone (CN)

Phenacyl chloride, alpha-chloracetephenone, omega-chloroacetephenone, chloromethyl phenyl ketone, and phenyl chloromethyl ketone

Orthochlorobenzalmalonitrile (CS)

O-chlorobenzylidene malonitrile, OCMB, and military tear gas

Oleoresin capsicum (OC)

Pepper spray, civilian tear gas

---

lasting injury. Civilian use of these agents includes riot control and self-protection. Chemical antipersonnel weapons have gained popularity with both the general public and law enforcement because they can be used to subdue persons without the use of extraordinary physical force. These chemical agents, which can be dispersed in a number of ways, offer a nonlethal form of protection that causes temporary extreme discomfort. Generally, there are three versions of these agents available: chloracetephenone (CN), orthochlorobenzalmalonitrile (CS), and the most popular civilian agent, oleoresin capsicum (OC).

CN and CS

The effects from CN begin in one to three seconds and are characterized by extreme irritation to the eyes, causing burning and tearing. Irritation to the skin is also common because the crystals stick to moist skin, causing burning and itching at the point of contact. CN also causes upper respiratory irritation. These effects last 10–30 minutes.

The symptoms of CS start in about three to seven seconds and last 10–30 minutes. The effects reported are stinging of the skin, especially in moist areas, and intense eye irritation with profuse tearing and blepharospasms. The burning also affects the nose and upper respiratory system. Some victims panic due to shortness of breath and chest tightness; they describe its effects as being ten times worse than CN. Police agencies still use this irritant for crowd dispersal.

Both CN and CS are submicron (less than 1 micron) particles. They are extremely light and are carried to the target area in a carrier solution that evaporates quickly, dispersing the agent. Because of their light, fine particles, both of these chemicals are capable of cross-contamination between the victim and emergency response personnel. Their submicron size may allow these irritating particles to gain access into the lungs, fine bronchioles, and alveoli, causing irritations and injury deep in the lungs.

OC

OC has become the safest and most popular of the chemical antipersonnel agents. It is found in police aerosol sprays and over-the-counter agents. It is a non-water-soluble agent prepared from an extract of the cayenne pepper plant. When contact with the eyes occurs, the effects of OC are almost immediately apparent. OC is not a submicron particle, so access to the lower respiratory system is limited. Contact with OC causes immediate nerve-ending stimulation but no irritation. The condition lasts 10–30 minutes and usually has no lasting effect.

Decontamination

Decontamination should consist of removing the clothing and washing exposed skin areas with soap and water. In the case of OC exposure, the agent is non-water-soluble, so its effects on mucous membranes will last until it is detoxified by the body as eye irrigation using water is not sufficient in removing the agent. The pain caused by these lacrimators generally lasts about 30 minutes, which is the duration of anesthesia provided by the eye numbing agents such as tetracaine or proparacaine.

Treatment

Less than 1% of the exposed victims will have symptoms severe enough to need medical care. As mentioned earlier, the symptoms generally last about 30 minutes and disappear without lasting injury. There is no known antidote for these irritants, so medical care is centered on the relief of the symptoms. Irrigation using a fat-containing substance such as whole milk will help remove the fat-soluble chemical from the eye. In the case of eye irritation caused by exposure, relief from symptoms can be accomplished with a topical anesthetic agent like tetracaine or proparacaine. If sensitivity to these chemicals causes bronchospasms, typical bronchodilating drugs may be used. Updrafts of albuterol will usually control bronchospasms caused by exposure to these agents.

Terrorism Using Biological Agents

In today's world of pandemics and endemics, it is easy to see how biological terrorism could be seriously considered. We have all witnessed how a pandemic can affect the economy of the world and cause great distrust in government. This is one of the goals of an act of terrorism. The thought of being infected by a deadly or debilitating disease or a biological toxin is truly a scary thought. Unfortunately, these agents are not difficult to cultivate; they are surprisingly easy for someone with a very limited knowledge of microbiology to produce. An attack using biological materials would bring international attention to a radical organization.

Research into the use of biological weapons gained momentum during World War II. Prior to this time, biological agents were used to inflict harm on individuals, but were not, for the most part, cultivated or produced in mass quantities for war. For example, plague-ridden corpses were flung over fortress walls in the fourteenth century and smallpox-ridden blankets were given to the American Indians during the French and Indian War of 1754. True biological warfare research, however, did not take place until much later.

Biological agents are made from a variety of microorganisms and biological toxins. Biological toxins are chemical compounds poisonous to humans produced by plants, animals, or microbes. Microorganisms are generally living viruses or bacteria that have the ability to establish debilitating or deadly infections in their victims. Although many of the organisms discussed in this section are considered to be military-type weapons, many others can be cultivated and introduced into the environment with the intent of inflicting harm on a targeted civilian population.

Bacteria

Bacteria are single-celled microorganisms that are plant-like in structure. They vary in size from approximately one half of a micron to ten microns. They can be either spherical (cocci) or rod shaped (bacilli). Bacterial agents include living cells of *Bacillus anthracis* (anthrax), cholera, *Yersinia pestis* (plague), *Francisella tularensis* (tularemia), Q fever, and salmonellae. These microorganisms can be grown in artificial media; many have the ability to spore (become seed-like) and live for long periods of time without infecting tissue.

Viruses

Viruses are smaller than most bacteria and live on or within other cells, using the host cells metabolic process to grow and replicate. A viral infection is the result of destruction to the host cell by the intracellular parasitic action of the virus. Viruses cannot be grown on artificial media. It can only be grown in media that contains living host cells. Each virus needs a particular type of host cells, making the production of viruses for warfare or terrorist use complicated and expensive. For this reason, it is problematic for low-budget terrorist organizations or private individuals who are not associated with larger groups or government entities.

Governments from around the world, including the United States, have experimented with virus use for weapons. This section will provide some basic information of those viruses considered most threatening, including variola virus (smallpox), Venezuelan equine encephalitis (VEE), and viral hemorrhagic fever (VHF).

Biological Toxins

Biological toxins are toxic substances originating in animals, plants, or microbes, and are more toxic than most chemicals used and produced in industry. Since these toxins are not volatile, they are generally not suitable for the battlefield. Instead, they are used for contaminating food sources, water supplies, and specifically targeted individuals. Toxins have been considered for military use including botulinum toxins (botulism), staphylococcal enterotoxin B (SEB), ricin, and trichothecene mycotoxins (T2s).

Bacterial Agents

Anthrax (*B. anthracis*)

Anthrax (splenic fever, woolsorters' disease, malignant pustule) has been the biological weapon of choice. It was first prepared as a weapon in the United States in the 1950s and continued to be produced until the biological weapon program was terminated. Anthrax is easily grown and can be kept almost indefinitely in the spore (dormant) form under the proper conditions. Although the United States no longer cultivates anthrax as a weapon, it is suspected that many other countries have biological weapons using this agent. It is reasonable to expect that terrorist organizations are either developing a weapon of this type or already have one at their disposal.

Anthrax can be delivered to a target as dust that can be inhaled. This was seen in 2001 here in the United States, when numerous letters containing anthrax were sent to politicians and the news media. Furthermore, it can contaminate the environment and drinking water, causing disease days or weeks after it is disseminated.

Once anthrax bacteria are inhaled or ingested, they will incubate for approximately one to six days in the victim. Signs of anthrax infection include chest pain, cough, fatigue, and fever. More serious symptoms develop as the infection becomes worse and include shortness of breath, diaphoresis, and cyanosis, leading to shock and death.

As anthrax infection progresses, the bacteria release a toxin that further poisons the body. Once this stage is reached in the infection, antibiotics may kill the bacteria but has little effect on the toxin and death is usually related to the effects of the toxin and not from the infectious process.

Treatment includes high-dose antibiotics and many only be successful if the infection is caught in the early stages. Other supportive therapy, such as intubation and ventilation, may also be necessary. If an anthrax infection is suspected, precautions by responders should include HEPA-style masks and blood and body fluid protection.

Cholera (*Vibrio cholerae*)

The bacteria responsible for the cholera infection gains entrance into the body through contaminated food and water sources. It is a disease that spreads rapidly when proper sewage precautions are not taken. For this reason, it has been investigated as a wartime biological agent. If an infection can be established in a field camp where sewage precautions are not taken, the disease can infect a high percentage of the troops. In countries where sewage disposal is not carefully monitored, cholera has infected and killed thousands of people.

The cholera bacterium attaches to the tissue of the small intestine, causing over secretion of fluid. This overwhelms the large intestine's ability to absorb the fluid and leads to diarrhea and severe hypovolemia.

Signs of cholera begin within 12–72 hours of exposure and include vomiting, intestinal cramping, and headache. 5–10 l of fluid loss can be expected per day. If not aggressively treated, the fluid loss will lead to shock and death.

Treatment includes fluid replacement with electrolyte therapy. Antibiotics will shorten the duration of the infection and kill the infecting microorganisms. It is difficult to become infected through direct contact with an infected person, but precautions should be taken to reduce contact with body fluids. Hypochlorite solutions should be used as a decontaminating agent for material or equipment that has been in contact with body fluids.

Pneumonic/Bubonic Plague (*Y. pestis*)

Both pneumonic and bubonic plague are caused by the same bacteria, but their route of entry and symptoms are greatly different. Normally this bacterium causing bubonic plague is spread by rodents or fleas. Pneumonic plague occurs when a victim develops the infection in the lungs and spreads the disease through aerosolization. During a terrorist event, the organism could be introduced into the environment through aerosolized bacteria. Japan has researched the feasibility of releasing plague-infected fleas as a means of dissemination.

Swollen lymph nodes (buboes) and fever characterize bubonic plague. The affected lymph nodes are most often those found in the groin, since infected fleas often bit the leg areas. The incubation period for bubonic infection is between 2 and 10 days.

Pneumonic plague's onset is usually faster and follows an incubation period of two to three days. Infection results from a respiratory exposure to *Y. pestis* spread from person to person and caused by an infected host coughing or sneezing. Symptoms of pneumonic plague include fever, chills, coughing, bloody sputum, and dyspnea and cyanosis.

Treatment of plague involves early diagnosis and administration of antibiotics. Other supportive measures may be needed if breathing is impaired or other organ systems are involved. Personnel protection is necessary during treatment of these patients, since secretions and body fluids may be infectious if bubonic plague is the cause of illness. If pneumonic plague is the cause, strict respiratory protection is necessary, along with clothing protection, to prevent secondary contamination. All reusable equipment must be decontaminated using hypochlorite solution.

Tularemia (*F. tularensis*)

Tularemia was prepared as a weapon by the United States during the 1950s but was discontinued with the termination of the biological weapons program. It is reasonable to expect that other countries are still cultivating tularemia as a weapon.

Tularemia is also known as rabbit fever or deer fly fever. Blood and body fluids of an infected person or animal, or the bite of an infected deer fly, tick, or mosquito, transmit the disease. Inhalation of aerosolized bacteria would initiate a typhoidal tularemia infection with respiratory symptoms that appeared in 2–10 days. Tularemia persists for weeks in water, soil, or animal hides. Because the bacteria are resistant to freezing, it can persist in frozen rabbit meat for years.

Signs and symptoms of an infection include local ulcerations, swollen lymph nodes, fever, chills, and headache. Typhoidal symptoms include fever, headache, substernal discomfort, and nonproductive cough. Even untreated, the death rate for this infection is about 5%.

Excellent results have been obtained with treatment that includes antibiotic therapy. Secondary infections unusual, so strict isolation is not needed. Only typical personal protection against secretion and lesions is required.

Q Fever (*Coxiella burnetii* rickettsia)

Q fever (query fever) is another weapon previously kept in US arsenals. It occurs naturally as an infection found in sheep, cattle, and goats, and is an occupational hazard in industries involving these livestock. Q fever is a rickettsial agent disseminated through inhalation of infected aerosolized material, even as little as one organism, according to one report. The ease with which this microorganism can be harvested and its infectious properties make it an agent that terrorists could efficiently use.

Symptom onset begins at 10–20 days and is usually self-limiting. Q fever disables and could cause panic but death is not usually the result of an infection. Q fever usually lasts two days to two weeks and is characterized by fever, headache, fatigue, and, in some cases, pneumonia and pleuritic chest pain.

Treatment involves antibiotic therapy and supportive care. When untreated, the illness is incapacitant but does not usually cause death. In rare cases the infection may cause endocarditis, which, if untreated, may be fatal. Caregivers need only protect themselves from a contaminated environment. Decontamination is accomplished with soap and water or a mild hypochlorite solution.

Salmonellae (*Salmonella typhimurium*)

Salmonella causes one of the most common types of food poisoning. Although this bacterium has not been used as a military biological weapon, it has been used for domestic terrorism. Naturally occurring infections of salmonellosis are caused by ingesting the organism in food contaminated with infected feces.

Terrorist use of salmonella to harm a target group would be simple. Once a food source has been identified as contaminated, it could be easily mixed with uncontaminated food and distributed to victims. Since the bacteria is prevalent in meat products and poultry, using meat by-products to cultivate it is a reasonable way to obtain and produce enough of the microbe to cause harm to a large number of people. Spreading the bacteria on foods that are not eaten cooked is one way to ensure that it will not be destroyed prior to ingestion.

After ingestion of contaminated food, symptoms begin within 8–48 hours. The victim experiences fever, headache, abdominal pain, and watery diarrhea that may contain blood and mucous. If the infection localizes, the results may be endocarditis, meningitis, pericarditis, pneumonia, or abscesses.

Infected victims can cause secondary infection if caretakers do not protect themselves from body fluids. The bacteria are killed with heat or hypochlorite solutions.

Viral Agents

Smallpox (Variola virus)

Smallpox was eradicated by 1980. Elimination of the disease was accomplished through strict quarantines, and extensive vaccinations. To this date, there have been no further outbreaks of small pox reported in the world. However, there are two known deposits of the variola virus (smallpox virus) remaining, one at the Centers for Disease Control in Atlanta, Georgia, and on other at Vector in Novosibirsk, Russia. The World Health Organization (WHO) has recommended that existing viruses be destroyed by 1999 but that has not occurred and both maintain these viruses for vaccine purposes in case there is a future outbreak.

Because the disease has been eradicated does not mean that it no longer poses a threat in the United States. Other countries have experimented with the smallpox virus for biological warfare, and it is unclear whether other governments or organization may still possess the virus for cultivation. The last vaccination given in the United States general public was in 1980; these vaccinations are no longer effective. The exposure of a segment of the US population to the virus would be devastating.

The ease with which this virus is produced and its extreme aerosolized virulence (toxicity) make it an ideal weapon. Terrorists able to secure even a small portion of the virus could, in a short time, cultivate enough to infect to thousands of people. An attack of this nature would overwhelm the healthcare system and simulate a panic that would be felt nationwide and probably worldwide.

Once exposed, a victim develops a rash, which is followed by red, raised papules that eventually form pustular vesicles. The vesicles are more common on the extremities and face, and are the opposite of those produced in chicken pox, which concentrate on the chest and abdomen. The patient remains extremely contagious for up to 14 days after the onset of symptoms and should be isolated until the scabs separate. Caregivers must exercise complete blood, body fluid, and respiratory protection.

Venezuelan Equine Encephalitis (VEE)

VEE causes disease in horses, mules, burros, and donkeys. It is typically found in South and Central America, Mexico, Trinidad, and Florida. It occurs as a result of a bite from a mosquito carrying blood from an infected animal.

When the disease occurs naturally, there always is an outbreak among the Equidae (horses, mules, etc.) population before humans are affected. If the virus is spread by unnatural means, an outbreak will not initially involve Equidae during its early stages. The occurrence of the disease outside of its natural geographical area is another clue that the outbreak is not of natural origin.

VEE was prepared as a weapon by the United States during the 1950s but was later destroyed when the biological weapons program was terminated. Other countries may have experimented with VEE; the extent to which those countries have been successful is unknown. It is reasonable to expect, however, that since this virus is easy to acquire, cultivation by terrorists presents a threat.

Once a victim is exposed, an incubation period of between one and five days is followed by a sudden onset of symptoms. VEE generally establishes an infection in the meninges of the brain and within the brain itself, so symptoms will coincide with this inflammation. Initial symptoms include generalized weakness and numbness of the legs, photophobia, and severe headache. As the infection progresses, the symptoms advance to nausea, vomiting, and diarrhea. Children can have more severe central nervous system symptoms, including coma, seizures, and paralysis, leading to a 20% chance of death. Unborn fetuses are endangered by encephalitis, placental damage, spontaneous abortion, or congenital anomalies.

No specific treatment exists except for supportive care that may include analgesic medicines. If seizures result from the infection, anticonvulsant medications may be required. The most severe phase of the infection lasts between 24 and 72 hours, and may be followed by lingering or permanent paralysis or lethargy.

Blood and body fluid precautions should be exercised when caring for and infected individual. Decontamination of equipment can be accomplished using a hypochlorite solution or by destroying the virus with heat (80 °C for 30 minutes).

Viral Hemorrhagic Fevers (VHFs)

VHFs are a group of viruses that cause uncontrollable external and internal bleeding. Since Ebola has been seen in the news many times in recent years, these types of infection have the potential to cause hysteria in a population. Terrorists would most likely trigger this hysteria through false threats and hoaxes rather than through using one of these viruses. But the possibility of obtaining possession of these viruses and then using them always exists.

VHFs are a group of illnesses caused by ribonucleic acid (RNA) viruses of several families, including Ebola and Marburg virus (Filoviridae family), Lassa fever, Argentine and Bolivian fever (Arenaviridae), Hantavirus genus, Congo-Crimean hemorrhagic fever virus (Bunyaviridae family), and yellow fever virus and dengue hemorrhagic fever virus (Flaviviridae family).

These viruses, in general, cause permeability and loss of intravascular fluid in the vasculature. Early signs of infection are fever, pain, and conjunctival subcutaneous bleeding. As the infection progresses, mucous membrane hemorrhaging, pulmonary involvement, and shock result.

Depending on the extent of the infection, and the virus responsible, the mortality rate can be as high as 90%. Ebola and Lassa have the highest mortality rate and rapid onset of symptoms.

These viruses are transmitted in a variety of ways, but the most dangerous is by aerosolizing the agent. It is conceivable that these viruses could be used as a warfare agent, but the difficulty in obtaining and cultivating such organisms makes them an unlikely weapon for terrorist organizations.

Biological Toxins

Botulinum Toxin

Botulism is a serious and occasionally fatal disease caused by a toxin produced from an anerobic bacterium, *Clostridium botulinum*. The bacteria are fond in poorly handled food and account for many cases of food poisoning (botulism) in the United States. This toxin could be easily produced and spread over a targeted area as an aerosolized particle. The effects of inhaling it are similar to those from ingesting it, except that the onset of symptoms may actually be slower.

Botulinum toxins are generally believed to be the most poisonous substances yet discovered. Toxic testing has revealed that it takes less than 0.001 µg/kg of body weight to kill 50% of the test population. That can be calculated to a lethal dose for a 220-lb man to be as little as 0.1 μg. As few as 950 molecules of toxin can be lethal for white mice. When these compounds are compared to other antipersonnel chemical weapons, their toxicity is placed in perspective. For example, botulinum toxin is 15,000 times more toxic than VX, the most toxic of known nerve agents.

Botulinum causes injury to the body by bonding to the nerve endings at the synaptic junction. This action prevents the nerves' terminal ends from releasing acetylcholine that is needed for nerve impulse transmission. The effect is exactly opposite that of organophosphate nerve agent poisoning. During organophosphate poisoning, the synaptic junction is filled with acetylcholine because the organophosphate binds with the enzyme acetylcholinesterase, causing a continuous transmission of nerve impulses.

The inability to transmit nerve impulses to the skeletal muscles and parasympathetic system causing generalized weakness, dry mucous membranes, and eventually respiratory failure. If the concentration of toxin is great enough, as seen in ingestion injury or concentrated inhalation of aerosolized toxin, the onset of symptoms occurs as early as 8–36 hours. Lower doses may cause an onset of symptoms days after exposure.

Signs and symptoms include dry mouth with difficulty speaking and swallowing; severe cases may cause the loos of gag reflex. Vision may be blurred and include double vision, photophobia, and dilated pupils. As the poisoning continues, the muscles become flaccid and develop paralysis, leading to respiratory failure and death. During the 1940s and 1950s, the mortality rate for botulism disease approached 60%, but due to improved respiratory care, the mortality rate is now below 20%.

Emergency treatment includes securing a good airway or endotracheal intubation, if necessary, and positive pressure ventilations. Hypotension may also result, so ensuring that blood pressure is also maintained may be an additional consideration. Current treatments include equine serum trivalent botulism antitoxin, human-derived botulinum immune globulin intravenous (BIG-IV), plasma exchange, 3,4-diaminopyridine, and guanidine. Definitive care is focused on supportive measures and can continue for weeks to months, depending on the patient and level of exposure.

Staphylococcal Enterotoxin B (SEB)

SEB represents another form of food poisoning that incapacitates victims but rarely causes death. Clinical effects are caused from toxins produced by the bacterium *Staphylococcus aureus*. Illness caused by SEB is much more common than that caused by botulism. Natural outbreaks are usually clustered and can be traced to one exposure source, such as a restaurant or picnic. Its usefulness in combat is related to its ability to render up to 80% of exposed victims incapacitated and unable to function. Terrorists may find SEB simple to produce and easily used to contaminate open food sources and water supplies. This exposure could affect hundreds or thousands of persons, taxing local healthcare facilities and generating desired publicity. The effects of the exposure can last several weeks, causing a long-term strain on the healthcare infrastructure.

Depending on the route of exposure, the effects of SEB are different. Inhaled SEB causes systemic injury that can lead to septic shock. Ingestion of SEB causes a slower onset of symptoms that are generally less dramatic and less serious. The mechanisms of toxicity are complex and involve the use of the body's own antibody response to cause injury.

The onset of symptoms, from inhalation exposure, is from 3 to 12 hours and is characterized by fever, headache, muscle pain, shortness of breath, and occasionally, chest pain. A gastrointestinal exposure may also include nausea, vomiting, and diarrhea.

Treatment is supportive in nature. Emergency responders must exercise good airway management using supplied oxygen. Severe cases may cause respiratory depression or arrest, so endotracheal intubation or other form of advanced airway is recommended and aggressive positive pressure ventilations should be performed.

Decontamination of exposed equipment can be accomplished using a hypochlorite solution. Vigorous hand washing is recommended and care must be taken not to eat any food that may be contaminated with the toxin.

Ricin

Ricin is a biological toxin formed in the seed of a castor plant (*Ricinus communis*). Castor plants grow throughout the world and are unregulated, so this toxin is important to review. Castor beans are used in the production of castor oil, which is produced by cold crushing of the bean. Castor oil is nontoxic and used as a cathartic; however, the remaining mash contains toxic levels of ricin. There is enough ricin in the ingestion of two to four beans by an adult or one to three beans by a child to cause poisoning and death. For military or terrorist use, ricin can be produced as a solid and aerosolized to affect large numbers of victims. Once the ricin gains access to the body, the globular glycoprotein attaches to the cell wall and is eventually internalized by the cell. Ricin then blocks protein synthesis, thus killing the cell.

When inhaled, ricin causes cellular damage in the respiratory system, leading to necrosis of tissue, bronchitis, interstitial pneumonia, and pulmonary edema. The onset is rapid, with symptoms seen in as little as three hours, but more commonly in 8–24 hours. When ingested, ricin damage is seen in the gastrointestinal system and is evidenced through gastrointestinal bleeding. These symptoms are followed by cardiovascular and systemic injury, including vascular collapse, hepatic damage, renal necrosis, and death.

Emergency medical treatment is supportive and involves good airway management and fluid replacement, if necessary. There is no antidote for ricin toxicity. If an exposure to this toxin is suspected or known, proper protective measures must be taken by the rescuer, including skin and respiratory protection. Decontamination of the patient can be accomplished by washing with a hypochlorite solution.

Trichothecene Mycotoxins (T2)

T2s are chemically complex toxins that are naturally produced by fungi. There are more than 40y such mycotoxins capable of inhibiting protein synthesis and destroying the integrity of the cell membrane. The toxin targets the most rapidly reproducing cells, so those found in the skin, mucous membranes, and bone marrow are most at risk.

T2 can enter through all routes causing injury along whichever ones are exposed. Symptoms appear within minutes and are evidenced by burning, itching, reddened skin, burning in the nose and throat, sneezing, burning of the eyes, and conjunctivitis. Burning and reddening rapidly advances to blackened necrosed tissue. Gastrointestinal symptoms include nausea, vomiting, diarrhea, and abdominal pain.

Rescuers must wear chemical protective clothing. All clothing worn by a victim must be removed to lessen exposure and prevent secondary exposure and prevent secondary exposure of the medical provider. This toxin is extremely hearty, requiring a heat source of 500 °F for 30 minutes to destroy. The most efficient detoxifying agent is sodium hypochlorite, which can eliminate the toxic activity of T2. The patient should then be washed with soap and water to remove residual material. T2 is non-water-soluble, so complete decontamination will require careful, complete washing.

There is no specific antidote to T2 poisoning and supportive care has limited benefit. If T2 is ingested, activated charcoal should be given.

If a terrorist chooses to use a biological agent to inflict harm on a population, emergency responders and emergency room staff will not only be in great danger from the product but also at a loss to immediately detect and determine what agent is used. For the terrorist to complete the goal of causing fear within the targeted population, they will, most likely, take responsibility for the attack and advise victims of their exposure. The only other clue will come from multiple patients complaining of similar or identical symptoms. In this case, the emergency responders' familiarity with the symptoms of each toxin may guide them in the development of a differential diagnosis, a proper response and emergency care.

Explosives and Incendiary Devices

It is easy to see why bombs have been the weapon of choice for terrorists. Bombings enable terrorists to make political statements through the media and cause a fear response associated with a specific target. In short, bombs work very well to fulfill terrorist goals. They are inexpensive, simple to build, easy to place, and can be assembled from commonly obtained materials. Bombings become high-profile media events with sudden impact, resulting in the destruction of political confidence.

Bomb Incidents

Bombs and incendiary devices cause a majority of terrorist incidents. Bombings are responsible for approximately 70% of the historical occurrences of terrorist activity. They are not only tools used by these criminals; they are just used more often.

The Department of Transportation (DOT), for the purposes of transportation, defines an explosive as any substance or article, including a device, which is designated to function by rapidly releasing gas and heat. DOT further states that an explosive may include a chemical reaction that is rapidly releasing gas and heat unless the substance or article is otherwise classified. These definitions are used to define placarding, labeling, and compatibility standards for transport. Under most conditions, the law-abiding transporter will comply with the transportation criteria defined by DOT. Obviously, terrorists are not concerned with the law and will go to great lengths not to bring attention to what they may be transporting. Therefore, responders may encounter any type of device or material being illegally transported by those intending harm. Improvised munitions mixed or assembled haphazardly add a much higher degree of hazard to an already dangerous response. Timing devices, poor-quality precursor chemicals, and unpredictable chemical reactions are examples of extreme hazards to expect from improvised munitions.

Anatomy of Explosives

Improvised explosive devices may contain filler material rated as high or low filler explosive material. High-filler materials are further described as high-order or low-order detonation substances. The difference between these materials is the degree of consumption during the explosion. High-order materials detonate using all of the filler material at a burning rate of 3300 ft/s (speed of sound), destroying the target by shattering structural materials. High order produces a supersonic over-pressurization shock wave. Examples of high order include TNT, C-4, Semtex, nitroglycerine, dynamite, and ammonium nitrate fuel oil (ANFO).

Low-order materials deflagrate (rapid burn) at a rate less than 3300 ft/s (1100 fps is the overpressure blast wave) and may not completely consume the filler, resulting in incomplete detonation. Low-order filler destroys targets by a push–pull–shove effect, weakening structural integrity. Low-order detonations are hazardous for the first responder because of not fully consumed; they leave explosive and even shock-sensitive materials scattered around the scene. Examples of low-order explosives are pipe bombs, gunpowder, and pure petroleum-based bombs such as Molotov cocktails or aircraft improvised as guided missiles. The pressure waves for high-order detonation reach 50,000–4,000,000 psi, while low-order detonation pressure waves reach up to 50,000 psi.

High-order filler explosives follow a sequence of events to detonate. This sequenced set of events is called an explosive train and has three separate segments: ignition source, initiating explosive, and main charge. Between the initiating explosive and main charge, a booster charge or delay charge may be used to ensure complete consumption.

Primary injuries expected from high-order explosions related to the over-pressurization wave include head injuries which may or may not include a history of a loss of consciousness and may be associated with:

- Headache, seizures, dizziness, memory problems
- Gait/balance problems, nausea/vomiting, difficulty concentrating
- Visual disturbances, tinnitus, slurred speech
- Disoriented irritability, confusion
- Extremity weakness or numbness

Tympanic membrane – ear injuries which include:

- Impaired hearing which can complicate the triage process
- Perforated tympanic membranes
- Spontaneous healing of the tympanic membrane occurs in 50–80% of all patients suffering from perforations.

Abdominal injuries including perforations with signs and symptoms that may develop 24–48 hours after the blast. Peritonitis resulting from the blast trauma, may manifest as long as days after the blast.

Expected Effects from Explosions

Potential injury	Pressure (psi)	Structural effects
Loss of balance/rupture of eardrums, internal organ damage	0.5–3	Glass shatters; façade failure
Internal organ damage	5–6	Cinderblock shatters; steel structures fail; containers collapse; utility poles fail
Pressure causes multisystem trauma	15	Structural failure of typical construction
Lung collapse	30	Reinforced construction
Fatal injury	100	Structural failure

Physiology of Blast Effects

When a detonation occurs, a wave of pressure moves from a detonation point outward in all directions. The initial shock wave is dependent on the type of explosive, confinement of the material, and the oxidizers present. The pressure continues in every direction until the released energy has been equalized. Depending on the level of explosive order, a pressure wave can reach above 50,000 psi. Think of this wave as a locomotive hauling freight that strikes objects while moving at 15,000 mph? Atomization and total disintegration of material, including all biological material, will occur close to the detonation.

The primary pressure wave produces a shearing effect in such organs as the gastrointestinal tract, eardrum, and surrounding bones, lungs, and central nervous system. The secondary wave causes the material to fly around, striking victims, and causing them traumatic injury. Blunt trauma, lacerations, abrasions, puncture wounds, penetrating injuries, and incisions are all common. The tertiary effects of the blast are due to deceleration injuries. Once the explosion occurs, the victim is thrown in the direction of the initial blast wave. See Figure 8.1 for the evacuation distances based on the size of the explosive device.

Figure 8.2 gives the explosive potential using cylinders of common industrial fuels (butane and propane) as explosive devices.

Blast Effects

The effects of a sudden release of energy from an explosion can be devastating. If a victim is far enough away from the center so that disintegration does not occur, one of the following four general categories of injuries will be seen:

1) The first category of injury is caused by the initial blast wave, and they involve the hollow organs and their interfaces. The gas-filled organs are compressed and develop a gas-pressure exchange, causing implosion injury. These injuries have a high mortality rate.

Explosives (DHS/FBI bomb threat stand-off card)				
Threat	Explosives capacity	Mandatory evacuation distance	Shelter in place zone	Preferred evacuation distance
Pipe bomb	5 lb	70 ft	71–1119 ft	+1200 ft
Suicide bomb	20 lb	110 ft	111–1699 ft	+1700 ft
Briefcase	50 lb	150 ft	151–1849 ft	+1850 ft
Car	500 lb	320 ft	321–1899 ft	+1900 ft
SUV/VAN	1000 lb	400 ft	401–2399 ft	+2400 ft
Small delivery truck	4000 lb	640 ft	641–3799 ft	+3800 ft
Container/ water truck	10,000 lb	860 ft	861–5099 ft	+5100 ft
Semi trailer	60,000 lb	1,570 ft	1571–9299 ft	+9300 ft

Figure 8.1 FBI's shelter and evacuation chart based on the size of the suspected vehicle bomb.

Liquified petroleum gas: LPG (butane or propane)			
Threat	LPG/mass/volume[1]	Fireball diameter[2]	Safe distance[3]
Small LPG cylinder	20 lb/5 gal 9 kg/19 l	40 ft/12 M	160 ft/40 M
Large LPG cylinder	100 lb/55 gal 45 kg/95 l	69 ft/21 M	276 ft/84 M
Commercial/ residential LPG cylinder	2000 lb/500 gal 907 kg/1893 l	184 ft/56 M	736 ft/224 M
Small LPG truck	8000 lb/2000 gal 3630 kg/7570 l	292 ft/89 M	1168 ft/356 M
Semi-tanker LPG	40,000 lb/10,000 gal 18,144 kg/37,840 l	499 ft/152 M	1996 ft/608 M

[1]Based on the maximum amount of material that could reasonably fit into a container or vehicle, variations are possible. [2]Assuming efficient mixing of the flammable gas with ambient air. [3]Determined by US firefighting practices wherein safe distances are approximately four times the flame height. Note that an LPG cylinder filled with high explosives would require a significantly greater standoff distance than if it were filled with LPG.

Figure 8.2 FBI's evacuation chart based on an improvised explosive device involving a known propane cylinder.

2) The second type of injury is spalling. It occurs when the pressure goes through a high-density organ then strikes a low-density organ, resulting in tissue violently spalling (tearing away from) off low-density tissue. These injuries have a high mortality rate.

3) The third type of injury involves organs that are attached or are in close proximity to one another. Target organs from this blast are the abdominal viscera, lungs, nervous system, and skin. Moderate to high mortality rates result and survivability depends on the distance from the center of the blast.

4) The fourth category of injury is due to the toxic gases, dust, and fire created by the explosion. Similar to radiation emergencies, distance and shielding provide the best protection.

Medical management should be geared toward the treatment of multiple systems trauma and deceleration injuries. If patients are contaminated, emergency medical treatment will be very difficult until decontamination is completed. If nerve agents or biologicals are used, technical decontamination may be nearly impossible. An explosion that occurs inside a sealed structure creates more trauma to victims compared to outside explosions. Outside explosions can disperse the energy easier and equalize the pressure more rapidly.

Multisystem trauma should be anticipated, and treatment modalities should consist of volume replacement and shock management. Shock should be managed with high-flow oxygen, fluid replacement, positions, and rapid transport. Normal post-explosion evaluations may include a 24-hour hospitalization, even if there is no evidence of internal injury. Even a simple earache or headache may indicate deeper injury.

Summary

This chapter focused on chemical and biological terrorism when in reality the majority of terrorist activity has been bombings and armed attacks. In the future, this will not change as terrorist get the most immediate recognition from these acts. Chemical and biological terrorism role out much slower and may immediately go unnoticed. But there seems to be greater fear from something that cannot be seen or, may not have immediate effects. Chemicals and especially biological agents have those features and, when used, do generate a greater amount of fear.

There have been isolated events that have caused local responders to re-evaluate their response plans. There have been cases of biological substances being produced with the sole purpose of sell it, killing one or many persons, assassinations, and other nefarious reasons. There have been a number of incidents where very talented young adults have synthesized biologicals just for the satisfaction of doing it. There have been a number of cases where both pipe bombs and ricin have been made just for the "heck of it"! At times the amount of ricin produced was enough to kill thousands of people. Although, these discoveries by law enforcement are kept out of the news, there have been a number of events across the country that could have created numerous deaths and injuries if they had been used. Here are just a few of the cases that have been found:

January 2006, Richmond, Virginia
February 2007, Orlando, Florida
February 2008, Las Vegas, Nevada
January 2009, Seattle, Washington
June 4, 2009, Everett, Washington
January 2011, Akron Ohio
November 2011, Gainesville, Georgia
April 16, 2013, Washington, DC
May 2013, New Boston, Texas
March 14, 2014, Georgetown, Virginia
March 21, 2014, Hatboro, Pennsylvania
May 2014, Spokane, Washington
October 2014, Oshkosh, Wisconsin
September 2020, Little Rock, Arkansas

Terrorism is not a new issue nor is it just an issue of the past. International terrorism has been a threat for decades. France, England, Ireland, Italy, Israel, Bosnia, Libya, Iran, and Iraq, to mention a few, have been dealing with terroristic acts for many years. Political unrest, different religious views, and/or indifferences have created these groups whose mission is to change their environment of political or religious climate they live under. Here in America, we too, have a long tradition of freedom fighters. A study of the growth of this country during the post-Civil War era reveals a list of militia groups that truly saw their cause as a right way and were willing to fight to change their future. Some of these same groups still exist today, albeit underground.

Terrorism has been redefined within the last decade. Recent Untied States target include:

- 1st attack on the World Trade Center in New York City (February 1993, a truck bomb was driven into the underground parking facility in an attempt to bring down the building. This attack was linked to Middle Easter factions).

- The Alfred P. Murrah Federal Building in Oklahoma City (April 1995, linked to American sovereign citizens dissatisfied with our government's handling of the Waco, Texas incident a year before. A truck bomb parked in front of the building detonated killing 168 and injuring another 680 people).
- The Summer Olympics in Atlanta (1996 an act of domestic terrorism from the same bomber; Eric Rudolph, who earlier had bombed a number of abortion clinics in the Atlanta area. The Olympic backpack bomb killed 2 and injured 111 others).
- 2nd attack on the World Trade Center and the Pentagon (September 11, 2001, where hijacked civilian aircraft flew into both the World Trade Center and the Pentagon killing 2750 at the WTC and another 184 at the Pentagon, linked to Middle Easter faction, Osama Bin Laden). A 3rd hijacked jet crashed in Pennsylvania before reaching its target, killing another 40 people.
- The Boston Marathon bombing (took place on April 15, 2013 conducted by two brothers, Dzhokhar Tsarnaev and Tamerian Tsarnaev, utilizing two homemade pressure cooker bombs near the finish of the 26-mi foot race. The bombs killed three victims and injured hundreds of others. In total, 17 victims lost one or more of their limbs. The brothers were Muslim and conducted the attacks over their religious beliefs.
- Mass shooting that occurred at the Pulse Nightclub in Orlando, Florida (June 12, 2016, the shooter, Omar Mir Seddique Mateen of Afghani decent and was Muslim, pledged allegiance to the Islamic State of Iraq upon carrying out the shooting killing 50 and wounding 53 others).
- Mass shooting that occurred during Harvest Music Festival in Las Vegas (October 1, 2017 killing 60 people and wounding 411 others. A total of 867 people were injured, many due to the panic caused from the shooting. The armed attack was conducted from the 32nd floor of a neighboring hotel where Stephen Paddock fired over 1000 bullets into the crowd and eventually killed himself with a self-inflicted gunshot before a motive for the attack could be determined).

The nature of the terrorism act is disturbing. These acts are unanticipated, premeditated, intimidating, politically motivated, and displays a total disregard for human life. It is not an act of war but instead an attack on uninvolved people to create the biggest outcry from citizens and put their agenda and causes boldly in front of them. It is, as described in an article produced by the New World Liberation Front, "an action that the urban guerilla must execute with the greatest cold-bloodedness, calmness, and decision." It is a crime that erodes the infrastructure of modern society with the violence of rape and the brutality of murder.

Epilogue

The subject of hazardous materials medicine is a broad one. It covers a vast number of disciplines that are intertwined. Along the way, the authors, who have gone through many changes in their own philosophy, have found that several things will always hold true.

1) Understanding the physiology of the poisoning is more important than the treatment because without the understanding of the physiology, an efficient and effective treatment cannot be provided.
2) Regardless of the event taking place, the safety of the responders is one of the most important concepts. This book has reviewed aspects of the response from the beginning, even before the response takes place, during the response, and after the response, and what needs to occur to keep the responder safe and productive throughout the response.
3) Emergency medical response to any hazardous materials incident involves a team effort, education, and resources to make the response a success.

With training and education comes insight. Understanding chemistry, biology, toxicology, and physiology, then applying this knowledge to an incident involving chemical exposures, a healthcare provider can move forward in effective assessment, presumptive diagnosis, and treatment.

When failure exists involving exposure, it is always related back to the lack of awareness, training, equipment, or staffing. All portions of the system, from the education to application must be evaluated regularly and an improvement plan developed. This book provides the guidance for both.

The authors know that the development of a hazardous materials medical program is not an easy thing to accomplish. After a combined 80+ years of response, many years of educating responders, and developing medical programs, the authors still find themselves learning cause and effect. It was our intent that these 80 years of research, running calls, teaching, and most important talking with responders from all over the country, would be used to improve the overall response to hazardous materials exposures and to provide early intervention to both save lives and to reduce the long-term effects of exposures. It is our hope that the next generation of responders take the initial steps found in this book to embrace hazmat medicine in the future.

This is not the end but rather the beginning; this subject must be taught at every level of emergency medical care from the basic EMT who responds on the initial response to the physician providing definitive care after an exposure. All need a level of education in order to properly assist the public when the public need help after a chemical emergency.

Hazardous Materials Medicine: Treating the Chemically Injured Patient, First Edition.
Richard Stilp and Armando Bevelacqua.
© 2023 John Wiley & Sons, Inc. Published 2023 by John Wiley & Sons, Inc.

Index

Hazardous Materials Medicine: Treating the Chemically Injured Patient, First Edition.
Richard Stilp and Armando Bevelacqua.
© 2023 John Wiley & Sons, Inc. Published 2023 by John Wiley & Sons, Inc.